Lecture Notes in Computer Science

Lecture Notes in Artificial Intelligence   14670

Founding Editor

Jörg Siekmann

Series Editors

Randy Goebel, *University of Alberta, Edmonton, Canada*
Wolfgang Wahlster, *DFKI, Berlin, Germany*
Zhi-Hua Zhou, *Nanjing University, Nanjing, China*

The series Lecture Notes in Artificial Intelligence (LNAI) was established in 1988 as a topical subseries of LNCS devoted to artificial intelligence.

The series publishes state-of-the-art research results at a high level. As with the LNCS mother series, the mission of the series is to serve the international R & D community by providing an invaluable service, mainly focused on the publication of conference and workshop proceedings and postproceedings.

Annalisa Appice · Hanane Azzag ·
Mohand-Said Hacid · Allel Hadjali ·
Zbigniew Ras
Editors

Foundations
of Intelligent Systems

27th International Symposium, ISMIS 2024
Poitiers, France, June 17–19, 2024
Proceedings

 Springer

Editors
Annalisa Appice ⓘ
University of Bari Aldo Moro
Bari, Italy

Hanane Azzag ⓘ
University of Sorbonne Paris Nord
Villetaneuse, France

Mohand-Said Hacid ⓘ
Claude Bernard University Lyon 1
Villeurbanne Cedex, France

Allel Hadjali ⓘ
LIAS/ENSMA, Poitiers
Cedex, France

Zbigniew Ras ⓘ
University of North Carolina
Charlotte, NC, USA

ISSN 0302-9743   ISSN 1611-3349 (electronic)
Lecture Notes in Artificial Intelligence
ISBN 978-3-031-62699-9   ISBN 978-3-031-62700-2 (eBook)
https://doi.org/10.1007/978-3-031-62700-2

LNCS Sublibrary: SL7 – Artificial Intelligence

This Springer imprint is published by the registered company Springer Nature Switzerland AG
The registered company address is: Gewerbestrasse 11, 6330 Cham, Switzerland

If disposing of this product, please recycle the paper.

Preface

This volume contains the papers selected for presentation at the 27th International Symposium on Methodologies for Intelligent Systems (ISMIS 2024), which took place in Poitiers, France, during June 17–19, 2024. The symposium was organized by the LIAS Laboratory of the Engineer School ISAE-ENSMA, Poitiers, in cooperation with the Computer Science Department at the University of Bari Aldo Moro, Bari (Italy), the LIPN Laboratory, Sorbonne Paris Nord, Villetaneuse (France) and the LIRIS Laboratory, Claude Bernard University, Lyon (France).

ISMIS is a symposium series that started in 1986. Held twice every three years, ISMIS continues its tradition as the unique venue for the latest advances in the development and analysis of methods for building intelligent systems. This year's symposium theme spans two main directions: "concepts and metrics to assess the explainability of AI systems" and "the principles of responsible AI". With this focus, ISMIS 2024 aims to contribute to emerging challenges relative to the development of a new generation of fair and ethical AI systems capable of both "understanding" and "being understandable by" humans, adapting to complex real-world environments, appropriately interacting in complex social settings, mitigating possible cyber-threats and vulnerabilities, and aligning with Environmental, Social and Governance principles. The increased relevance of areas such as Autonomous Systems, Financial Services and Cybersecurity, and sensitive application domains such as Medicine, Manufacturing, Education, Earth Observation, Intelligent Maintenance and Defense require systems with high accuracy, robustness, accountability, fairness, privacy/security, transparency, interpretability, and ethics. This year, we also organised an "Industry Session" devoted to creating a forum for exchanging of ideas between the leading researchers and industry practitioners for effective applications of various intelligent methods, algorithms and tools related to Artificial Intelligence and Big Data in industry.

ISMIS 2024 received 46 international submissions that were carefully reviewed by at least three Program Committee (PC) members or external reviewers. Papers submitted to the "Industry Session" were subject to the same reviewing procedure as those submitted to the regular session. After a rigorous reviewing process, 18 regular papers, 6 short papers and 5 industrial papers were accepted for presentation at the conference and publication in the ISMIS 2024 proceedings volume.

The conference program included two invited keynotes. Eyke Hüllermeier from the Institute of Informatics at LMU Munich, Germany, contributed a talk titled "Uncertainty Quantification in Machine Learning - From Aleatoric to Epistemic". Leila Amgoud from the French National Center for Scientific Research (CNRS) at Toulouse - Paul Sabatier University, France, contributed a talk titled "Argumentation Theory - Foundations and Applications". Abstracts of the talks of the invited speakers are included in this volume.

We would like to sincerely thank all people who helped this volume come into being and made ISMIS 2024 a successful and exciting event. In particular, we would like to express our appreciation for the work of the ISMIS 2024 Program Committee members and external reviewers who helped assure the high standard of accepted papers. We would like to thank all authors of ISMIS 2024, without whose high-quality contributions it would not have been possible to organise the conference. We are grateful to the Steering Committee Chair, Zbigniew Ras, and the whole Steering Committee for their support in critical decisions concerning the event plan. We wish to express our thanks to the Local Organizing Committee Chair, Stéphane Jean, and the whole organisation team for their support and incredible work. We would also thank the Principal Secretary, Bénédicte Boinot, for her professional work and the Web and Social Media Chair, Mickael Baron, for his patience and availability to respond to all requests. We would like to express our deepest gratitude to organizers of the "Industry Session" at ISMIS 2024: Brice Chardin and Yannig Goude. We wish to thank the Publication Chairs, Karam Bou Chaaya and Richard Chbeir for their work on the proceedings. Our thanks are also due to the staff of Springer for their continuous support and work on the proceedings. For ISMIS 2024, Springer also supported both the Best Paper Award and the Best Student Paper Award.

Finally, we would like to thank our sponsors "Region Nouvelle Aquitaine", "Grand Poitiers Communauté Urbaine" and "Ville de Poitiers" for their contributions and supports.

We believe that the proceedings of ISMIS 2024 will become a valuable source of references for your ongoing and future research activities.

April 2024

Annalisa Appice
Hanane Azzag
Mohand-Said Hacid
Allel Hadjali
Zbigniew Ras

Organization

General Symposium Chair

Allel Hadjali   LIAS, ISAE-ENSMA, Poitiers, France

Program Committee Co-chairs

Annalisa Appice   Università degli Studi di Bari, Italy
Hanane Azzag   University of Sorbonne Paris Nord, France
Mohand-Said Hacid   Claude Bernard University - Lyon 1, France

Industry Session Co-chairs

Brice Chardin   LIAS, ISAE-ENSMA, France
Yannig Goude   EDF Lab Saclay, France

Steering Committee Chair

Zbigniew Ras   UNC-Charlotte, USA & Polish-Japanese
                                       Academy of IT, Poland

Steering Committee Members

Annalisa Appice   Università degli Studi di Bari, Italy
Michelangelo Ceci   Università degli Studi di Bari, Italy
Alexander Felfernig   Graz University of Technology, Austria
Mohand-Said Hacid   Université Claude Bernard Lyon 1, France
Denis Helic   Graz University of Technology, Austria
Nathalie Japkowicz   American University, USA
Marzena Kryszkiewicz   Warsaw University of Technology, Poland
Gerhard Leitner   University of Klagenfurt, Austria
Jiming Liu   Hong Kong Baptist University, China
Henryk Rybinski   Warsaw University of Technology, Poland
Andrzej Skowron   Polish Academy of Sciences, Warsaw, Poland

Dominik Slezak   University of Warsaw, Poland
Martin Stettinger   Graz University of Technology, Austria

Proceedings Co-chairs

Karam Bou Chaaya   EXPLEO Group, France
Richard Chbeir   Università UPPA - Pau, France

Program Committee

Giuseppina Andresini   University of Bari "Aldo Moro", Italy
Fabrizio Angiulli   University of Calabria, Italy
Arunkumar Bagavathi   Oklahoma State University, USA
Sylvio Barbon Junior   University of Trieste, Italy
Ladjel Bellatreche   ISAE-ENSMA, France
Mario Luca Bernardi   University of Sannio, Italy
Marenglen Biba   University of New York Tirana, Albania
Jose Borges   University of Porto, Portugal
Luciano Caroprese   University "G. d'Annunzio" of Chieti-Pescara,
                                        Italy
Michelangelo Ceci   University of Bari "Aldo Moro", Italy
Brice Chardin   ISAE-ENSMA, France
Zaineb Chelly Dagdia   University Paris-Saclay, France
Henning Christiansen   Roskilde University, Denmark
Marta Cimitile   University of Rome Unitelma Sapienza, Italy
Roberto Corizzo   American University, USA
Alfredo Cuzzocrea   University of Calabria, Italy
Graziella De Martino   University of Bari "Aldo Moro", Italy
Sarah Jane Delany   Technological University Dublin, Ireland
Claudio Di Ciccio   Utrecht University, The Netherlands
Habiba Drias   USTHB University, Algeria
Haytham Elghazel   University of Lyon 1, France
Naoki Fukuta   Shizuoka University, Japan
Paolo Garza   Politecnico di Torino, Italy
Martin Gebser   University of Klagenfurt, Austria
Massimo Guarascio   ICAR-CNR, Italy
Slimane Hammoudi   ESEO, Angers, France
Dino Ienco   INRAE, Montpellier, France
Angelo Impedovo   Università degli Studi di Bari, Italy
Roberto Interdonato   CIRAD - UMR TETIS, France

Stéphane Jean	University of Poitiers, France
Dragi Kocev	Jožef Stefan Institute, Slovenia
Grazina Korvel	Vilnius University, Lithuania
Mieczysław Kłopotek	ICS, Polish Academy of Sciences, Poland
Gerhard Leitner	University of Klagenfurt, Austria
Marie-Jeanne Lesot	LIP6, Sorbonne University, France
Rory Lewis	University of Colorado Colorado Springs, USA
Corrado Loglisci	University of Bari "Aldo Moro", Italy
Donato Malerba	University of Bari "Aldo Moro", Italy
Giuseppe Manco	ICAR-CNR, Italy
Claudia Marinica	Polytech Nantes, France
Mamoun Mardini	University of Florida, USA
Gabriel Marques Tavares	Università degli Studi di Milano, Italy
Elio Masciari	University of Naples "Federico II", Italy
Amin Mesmoudi	University of Poitiers, France
Paolo Mignone	University of Bari "Aldo Moro", Italy
Mikhail Moshkov	KAUST, Saudi Arabia
Agnieszka Mykowiecka	Polish Academy of Sciences, Poland
Mirco Nanni	KDD-Lab ISTI-CNR Pisa, Italy
Amedeo Napoli	LORIA Nancy, France
Louise Parkin	University of Tours, France
Vincenzo Pasquadibisceglie	University of Bari "Aldo Moro", Italy
Ruggero G. Pensa	University of Torino, Italy
Jean-Marc Petit	University of Lyon 1, France
Fabien Picarougne	University of Nantes, France
Gianvito Pio	University of Bari "Aldo Moro", Italy
Luca Piovesan	Università del Piemonte Orientale, Italy
Luigi Portinale	Università del Piemonte Orientale, Italy
Celine Robardet	INSA Lyon, France
Hiroshi Sakai	Kyushu Institute of Technology, Japan
Monika Seisenberger	Swansea University, UK
Giancarlo Sperli	University of Naples "Federico II", Italy
Paolo Terenziani	Università del Piemonte Orientale, Italy
Herna Viktor	University of Ottawa, Canada
Alicja Wieczorkowska	Polish-Japanese Academy of IT, Poland
David Wilson	University of North Carolina, USA
Yiyu Yao	University of Regina, Canada

Additional Reviewers

Sean Walton
Francesco Benedetti
Alice Tarzariol
Jay Morgan
Vivien Leonard
Widad Hassina Belkadi
Mohammed El-Kholany
Paolo Zicari

Stefano Polimena
Massimiliano Altieri
Daniela Gallo
Ramsha Ali
Galileo Sartor
Ilyes Khennak
Tommaso Ruga

Invited Talks

Uncertainty Quantification in Machine Learning - From Aleatoric to Epistemic

Eyke Hüllermeier

Institute of Informatics at LMU Munich, Germany

Abstract. Due to the steadily increasing relevance of machine learning for practical applications, many of which are coming with safety requirements, the notion of uncertainty has received increasing attention in machine learning research in the recent past. This talk will address questions regarding the representation and adequate handling of (predictive) uncertainty in (supervised) machine learning. A particular focus will be put on the distinction between two important types of uncertainty, often referred to as aleatoric and epistemic, and how to quantify these uncertainties in terms of appropriate numerical measures. Roughly speaking, while aleatoric uncertainty is due to the randomness inherent in the data generating process, epistemic uncertainty is caused by the learner's ignorance of the true underlying model.

Argumentation Theory - Foundations and Applications

Leila Amgoud

French National Centre for Scientific Research (CNRS),
and IRIT Lab in Toulouse

Abstract. Argumentation is a reasoning approach based on the justification of claims by arguments, i.e. reasons for accepting claims. Due to its explanatory power, it has been used for solving various AI problems including inconsistency handling and decision making. Whatever the problem to be solved, an argumentation process follows generally four main steps: it justifies claims by arguments, identifies (attack, support) relations between arguments, evaluates the arguments, and defines an output. Evaluation of arguments is crucial as it impacts the outcomes of argument-based systems. In this talk, I will introduce abstract argumentation frameworks, their formal foundations, various evaluation methods, and an example of a paraconsistent logic based on argumentation.

Contents

Neural Network and Data Mining

Explainability in AI

Industry Session

Learning with Complex Data

Recommendation Systems and Prediction

Classification and Clustering

Improving the Robustness to Color Perturbations of Classification and Regression Models in the Visual Evaluation of Fruits and Vegetables

Stefano Polimena[1,2] , Gianvito Pio[1,2(✉)] , Giovanni Attolico[3] ,
and Michelangelo Ceci[1,2,4]

[1] Department of Computer Science, University of Bari Aldo Moro, Bari, Italy
gianvito.pio@uniba.it
[2] Big Data Lab., National Interuniversity Consortium for Informatics, Rome, Italy
[3] Institute of Intelligent Industrial Technologies and Systems for Advanced
Manufacturing, National Research Council of Italy, Bari, Italy
[4] Department of Knowledge Technologies, Jožef Stefan Institute, Ljubljana, Slovenia

Abstract. The evaluation of the quality of fruits and vegetables not only influences customer choices but also plays a crucial role for companies to monitor the production and distribution processes. In recent years, there has been a growing interest on the adoption of non-destructive techniques to automatically assess the product quality along the agroalimentary supply chain, also to be adopted in live environments on devices with low computational resources (e.g., fridges).

In this paper, we present a solution that leverages the color distribution to mitigate the sensitivity of machine learning models to light/color variations. Specifically, we extract the color histogram and create multiple color aggregations to reduce the impact of perturbations on the model output. Our experiments, conducted on two real-world datasets and across two distinct learning tasks, demonstrated the effectiveness of the proposed method, also compared to state-of-the-art approaches based on complex neural network architectures.

Keywords: Machine Learning · Quality assessment · Agrifood

1 Introduction

The research of contactless and non-destructive quality assessment approaches for fruits and vegetables is of paramount importance in ensuring the integrity, safety, and reliability of products across the agroalimentary supply chain. It is crucial to know the quality of a product since it can provide benefits for both consumers, who are looking for healthy foods, and companies, to promote the design and distribution of products and services that minimize the environmental impact, preserve resources, and prioritize waste reduction.

The traditional manual inspection is often destructive, labor-intensive and also requires specialized operators which may introduce inconsistencies and

A. Appice et al. (Eds.): ISMIS 2024, LNAI 14670, pp. 3–13, 2024.
https://doi.org/10.1007/978-3-031-62700-2_1

subjectiveness in the evaluation. Recently, non-destructive computational methods have been proposed to replace the perception of human senses in order to estimate internal and external characteristics of the product.

In the literature, we can find several works where hyperspectral images (HSI) are analyzed through machine learning/deep learning techniques for classification and regression purposes in the agricultural domain. For example, in [11], the authors proposed an approach to discriminate among different types of fruits. They combined HSIs and state-of-art CNN models (e.g., ResNet [7]) with two additional convolutional layers to compress and simulate standard RGB channels. In [13], the authors proposed the adoption of a wavelet transform for a multi-scale transformation of the images, and a stacked convolution auto-encoder for the spectral feature extraction. Finally, they trained a Support Vector Regressor (SVR) on the obtained features in order to predict the compound metals content of lettuce leaves. In [6], the authors proposed a CNN in order to predict the ripening time of mature avocado fruit, both mature and unripe. However, it is noteworthy that hyperspectral imaging requires huge computational resources, since it generates highly dimensional data that are computationally heavy to analyze. This issue makes them currently unsuitable in some real-life environments, where classification and regression tasks need to be solved through limited computational resources (e.g., in small devices installed in supermarket fridges).

On the other hand, the adoption of computer vision systems (CVS) in agriculture has grown significantly in recent years. This is due to their ability to provide valuable insights about agricultural products, reduce operational costs, maintain consistent quality standards, and provide real-time information, estimations or predictions. CVS have been used to evaluate the quality of different products, such as apple [12], table grapes [3], lettuce [2] and, more recently, rocket leaves [10]. The specific interest for rocket comes from the fact that it contains a lot of antioxidants and other beneficial compounds that are essential for human health. In particular, the amount of chlorophyll is an important indicator of the quality of rocket leaves, because it is directly related to the plant nutritional values, freshness and color [10]: rocket leaves with high amounts of chlorophyll are typically more nutritious and have a longer shelf life than rocket leaves with low amounts of chlorophyll. Fresh rocket leaves tend to be bright green, with a uniform color and a crisp texture. In contrast, yellow and wilted leaves are undesirable because they are old and no longer edible.

Despite the rocket analysis being increasingly popular within the scientific community, there are only few works regarding the computational estimation of the amount of chlorophyll from rocket images. In [10], the authors proposed an approach for the contactless and non-destructive evaluation of fresh-cut rocket leaves, based on the estimation of the amount of chlorophyll and ammonia (considered as a regression task) and the of visual quality level (considered as a classification task). In [9] the authors identified the regions of the product in the a^*b^* plane of the CIE $L^*a^*b^*$ color space which correlates with properties of interest of rocket leaves. They also proposed an algorithm that distinguishes between marketable and non-marketable rocket leaves while also providing estimates for their ammonia and chlorophyll contents. In all these works, a histogram of 68,121 colors (261 values for the a channel $\times$ values for the b channel) is used, where each feature represents the value of a specific color. However, all the experiments were performed in ideal conditions, with no background and perfectly balanced colors (i.e., with no light/color perturbation). The fine-grained representation of colors adopted by the above-mentioned approaches can make the model highly sensitive to small variations, leading to inaccurate estimations.

Table grapes are also facing an increasing interest. Typically, a green color in the grapes' skin is an indicator of freshness, while a brown grape may push the consumer towards rejection. In [3], the authors applied a Random Forest to identify the level of quality of two different table grape cultivars, based on the histogram of colors. In [4] the assessment of table grapes has been carried out across different tasks by using a Neural Network. The authors evaluated the detection and segmentation of the fruit on the images with a limited data labeling, adopting a semi-supervised approach based on the Mask-RCNN architecture. In [8], a YOLO-grape network has been proposed for the identification of grapes. They added attention mechanisms and soft-NMS functions to detect grapes over complex backgrounds. These works solve classification tasks using neural networks. However, the adopted architectures are naturally able to recognize and exploit shapes, rather than color distribution, which can be more informative in some specific scenarios.

In this context, this paper explores the possibility to properly exploit the color distribution of images. Specifically, to alleviate the sensitiveness to possible light/color perturbations, we build multiple *color aggregations*. Following the work [10], we extract the 68,121-dimensional color histogram as descriptive features, where each feature represents a specific pair of (a, b) in the CIE $L^*a^*b^*$ color space. Subsequently, we evaluate different aggregations of these colors for enhancing robustness and reducing sensitivity to small variations in color. Our experimental evaluation, performed on two real-world datasets, in two distinct learning tasks, proves the effectiveness of the proposed approach.

The remaining of the paper is organized as follows: in Sect. 2, we describe the approach we adopt to collect and prepare data, along with detailing our feature aggregation step; in Sect. 3, we describe our experimental evaluation; finally, in Sect. 4, we draw some conclusions and outline possible future work.

2   Methodology

The learner we adopt to solve machine learning tasks is based on Random Forests, since it proved to be suitable for the application domain at hand in previous works [10]. However, as we emphasized in the introduction, providing it with fine-grained features may lead to models that are sensitive to even slight light/color variations. In this section, we describe the feature construction process that we adopt to alleviate this issue. We first start with a brief overview of the approach adopted in [10] to extract fine-grained features. Then, we illustrate how these features can be combined and employed during the training. The general workflow is depicted in Fig. 1.

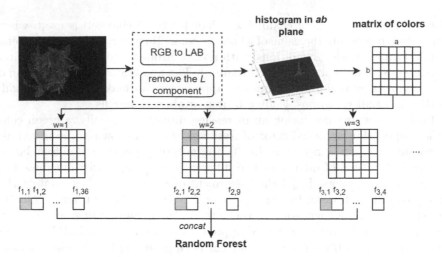

Fig. 1. An overview of the proposed approach. The input image is converted into the CIELab color space, the L component is removed, and the 261×261-dimensional color histogram is extracted. Different color aggregations are then computed and concatenated to obtain the feature space to train Random Forest. (Color figure online)

Given an input image, we first extract the color distribution. We perform color analysis only on the part of the image that is actually associated with the product (foreground). This portion is separated from the background by applying a multi-threshold approach, based on the Otsu algorithm, to the Hue component of the image converted in the HSV color space as described in [1]. A linear transformation is then used for color correction, assuming to have the expected and the measured RGB values of a color chart of k reference colors. This approach is *effective*, i.e., it provides consistent color measurements, and *efficient*, since it is computationally suitable for real applications along the supply chain. However, in our evaluation, we also made experiments in different color perturbation scenarios, starting from the corrected images (see Sect. 3).

More formally, given $[r_e^i g_e^i b_e^i]^T$ and $[r_m^i g_m^i b_m^i]^T$ the expected and the measured RGB values, respectively, for the i-th patch of the reference color chart, with $i = 1, 2, \ldots, k$, a correction matrix is computed through a least-square approach. Such a matrix is then used to reduce the distance between the expected and the measured values, using the following equation:

$$\begin{bmatrix} r_c \\ g_c \\ b_c \end{bmatrix} = \begin{pmatrix} m_{11} & m_{12} & m_{13} \\ m_{21} & m_{22} & m_{23} \\ m_{31} & m_{42} & m_{33} \end{pmatrix} \begin{bmatrix} r_m \\ g_m \\ b_m \end{bmatrix} \tag{1}$$

where $[r_c g_c b_c]^T$ are the colors corrected using the matrix.

To perform the analysis, we convert images from the RGB color space to the device-independent and perceptually-uniform CIE $L^* a^* b^*$ color space. The L^* component is then discarded, since it is too sensible to non-uniform illumination across the scene. As initial features, we consider the color histogram of the foreground pixels. In particular, for each color in the a^*b^* plane, we count the number of pixels in the image. Each axis (a^* and b^*) of the continuous (a^*, b^*) plane is discretized into 261 integer values (in the range $[-130;130]$), leading to a color matrix $H \in \mathbb{N}^{261 \times 261}$, as in [10].

Considering the sensitivity of the color histogram to even the slightest variations in color, we adopt a technique for aggregating nearby colors. Given the standard matrix of colors $H \in \mathbb{N}^{261 \times 261}$ and an integer w, we aim to define groups of w nearest colors along each channel. Specifically, we construct a matrix $\tilde{H}_w \in \mathbb{N}^{261/w \times 261/w}$ by aggregating blocks of $w \times w$ cells through summation, with no overlaps among blocks. The resulting matrix $\tilde{H}_w$ is then linearized. This process can be performed with different values of w, obtaining color histograms with different levels of granularity. Note that low values of w lead to fine-grained representations (with the lowest value $w = 1$ leading to the original matrix H, namely, $\tilde{H}_1 = H$), that are more precise but more sensitive to color variations. On the other hand, high values of w lead to coarser but less sensitive representations. Therefore, our approach consists in extracting multiple representations, with different values of w, and in concatenating all the identified features. We argue that this solution is able to generalize from specific colors to *groups of colors*, making the Random Forest aware of different color distributions at different levels of granularity, and leading to models that are more robust to slight differences in terms of colors. The workflow is depicted in Fig. 1, with an example extracting two different aggregations, with $w = 2$ and $w = 3$, that are concatenated together with the original feature representation ($w = 1$).

3 Experiments

In this section, we first provide a description of the considered datasets and of the experimental setting. Then, we report and discuss the obtained results.
In our empirical evaluation, we considered two real-world datasets:

- *Rocket*, that consists of 1191 images of rocket leaves, each associated with the amount of chlorophyll. In this case, the considered task is regression.
- *Grapes*, that consists of 411 images of table grapes, each associated with a one of the 5 possible classes representing the quality of the product: very good, good, limit of acceptability or marketability, poor, very poor (unacceptable). Therefore, in this case, the considered task is classification.

For both datasets, photos were acquired using a 3CCD digital camera (having a dedicated Charged Coupled Device for each color channel) with a resolution of 1024×768 pixels, that covers an area of 32×24 cm. A small X-Rite color-chart with 24 patches of known colors was placed in the scene to measure color variations due to environmental conditions. The colors in the color-chart were employed to estimate the linear transformation used to correct the images (see Equation (1)). This approach was necessary to have a neutral starting condition and introduce controlled color perturbations during our experiments. In this respect, in order to simulate color perturbations in a controlled manner, we applied a constant perturbation of ±10 on each RGB channel, separately, generating 6 different settings, called **CC+10r**, **CC-10r**, **CC+10g**, **CC-10g**, **CC+10b**, and **CC-10b**, together with the setting with no perturbation (henceforth indicated as **NoPert**). These settings simulate multiple possible conditions in real environments where the color temperature of the lights can perturbate the colors of the products during the image acquisition. In this way, we can evaluate how much our approach and the competitors are affected by such perturbations.

All the experiments were performed using the 10-fold cross validation, while as evaluation measures, we considered the Mean Squared Error (MSE) and the Relative Squared Error (RSE) for the regression task on the *Rocket* dataset, and the accuracy for the classification task on the *Grapes* dataset. We run our method (henceforth denoted with **our approach**) by concatenating the features extracted considering $w \in \{1, 4, 8, 10\}$, and adopting the default parameter setting of Random Forest in the Python scikit-learn library.

We compared the results with those achieved by some approaches based on state-of-the-art neural network architectures: Residual Network [7] (**resnet50**) and a Vision Trasformer [5] (**vit_b_16**). It is important to note that, unlike our approach, both architectures were directly trained from the original images, without extracting color histograms. This is because they should naturally be able to learn relevant features related to both colors and shapes from the images.

We finally assessed the contribution provided by the multiple granularities of color features. In this respect, we run the experiments considering the **original features** (that corresponds to $w = 1$), $w = 4$, $w = 8$ and $w = 10$, separately, and in combination (through feature concatenation).

3.1 Results

In Fig. 2 we report the results on the *Rocket* dataset in terms of MSE and RSE, obtained by our approach, by the model learned only from the original features (i.e., with $w = 1$), and by the other state-of-the-art competitors. From the figures, we can observe that our approach leads to better results (lower errors) in all the perturbation settings considered with respect to all the other methods. This implies that employing multiple color aggregations allows the model to generalize more effectively then only adopting fine-grained features. We can also observe that the competitor neural networks (vit_b_16 and resnet50) exhibit higher errors then the methods based on color histograms. This may be motivated by the fact that the estimation of the chlorophyll mainly depends on the distribution of colors, rather than on shapes captured by such architectures.

We can draw similar conclusions by examining the results in Fig. 3, where we report the classification performance in terms of accuracy for the *Grapes* dataset. Indeed, also in this case, the methods based on color histogram yield to higher accuracy. Specifically, although the setting based on original fine-grained features produces acceptable results, even compared with state-of-the-art neural networks, our approach provides interesting benefits in most cases.

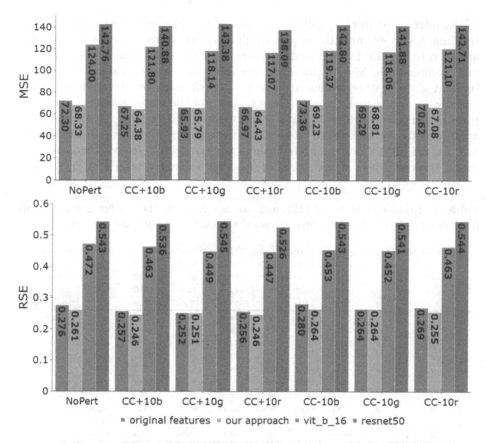

Fig. 2. MSE (on top) and RSE (in the bottom) results (lower is better) for the dataset *Rocket* obtained by our approach, the baseline based on the original features, and competitor systems. (Color figure online)

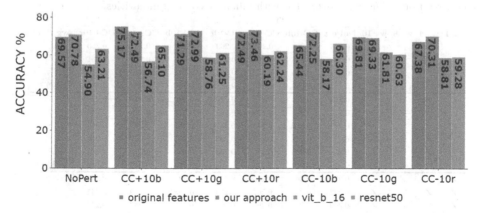

Fig. 3. Accuracy results (higher is better) for the dataset *Grapes* obtained by our approach, the baseline based on the original features, and competitor systems. (Color figure online)

As introduced in the previous subsection, we also conducted a comprehensive analysis to examine the impact of different color granularities. The results are reported in Tables 1, 2, and 3, where we also report the average rank of each combination. Specifically, each line represents the concatenation of features at given feature granularities (e.g., a check mark on the column $w = 1$ and $w = 4$ means that the model was trained on the concatenation of the features constructed using $w = 1$ and $w = 4$), where the last line is our proposed approach (the model learned from the concatenation of features obtained with $w \in \{1, 4, 8, 10\}$). Looking at the results, we can easily observe that our approach, based on the concatenations of the color extracted with $w \in \{1, 4, 8, 10\}$

Table 1. Results in terms of MSE and average rank on the *Rocket* dataset of all possible concatenations of features extracted with different color granularities.

w=1	w=4	w=8	w=10	NoPert	CC-10b	CC-10g	CC-10r	CC+10b	CC+10g	CC+10r	Avg Rank
✓				72.30	73.36	69.29	70.62	67.25	65.93	66.97	9.00
	✓			70.25	69.40	69.92	73.86	69.37	70.52	68.16	8.86
		✓		69.29	75.82	72.29	74.19	73.91	73.05	71.72	12.29
			✓	74.82	73.90	79.17	75.48	79.06	73.26	80.97	14.71
✓	✓			71.45	69.47	71.03	71.54	65.01	65.15	64.43	7.29
✓		✓		70.95	70.80	69.76	70.42	66.31	65.07	64.96	7.14
✓			✓	70.52	73.18	70.56	67.45	67.37	65.26	65.21	8.00
	✓	✓		69.16	69.38	69.74	73.39	71.25	69.71	68.35	8.29
	✓		✓	70.43	69.91	69.99	68.36	67.32	70.77	68.74	8.57
		✓	✓	72.11	76.34	72.63	70.51	70.47	72.52	73.32	12.86
✓	✓	✓		71.63	69.37	68.60	70.03	65.02	64.63	63.11	4.29
✓	✓		✓	68.84	69.03	70.05	66.79	64.83	65.39	63.08	3.43
✓		✓	✓	70.49	70.04	69.17	66.09	65.53	64.80	64.70	4.71
	✓	✓	✓	69.99	69.85	69.36	68.64	69.56	69.58	68.79	7.86
✓	✓	✓	✓	68.33	69.23	68.81	67.08	64.38	65.79	64.43	2.71

Table 2. Results in terms of RSE and average rank on the *Rocket* dataset of all possible concatenations of features extracted with different color granularities.

w=1	w=4	w=8	w=10	NoPert	CC-10b	CC-10g	CC-10r	CC+10b	CC+10g	CC+10r	Avg Rank
✓				0.276	0.280	0.264	0.269	0.257	0.252	0.256	8.86
	✓			0.268	0.264	0.266	0.282	0.266	0.270	0.261	8.57
		✓		0.264	0.289	0.277	0.283	0.282	0.280	0.274	12.43
			✓	0.286	0.280	0.302	0.288	0.303	0.280	0.309	14.57
✓	✓			0.273	0.264	0.271	0.273	0.249	0.249	0.246	7.14
✓		✓		0.271	0.270	0.267	0.269	0.253	0.248	0.248	7.71
✓			✓	0.269	0.279	0.269	0.257	0.257	0.249	0.249	7.86
	✓	✓		0.264	0.264	0.267	0.280	0.272	0.267	0.261	8.57
	✓		✓	0.269	0.266	0.267	0.260	0.258	0.270	0.263	8.29
		✓	✓	0.276	0.291	0.279	0.269	0.269	0.277	0.280	12.71
✓	✓	✓		0.274	0.264	0.263	0.267	0.248	0.247	0.241	4.29
✓	✓		✓	0.262	0.263	0.267	0.254	0.248	0.249	0.241	3.43
✓		✓	✓	0.269	0.267	0.265	0.252	0.250	0.247	0.247	4.86
	✓	✓	✓	0.267	0.266	0.266	0.261	0.266	0.266	0.263	8.00
✓	✓	✓	✓	0.261	0.264	0.264	0.255	0.246	0.251	0.246	2.71

Table 3. Results in terms of Accuracy and average rank on the *Grapes* dataset of all possible concatenations of features extracted with different color granularities.

w=1	w=4	w=8	w=10	NoPert	CC-10b	CC-10g	CC-10r	CC+10b	CC+10g	CC+10r	Avg Rank
✓				69.57	65.44	69.81	67.38	75.17	71.29	72.49	8.57
	✓			69.57	67.62	70.05	68.84	69.56	70.54	68.84	10.14
		✓		70.06	71.52	68.59	67.62	70.06	68.59	70.54	9.29
			✓	69.57	69.58	68.84	69.58	68.10	72.24	67.14	9.57
✓	✓			69.33	68.84	68.35	68.11	74.19	73.47	73.23	8.00
✓		✓		67.85	69.09	68.86	66.17	75.42	72.49	74.69	8.14
✓			✓	68.11	69.81	67.64	69.59	72.49	72.50	75.17	7.43
	✓	✓		67.63	71.28	68.59	71.28	71.26	71.03	70.54	8.86
	✓		✓	70.31	69.80	67.38	68.84	71.51	72.50	68.59	9.29
		✓	✓	71.03	72.00	70.05	69.08	73.46	73.70	69.08	4.29
✓	✓	✓		69.08	70.55	68.36	67.87	72.01	73.46	73.95	7.71
✓	✓		✓	69.09	69.07	67.14	69.09	72.98	72.98	73.94	8.43
✓		✓	✓	70.30	68.84	67.37	69.09	74.68	73.47	73.47	6.57
	✓	✓	✓	68.34	70.05	67.87	70.79	72.48	72.49	69.08	8.71
✓	✓	✓	✓	70.78	72.25	69.33	70.31	72.49	72.99	73.46	4.00

leads to the best results. We can also note that, in general, the features at the finest granularity ($w = 1$) are not enough if taken alone to properly model color variations. The same holds when considering the other color granularities alone (i.e., $w = 4$, $w = 8$, and $w = 10$, taken alone), as can be observed in the first 4 lines of the tables. In general, triplets provide good results (see $w \in \{1, 4, 8\}$, $w \in \{1, 4, 10\}$ and $w \in \{1, 8, 10\}$), except for the one that excludes $w = 1$ (i.e., $w \in \{4, 8, 10\}$). This means that the finest color granularity, although taken alone can lead the model to overfit to slight color variations, remains fundamental in the combination of granularities to achieve good predictive performances.

The results of our experiments outline that the proposed approach can support the contactless estimation of some characteristics of fruits and vegetables and that it is more robust to color variations in the environment with respect to other methods, also based on complex neural network architectures.

4 Conclusion

In this paper we proposed an approach for the contactless quality estimation of some characteristics of fruits and vegetables in the agroalimentary field. We employed a solution that leverages the color distribution to address the sensitivity of machine learning models to light or color variations. Specifically, the proposed method is based on the extraction of multiple color aggregations to represent the color distribution at different granularities and on feeding a Random Forest model with the concatenation of all the identified color distributions.

Our experiments on a dataset about rocket leaves (solving a regression task) and on a dataset about table grapes (solving a classification task) demonstrated the effectiveness of the proposed approach, also compared with complex models based on neural network architectures. This aspect makes the proposed method adoptable in live environments, where color perturbations can often happen due to different lightning, and where the computational resources of the devices is limited and does not allow the deployment of too complex models.

For future work, we will investigate the adoption of more advanced methods to identify color aggregations at different granularities, possibly directly learned from the data, aiming to construct *smart* color groups that would let the learning method to focus on relevant characteristics for the learning task at hand. Moreover, we will evaluate if the proposed approach may provide benefits also in other application domains, such as that of medical images.

Acknowledgments. This work has been partially supported by the European Union - NextGenerationEU through the Italian Ministry of University and Research, projects FAIR - Future AI Research (PE00000013), Spoke 6 - Symbiotic AI and PRIN 2022 "COCOWEARS" (A framework for COntinuum COmputing WEARable Systems), grant n. 2022T2XNJE, CUP: H53D23003650001. S. Polimena also acknowledges the financial support of his Ph.D. fellowship provided by the CNR STIIMA - "Institute of Intelligent Industrial Technologies and Systems for Advanced Manufacturing" of the National Research Council of Italy.

References

1. Cavallo, D.P., Cefola, M., Pace, B., et al.: Contactless and non-destructive chlorophyll content prediction by random forest regression: a case study on fresh-cut rocket leaves. Comput. Electron. Agric. **140**, 303–310 (2017)
2. Cavallo, D.P., Cefola, M., Pace, B., Logrieco, A.F., Attolico, G.: Non-destructive automatic quality evaluation of fresh-cut iceberg lettuce through packaging material. J. Food Eng. **223**, 46–52 (2018)
3. Cavallo, D.P., Cefola, M., Pace, B., Logrieco, A.F., Attolico, G.: Non-destructive and contactless quality evaluation of table grapes by a computer vision system. Comput. Electron. Agric. **156**, 558–564 (2019)
4. Ciarfuglia, T.A., Motoi, I.M., Saraceni, L., et al.: Weakly and semi-supervised detection, segmentation and tracking of table grapes with limited and noisy data. Comput. Electron. Agric. **205**, 107624 (2023)
5. Dosovitskiy, A., et al.: An image is worth 16x16 words: transformers for image recognition at scale. arXiv preprint: arXiv:2010.11929 (2020)
6. Han, Y., Bai, S.H., Trueman, S.J., Khoshelham, K., Kämper, W.: Predicting the ripening time of 'hass' and 'shepard' avocado fruit by hyperspectral imaging. Precision Agric. **24**, 1889–1905 (2023)
7. He, K., Zhang, X., Ren, S., Sun, J.: Deep residual learning for image recognition. In: 2016 IEEE Conference on CVPR, pp. 770–778 (2016)
8. Li, H., Li, C., Li, G., Chen, L.: A real-time table grape detection method based on improved YOLOv4-tiny network in complex background. Biosys. Eng. **212**, 347–359 (2021)
9. Palumbo, M., Cefola, M., Pace, B., Colelli, G., Attolico, G.: Machine learning for the identification of colour cues to estimate quality parameters of rocket leaves. J. Food Eng. **366**, 111850 (2024)
10. Palumbo, M., Pace, B., Cefola, M., et al.: Non-destructive and contactless estimation of chlorophyll and ammonia contents in packaged fresh-cut rocket leaves by a computer vision system. Postharvest Biol. Technol. **189**, 111910 (2022)

11. Steinbrener, J., Posch, K., Leitner, R.: Hyperspectral fruit and vegetable classification using convolutional neural networks. Comput. Electron. Agric. **162**, 364–372 (2019)
12. Wang, X., Kang, H., Zhou, H., Au, W., Chen, C.: Geometry-aware fruit grasping estimation for robotic harvesting in apple orchards. Comput. Electron. Agric. **193**, 106716 (2022)
13. Zhou, X., Sun, J., Tian, Y., Lu, B., Hang, Y., Chen, Q.: Hyperspectral technique combined with deep learning algorithm for detection of compound heavy metals in lettuce. Food Chem. **321**, 126503 (2020)

Clustering Under Radius Constraints
Using Minimum Dominating Sets

Quentin Haenn[✉], Brice Chardin, and Mickael Baron

ISAE-ENSMA, Université de Poitiers, Lias, France
{quentin.haenn,brice.chardin,mickael.baron}@ensma.fr

Abstract. In this paper, we evaluate the applicability of algorithms designed to solve the minimum dominating set problem to perform clustering. The associated clustering problem relies on user constraints, and more specifically on radius intra-cluster constraints. We adapt and evaluate implementations from the state of the art on classification datasets, to compare them with other exact or approximate radius-based clustering algorithms, namely equiwide clustering and hierarchical agglomerative clustering with minimax linkage. We consequently provide the benchmark tools and datasets used in this work.

Keywords: Constrained Clustering · Radius Based Clustering · Minimum Dominating Set

1 Introduction

One of the major benefits of clustering is the ability to find groups and patterns that lie under the data, particularly without much knowledge of it [12]. Clustering is used in many fields, such as data mining, machine learning, pattern recognition, image analysis, bioinformatics, etc. [2,8,12,14], and is a very active field of research.

Clustering under user constraints is a particular type of clustering, where users provide as input of the algorithm constraints that are coherent with their prior knowledge of the data [9]. Typical constraints are instance-based, i.e., users can either specify that two points must belong to the same cluster (must-link), or that two points must belong to different clusters (cannot-link). This has led to global, cluster-based constraints such as diameter and radius constraints. In this work, we focus on the clustering under radius constraints (CRC) problem, which is capable of finding clusters that satisfy a provided error bound and find an appropriate representative element for each cluster.

Since such constraints can lead to trivial solutions such as each points belonging to its own cluster, we add the global constraint that the number of clusters must be minimal. This is often desirable, especially when the user does not know how many clusters are originally present in the data.

Previous work has already introduced such a problem [1,6], but has mainly focused on diameter constraints. Consequently, we found that the minimum dominating set (MDS) has been linked to the CRC problem [1–3,11]. The MDS problem is a well-known NP-Hard graph problem, but recent advances in exact and approximate algorithms have made it possible to solve it efficiently [4,13]. As solving the MDS problem relies on heavy computations but theoretically offers optimal solutions to the CRC problem (i.e., the number of clusters is minimal), we propose to study the feasibility of this approach in comparison to state of the art clustering algorithms. If this approach is efficient, it could be a new way to solve the CRC problem, providing users a new tool to reduce the dimensionality of their data.

Industrial Use Case: We aim to apply this approach to the analysis of electrical consumers and producers along an electrical grid. This is an instance of dimension reduction, as we attempt to identify representative components on the grid. This step is part of a process to consistently reduce the computation time required for further modeling of the grid.

Paper Organization. In the remainder of this section we introduce necessary notions and definitions and formalize the clustering under radius constraint problem. Section 2 provides a brief overview of the CRC problem and the MDS problem. Section 3 presents a CRC algorithm based on MDS. We provide an experimental evaluation of state-of-the-art algorithms in Sect. 4. We conclude by discussing the feasibility of this approach.

1.1 Preliminaries

A clustering task is concerned with finding a partition $\mathcal{P} = \{C_1, C_2, \ldots, C_n\}$ of a population $\mathcal{S}$ such that each cluster C_i is a subset of $\mathcal{S}$, and $\bigcup_{i=1}^{n} C_i = \mathcal{S}$. Assigning a point to a cluster is based on dissimilarity measures between points. We note $d(a, b)$ the dissimilarity between points a and b. Upon this definition, some authors introduced various wideness measures of a cluster, such as the diameter, the radius, the average distance, etc. [1–3,6,7,9]. In this paper, we only consider the radius concept of wideness characterization of a subset of elements of a population.

The radius of a cluster C has been identified by Hubert [11] and applied by Ao et al. [2] as the minimum eccentricity within C: $R(C) = \min_{a \in C} \max_{b \in C} d(a, b)$. Based on this, the natural definition of the center (i.e. the representative point) of a cluster is the point a of C such that $a = \arg\min_{a \in C} \max_{b \in C} d(a, b)$. Intuitively, this center is the point with the best worst-case dissimilarity to any other point in the cluster: $\forall b \in C, d(a, b) \leq R(C)$.

Definition 1 (Clustering Under Radius Constraints Problem). *Let $\mathcal{S}$ be a population, d a dissimilarity function, and T a threshold. The clustering under radius constraints problem is the problem of finding a partition $\mathcal{P}$ of $\mathcal{S}$ such that:*

- $\forall C \in \mathcal{P}, R(C) \leq T$
- $\bigcup_{C \in \mathcal{P}} C = \mathcal{S}$

– $|\mathcal{P}|$ *is minimum, i.e., there is no partition* $\mathcal{P}'$ *such that* $|\mathcal{P}'| < |\mathcal{P}|$ *and* $\mathcal{P}'$ *is a solution to the initial CRC problem.*

Because we aim to apply graph based approaches to the CRC problem, we introduce related notions and definitions. A simple graph $G = (V, E)$ is a couple where V is a set of vertices and E is a set of edges. An edge $e = (u, v)$ is a pair of adjacent vertices u and v. A set of vertices $D \subseteq V$ is a dominating set of G if and only if $\forall v \in V \setminus D, \exists u \in D, (u, v) \in E$. The minimum dominating set problem is the problem of finding a dominating set $D \subseteq V$ of G such that $|D|$ is minimum.

2   Related Work

Radius-based approaches provide some benefits compared with other approaches, as they lead intuitively to a clustering that includes a representation of the data. This representative element is not available under diameter constraint, and sometimes not even computable [1]. This representation problem is an integral part of the clustering problem [12].

One approach to the CRC problem is the Equiwide Clustering (EQW) algorithm [1]. In this work, the authors propose an exact algorithm based on a linear programming (LP) formulation. Their experiments show that the algorithm is able to find the optimal solution in reasonable time on relatively small real world dataset (less than 2000 instances). However, this algorithm efficiency drops when applied to bigger datasets.

Another well-known approach is hierarchical agglomerative clustering (HAC) with the min max linkage criterion [2,3], but it is also well-known that HAC is not designed to find the optimal solution regarding the minimality of the partition under a given constraint. To date, the most efficient HAC algorithm using MinMax criterion identified in the literature is Protoclust [3,17].

The MDS problem has been proven linked to the CRC problem with a few adjustments [2,11]. Thus, the possibility of using MDS to solve CRC problems has already been mentioned [2,3] but not implemented due to the lack of efficient algorithms.

Lately, an approximate MDS algorithm has been proposed by Casado et al. [4]. The authors showed that the approximation is efficient on classical graph instances when compared to other state of the art approximate algorithms [5,15, 16], and concluded that it may be interesting to use it on different combinatorial optimization problems.

Jiang and Zheng [13] released an exact algorithm to solve the MDS problem, based on the idea that the MDS problem can be solved by a novel branch and bound algorithm and bounded the search space thanks to the 2-hop adjacency of the graph, that is to say, two vertices are 2-hop adjacent if and only if they are adjacent or if there exists a vertex that is adjacent to both of them. Considering this definition, if two vertices are not 2-hop adjacent, then they cannot be in the same dominating set. According to Jiang and Zheng, it is the first efficient Branch-and-Bound exact algorithm to solve the MDS problem to date.

Table 1. Selected state of the art algorithms

Algorithm	Paradigm	Minimal #Clusters	Language
Protoclust [3]	MinMax HAC	No	R
EQW-LP [1]	LP	Yes	Python
MDS-APPROX [4]	MDS	No	Python + Java
MDS-EXACT [13]	MDS	Yes	Python + C

Following the previous state of the art, we propose to study implementations of CRC algorithms built on top of both the exact and approximate MDS algorithms [4,13], and to compare their efficiency with two state of the art algorithms: Protoclust [3] and Equiwide Clustering [1].

Table 1 lists the characteristics of the algorithms included in the experimental evaluation of this study.

3   Minimum Dominating Set Based Clustering

A Ten-Point Example for MDS-Based Clustering. Let $\mathcal{S}$ be a set of points that we want to cluster into $\mathcal{P}$. Let d be a dissimilarity function. Let us consider that the user already analyzed the data and found that the maximum admissible dissimilarity between a point and its representative is 2. Thus, the user wants to cluster the data into clusters under a radius constraint, or threshold, $T = 2$. We illustrate the MDS-based clustering algorithm on a simple example, shown in Fig. 1. Figure 1a represents the input data transformed into an equivalent graph. Each vertex represents a point of the population, and each edge represents a dissimilarity between two points. The weight of the edge is the dissimilarity between the two points. For readability we did not represent each pairwise dissimilarity. Dark edges in Fig. 1a represent the edges whose weight exceeds the threshold ($T = 2$).

To convert the original clustering input, i.e: $\mathcal{S}$, d and T, into a suitable input for MDS, the initial equivalent graph is converted into a graph $G'(\mathcal{S}, E)$ where $E = \{(x_i, x_j) \mid d(x_i, x_j) \leq T\}$, i.e., over-weighted edges are removed. This graph is illustrated in Fig. 1b.

This graph is then provided as the input of MDS algorithms. Their output is a dominating set, i.e., a set of vertices. In the example represented in Fig. 1, the set $\{3, 7\}$ is the only MDS, because we cannot find a dominating set with less than two vertices. Using this MDS, we say that points 3 and 7 are representative elements, or centers, of their cluster.

To provide a clustering, this dominating set has to be transformed into a proper partition of the input population. Several solutions might be considered to assign a point to its cluster, as long as the affected point is dominated by the center of the cluster it is assigned to. The assignment selected for this evaluation is to group elements with the closest center, i.e., the center with which the dissimilarity is minimal. When a point is dominated by multiple representative

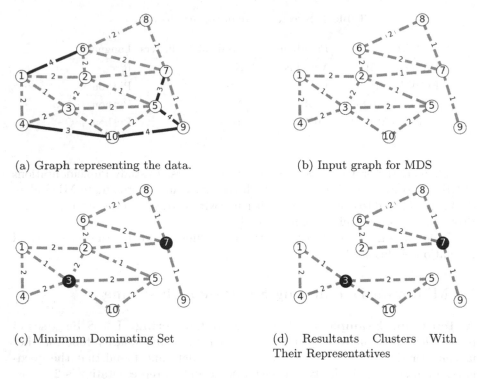

(a) Graph representing the data.

(b) Input graph for MDS

(c) Minimum Dominating Set

(d)   Resultants   Clusters   With
Their Representatives

Fig. 1. Illustration of the clustering process using the MDS approach

points, it is assigned to the cluster of the closest representative point. In our
example, 2 is dominated by both 3 and 7, as illustrated in Fig. 1c. Thus, it is
assigned to the cluster of 7, as $d(2,7) < d(2,3)$. An edge case can occur when a
point is dominated by multiple representative points with the same dissimilarity.
In this case, the point is assigned arbitrarily.

Once each vertex has been assigned to a cluster, the algorithm returns the
clusters and their representative points. The clustering result is illustrated in
Fig. 1d.

This example shows how the MDS problem solves the CRC problem.

4   Experiments

In this section we evaluate the algorithms previously listed in Table 1 on datasets
from OpenML [18]. Their characteristics are described in Table 2.

The experiments are run on a Linux computer running Debian 12 with an
Intel Core I5-10505 CPU at 3,20GHz, and 32GiB of DDR4 RAM. The Linux kernel is 6.1.0. The implementations of the algorithms based upon MDS approaches
and EQW-LP are in Python. The approximative algorithm provided by [4] is
implemented in Java. The exact algorithm, provided by [13] is implemented in
C. Protoclust [3] is implemented in R. The python interpreter is version 3.11, the

Table 2. Experimental Evaluation Datasets

Name	#Elements	#Dimensions	#Classes	R_{opt}
Iris	150	4	3	1.43
Wine	178	13	3	232.09
Glass Identification	214	9	6	3.94
Ionosphere	351	34	2	5.46
WDBC	569	30	2	1197.42
Synthetic Control	600	60	6	70.12
Vehicle	846	18	4	155.05
Yeast	1484	8	10	0.4235
Ozone	2534	72	2	245.59
Waveform	5000	40	3	10.74

R interpreter version is 4.2.2 and the java platform is OpenJDK 17.0.9. The C compiler is gcc 12.2.0. The LP solver is Gurobi 10.0.0. Experiments are run ten times to evaluate the variance of the results. All the experiments are available online for reproducibility purposes [10].

4.1 Radius Threshold and Dissimilarity

Clustering under radius constraints requires a radius threshold and a dissimilarity function in addition to the datasets. We used the Euclidean Distance as the dissimilarity, in line with previous experimental evaluations on similar datasets [1,6]. As for the radius threshold, it is not trivial to find an appropriate value without domain-specific knowledge. The same issue occurred in previous work for diameter constraints [6]. The authors proposed an iterative way of finding the optimal diameter from the number of classes taken as the desired number of clusters. Therefore, we adapted their work to find the optimal threshold to use in our experiments.

To do so, we performed a binary search on the threshold to use. The stopping criterion for that search is when the number of clusters returned by the CRC algorithm is the number of classes in the original dataset, and the immediate lower dissimilarity yields a greater number of clusters.

The optimal radius obtained from the binary search for each dataset are presented in the column R_{opt} of Table 2. Every radius is rounded up to 10^{-2} decimals for readability, except for the Yeast dataset due to its density. Nevertheless, the radius shown in Table 2 are the ones used in the experiments.

4.2 Results

Number of Clusters. The first metric we analyzed is the number of clusters found by the algorithms. The results are presented in Table 3. First, the number of

Table 3. Number of cluster found by the algorithms on the datasets using the radius found

	MDS-Approx	MDS-Exact	EQW-LP	Protoclust
Iris	**3**	**3**	**3**	4
Wine	4	**3**	**3**	4
Glass Identification	7	**6**	**6**	7
Ionosphere	**2**	**2**	**2**	5
WDBC	**2**	**2**	**2**	3
Synthetic Control	8	**6**	**6**	8
Vehicle	5	**4**	**4**	6
Yeast	**10**	**10**	**10**	13
Ozone	3	**2**	**2**	3
Waveform	**3**	**3**	**3**	6

cluster is constant on all runs of all algorithms on all datasets. This supports the idea that the algorithms are stable. Second, exact algorithms (MDS-Exact and EQW-LP) always identify the optimal number of clusters while approximate algorithms might not. In addition, the number of clusters found by MDS-Approx is either exact or close to the optimal number of clusters, with at most two clusters difference noticed on the Synthetic Control dataset. Protoclust never reaches the optimal number of clusters, but remains relatively close to it with at most a three clusters difference for the Yeast dataset.

Effective Radius. This metric allows to assess if the clusters found are valid, i.e., if the radius of the clusters is at most equal to the given constraint. Plus, this metric allows us to know if the clusters built are compact, i.e., if the effective maximal radius of the clusters is the minimal one that can be found under the constraint given. Results are displayed in Table 4 and optimal values are typeset in bold.

The effective radius is constant on all ten runs of all algorithms on all datasets. As the number of clusters before, this metric strengthens the idea that the algorithms are mostly stable.

The second thing we note is that, with MDS-Exact and EQW-LP, the effective radius is always equal to the constraint given. This confirms that, on one hand, the algorithms are optimal under this metric and, on the other hand, that the radius constraint is tight, namely that the MDS admits a single solution or, potentially, multiple solution but with the same resulting wideness. On the contrary, the effective radius given by Protoclust and MDS-Approx are sometimes smaller than the constraint given. This is directly linked to the number of clusters found by the implementation, because if the number of clusters is not optimal, the effective radius can be lower. This is always the case except on particular datasets such as Glass Identification, where the effective radius remains equal to the threshold despite the number of clusters being larger for MDS-Approx, and on Yeast dataset for Protoclust.

Table 4. Effective radius after clustering on the datasets

	MDS-APPROX	MDS-EXACT	EQW-LP	PROTOCLUST
Iris	1.43	1.43	1.43	1.24
Wine	220.05	232.08	232.08	181.35
Glass Identification	3.94	3.94	3.94	3.31
Ionosphere	5.45	5.45	5.45	5.35
WDBC	1197.42	1197.42	1197.42	907.10
Synthetic Control	66.59	70.11	70.11	68.27
Vehicle	150.87	155.05	155.05	120.97
Yeast	0.423	0.423	0.423	0.419
Ozone	235.77	245.58	245.58	194.89
Waveform	10.73	10.73	10.73	10.47

Lastly, we note that every algorithm satisfies the given radius constraint. This means that they are indeed all valid solutions to solve the CRC problem, although sometimes approximate w.r.t. the minimality criteria.

Execution Time. Based upon the results presented in Table 5 we can split the analysis into two parts: exact algorithms and approximate algorithms.

Among exact algorithms, the MDS-EXACT implementation is faster on small datasets, except on the ionosphere dataset, and is up to 4 times faster than EQW-LP, as observed on Synthetic Control. On the contrary, EQW-LP is faster on larger datasets, up to 100 times faster than MDS-EXACT, as observed on Yeast. Based on these experiments, we conclude that EQW-LP is to be preferred on average, since its execution time remains comparable on small datasets, but becomes largely preferable with larger datasets under this metric.

As for approximate algorithms, MDS-APPROX is faster than PROTOCLUST on all datasets except Vehicle and Ozone, where it is 3 times slower. However, both algorithms run in mostly comparable execution times. Thus, considering MDS-APPROX is either better or equivalent to PROTOCLUST on the other metrics, we conclude that MDS-APPROX is to be preferred on all datasets. By its design, we expected PROTOCLUST to be faster on all datasets, however, this is mostly not the case. This is partially due to an implementation limitation of PROTOCLUST because it always computes the entire dendrogram, to then cut it at the provided threshold. An improvement for this use case would be to stop the agglomerative clustering as soon as the threshold is reached.

Overall, the MDS-APPROX algorithm becomes the preferred solution among all four with larger datasets when the minimality of the number of clusters is not to be guaranteed, since it becomes up to five times faster than exact solutions.

Table 5. Execution time of the algorithms on the datasets in seconds

	MDS-APPROX	MDS-EXACT	EQW-LP	PROTOCLUST
Iris	0.062 ± 0.01	**0.009 ± 0.00**	0.018 ± 0.01	0.026 ± 0.00
Wine	0.029 ± 0.00	**0.010 ± 0.00**	0.014 ± 0.00	0.034 ± 0.00
Glass Identification	**0.015 ± 0.00**	0.020 ± 0.00	0.026 ± 0.00	0.046 ± 0.00
Ionosphere	**0.078 ± 0.01**	2.640 ± 0.05	0.104 ± 0.00	0.12 ± 0.00
WDBC	0.315 ± 0.01	**0.138 ± 0.00**	0.197 ± 0.01	0.402 ± 0.00
Synthetic Control	0.35 ± 0.03	**0.036 ± 0.00**	0.143 ± 0.01	0.489 ± 0.00
Vehicle	0.955 ± 0.04	**0.185 ± 0.00**	0.526 ± 0.01	0.830 ± 0.01
Yeast	**2.361 ± 0.03**	622.87 ± 0.30	6.718 ± 0.02	2.374 ± 0.08
Ozone	49.82 ± 1.18	1350.86 ± 1.5	26.86 ± 0.63	**15.32 ± 0.15**
Waveform	**48.01 ± 0.39**	5559.9 ± 15.3	233.9 ± 1.45	61.27 ± 0.08

5 Conclusion and Future Work

In this work, we studied MDS-based clustering under radius constraint. Despite the fact that MDS approaches were identified very early on as being suitable for CRC problems, the lack of efficient algorithms, the inherent complexity of the problem itself and the hardware capabilities of computers meant that it was never applied. We have showed through various experiments that those approaches can indeed be considered as a tangible alternative to various CRC algorithms, both in terms of execution time and quality of the results. Plus, we showed that both the exact and approximate variants of the algorithm are very efficient on small datasets, and that the approximate variant remains competitive on larger datasets. This work also may be considered as a first usage of Casado et al. [4] and Jiang and Zheng [13] algorithms on real combinatorial problem.

We conclude that the MDS approach seems promising and recent advances in this field mean this paradigm can be seen as a real alternative to classical algorithms or those requiring heavy and proprietary solvers such as linear programming.

Acknowledgments. This work was supported by the french region Nouvelle Aquitaine and the aLIENOR ANR LabCom (ANR-19-LCV2-0006). We also thank Jiang and Zheng for kindly providing us with the source code of their exact algorithm.

Disclosure of Interest. The authors declare that they have no conflict of interest.

References

1. Andersen, J., Chardin, B., Tribak, M.: Clustering to the fewest clusters under intra-cluster dissimilarity constraints. In: 2021 IEEE 33rd International Conference on Tools with Artificial Intelligence (ICTAI), pp. 209–216 (2021). https://doi.org/10.1109/ICTAI52525.2021.00036

2. Ao, S.I., et al.: CLUSTAG: hierarchical clustering and graph methods for selecting tag SNPs. Bioinformatics **21**(8), 1735–1736 (2005). https://doi.org/10.1093/bioinformatics/bti201

3. Bien, J., Tibshirani, R.: Hierarchical clustering with prototypes via minimax linkage. J. Am. Statist. Assoc. **106**(495), 1075–1084 (2011). https://doi.org/10.1198/jasa.2011.tm10183. ISSN 0162-1459

4. Casado, A., et al.: An iterated greedy algorithm for finding the minimum dominating set in graphs. Math. Comput. Simul. **207**, 41–58 (2023). https://doi.org/10.1016/j.matcom.2022.12.018. ISSN 0378-4754

5. Chalupa, D.: An order-based algorithm for minimum dominating set with application in graph mining. Inf. Sci. **426**, 101–116 (2018). https://doi.org/10.1016/j.ins.2017.10.033. ISSN 0020-0255

6. Dao, T.B.H., Duong, K.C., Vrain, C.: Constrained clustering by constraint programming. Artif. Intell. **244**, 70–94 (2017). https://doi.org/10.1016/j.artint.2015.05.006. ISSN 00043702

7. Dinler, D., Tural, M.K.: A Survey of Constrained Clustering. In: Celebi, M.E., Aydin, K. (eds.) Unsupervised Learning Algorithms, pp. 207–235. Springer, Cham (2016). https://doi.org/10.1007/978-3-319-24211-8_9 ISBN 978-3-319- 24211-8

8. Gao, Z., et al.: Multi-level aircraft feature representation and selection for aviation environmental impact analysis. Transp. Res. Part C: Emerg. Technol. **143**, 103824 (2022). https://doi.org/10.1016/j.trc.2022.103824. ISSN 0968-090X

9. Gordon, A.D.: A survey of constrained classification. Comput. Statist. Data Anal. **21**(1), 17–29 (1996). https://doi.org/10.1016/0167-9473(95)00005-4. ISSN 0167-9473

10. Haenn, Q., Chardin, B., Baron, M.: MDS clustering experiments. Source Code repository (2024). https://forge.lias-lab.fr/mds_clustering

11. Hubert, L.J.: Some applications of graph theory to clustering. Psychometrika **39**(3), 283–309 (1974). https://doi.org/10.1007/BF02291704. ISSN 1860-0980

12. Jain, A.K., Murty, M.N., Flynn, P.J.: Data clustering: a review. ACM Comput. Surv. **31**(3), 264–323 (1999). https://doi.org/10.1145/331499.331504. ISSN 0360-0300, 1557-7341

13. Jiang, H., Zheng, Z.: An exact algorithm for the minimum dominating set problem. In: Proceedings of the Thirty-Second International Joint Conference on Artificial Intelligence, pp. 5604–5612 (2023). https://doi.org/10.24963/ijcai.2023/622. ISBN 978-1-956792-03-4

14. Liu, Y., Sioshansi, R., Conejo, A.J.: Hierarchical clustering to find representative operating periods for capacity-expansion modeling. IEEE Trans. Power Syst. **33**(3), 3029–3039 (2018). https://doi.org/10.1109/TPWRS.2017.2746379. ISSN 1558-0679

15. Potluri, A., Singh, A.: Two hybrid meta-heuristic approaches for minimum dominating set problem. In: Panigrahi, B.K., Suganthan, P.N., Das, S., Satapathy, S.C. (eds.) SEMCCO 2011. LNCS, vol. 7077, pp. 97–104. Springer, Heidelberg (2011). https://doi.org/10.1007/978-3-642-27242-4_12 ISBN 978-3-642-27242-4

16. Potluri, A., Singh, A.: Hybrid metaheuristic algorithms for minimum weight dominating set. Appl. Soft Comput. **13**(1), 76–88 (2013). https://doi.org/10.1016/j.asoc.2012.07.009. ISSN 1568-4946

17. Tai, X.H., Frisoli, K.: Benchmarking minimax linkage (2019). arXiv:1906.03336 [cs, stat]

18. Vanschoren, J., van Rijn, J.N., Bischl, B., Torgo, L.: OpenMl: networked science in machine learning. SIGKDD Explor. **15**(2), 49–60 (2013). https://doi.org/10.1145/2641190.2641198

Learning Typicality Inclusions in a Probabilistic Description Logic for Concept Combination

Alberto Valese, Valentina Gliozzi[iD], and Gian Luca Pozzato[(⊠)][iD]

Dipartimento di Informatica, Università di Torino, Turin, Italy
{alberto.valese,valentina.gliozzi,gianluca.pozzato}@unito.it

Abstract. Our paper introduces an innovative automated system designed to extract logical rules using the $\mathbf{T}^{CL}$ logic from diverse datasets, with a particular emphasis on tabular data. Our starting point is the CN2 algorithm. Typically employed for classification tasks, we have adapted this algorithm to suit our descriptive objectives. We consider well-known datasets (such as iris and zoo) to illustrate our approach. Furthermore, we extend this analysis to intricate datasets, notably the GTZAN musical dataset and the "Adult" dataset. These examples showcase the algorithm's efficacy in generating descriptive rules across different data domains. We discuss the adaptability of the proposed approach across various data types, including images, sounds, and diverse heterogeneous structures.

1 Introduction

Generating new knowledge via conceptual combination concerns high-level capacities associated to creative thinking and problem solving. From an AI perspective, this requires the harmonization of two conflicting requirements: the need of a syntactic and semantic compositionality, on the one hand; the need of capturing typicality effects, on the other hand, that can be hardly accommodated in standard symbolic systems, including formal ontologies. According to a well-known argument [11], prototypes, are not compositional. Consider a concept like *pet fish*: it results from the composition of the concept *pet* and of the concept *fish*, however, the prototype of *pet fish* cannot result from the composition of the prototypes of a pet and a fish. For instance, a typical pet is furry, whereas a typical fish is grayish, but a typical pet fish is neither furry nor grayish (typically, it is red). Examples of such difficulties concern handling exceptions to attribute inheritance and handling the possible inconsistencies arising between conflicting properties of the concepts to be combined.

In order to tackle this problem, in [5] the authors have introduced the Description Logic $\mathbf{T}^{CL}$, where "typical" properties can be directly specified by means of a "typicality" operator $\mathbf{T}$ enriching the underlying DL, and a TBox can contain inclusions of the form $\mathbf{T}(C) \sqsubseteq D$ to represent that "typical Cs are also Ds". As a difference with standard DLs, in the logic $\mathbf{T}^{CL}$ one can consistently express exceptions and reason about defeasible inheritance as well. Typicality inclusions are also equipped by a real number $p \in (0, 1)$ representing the probability/degree of belief in such a typical property. For instance, a knowledge base (KB) can consistently express:

(1) $SeniorStudent \sqsubseteq Student$ (3) $0.95 :: \mathbf{T}(SeniorStudent) \sqsubseteq \neg GoParty$

(2) $0.70 :: \mathbf{T}(Student) \sqsubseteq GoParty$ (4) $0.85 :: \mathbf{T}(SeniorStudent) \sqsubseteq Married$

A. Appice et al. (Eds.): ISMIS 2024, LNAI 14670, pp. 24–32, 2024.
https://doi.org/10.1007/978-3-031-62700-2_3

Probabilities allow us to define a semantics inspired to the DISPONTE semantics [13], which in turn is used in order to describe different *scenarios* where only some typicality properties are considered; in the example, we consider eight different scenarios, representing all possible combinations of typicality inclusion, for instance $\{((2), 1), ((3), 1), ((4), 0)\}$ represents the scenario in which (2) and (3) hold, whereas (4) is not considered. The standard inclusion (1) holds in every scenario, representing a rigid property not admitting exceptions. Each scenario is equipped with a probability depending on those of the involved inclusions. Given a KB containing the description of two concepts C_H and C_M occurring in it, one can consider only *some* scenarios in order to define a revised knowledge base, enriched by typical properties of the combined concept $C \sqsubseteq C_H \sqcap C_M$ by also implementing a HEAD/MODIFIER heuristics coming from the cognitive semantics.

The logic $\mathbf{T}^{CL}$ has been recently exploited in order to build a goal-oriented framework for knowledge invention in the cognitive architecture of SOAR [4], as well as for the generation and the suggestion of novel editorial content in multimedia broadcasting [1] and in the artistic domain of paintings, poetic content [8], and museum items [6,9]. However, the mechanism implementing the concept combination is strongly related to the probabilities used to equip typicality inclusions, and at present, the problem of choosing the right probabilities/frequencies is left to the ontology engineer, whereas the goal of having an intelligent system able to automatically extract such probabilities from data is far from being solved.

In this work we introduce an innovative automated system designed to extract logical rules employing the logic $\mathbf{T}^{CL}$ from datasets of diverse forms and natures. Our focus lies predominantly on tabular data, where we offer concrete demonstrations using the iris and zoo datasets as working examples. To obtain the rules, our exploration led us to focus on "Rule Learning Algorithms" within machine learning. Our aim was to identify an appropriate solution that aligned with our objectives. While exploring various possibilities in the literature, considering our objective and without seeking unnecessarily complex solutions, we found an algorithm that met our needs: CN2 [2]. The basic version of this algorithm doesn't precisely meet our requirements; however, its capacity to generate rules in the 'if-then' format allows us to easily generate logical rules. Therefore, we adapt it to generete descriptive rules rather than classification rules.

2   The CN2 Algorithm

In the field of machine learning, induction algorithms are used to automatically generate rules that can be used to classify new data. One such algorithm is the CN2 algorithm, developed by Clark and Niblett in 1989 [2]. The algorithm is designed to handle noisy data (data containing uncertainties, errors, or inconsistencies), making it well-suited for real-world domains. It combines the efficiency and noise-handling capabilities of the ID3 algorithm [12] with the if-then rule form and flexible search strategy of the AQ family [10]. The reasons behind our choice of this algorithm primarily stem from its non-overcomplicated nature, allowing for adaptability through modifications tailored to our needs while preserving its simplicity. This ensures computational efficiency, making it viable even for large datasets. During rule construction, the CN2 algorithm makes use of a user-defined heuristic function to estimate the level of noise within the data. Typically, this involves a threshold set by the user based on an evaluation function determining the quality of a rule. When the best rule, evaluated by the evaluation function, falls below a certain goodness threshold, the algorithm terminates the search process. This approach enables the algorithm to conclude, generating rules that may not perfectly classify all training examples but exhibit good generalization performance when applied to new data. Upon application of the CN2 Algorithm, a set of classification rules is generated. The algorithm systematically identifies the optimal complex (i.e., the most favorable rule) through iterative processes. This involves the selection of the best rule, determination of instances adhering to the rule's antecedent (i.e., satisfying the rule), and assignment of the rule to the class satisfied by the highest number of examples, designating it as the consequent. Specifically, in reference to CN2, the algorithm invokes an auxiliary function to acquire the best rule and subsequently removes the examples covered by that rule from the set of examples. Following this, it appends the best rule to the set of rules and repeats the process until either the best rule is nil or the set of examples becomes empty.

This algorithm deals with only discrete domains, but the obvious solution to this limitation is given as a preprocessing phase consisting in the creation of intervals to discretize the input data. Furthermore, the definition of rule evaluation criteria, such as the evaluation function and the threshold, is left to the user, thus allowing freedom in choosing the right function. In the next section, we will explain how we adapt the algorithm to our goal, which is to *describe* a dataset rather than *classify* it.

3   Descriptive Rule Extraction Using the Modified CN2 Algorithm

In order to extract typicality inclusions with probabilities in the logic $\mathbf{T}^{CL}$ we have to modify the original CN2 algorithm that, as said, was oriented towards classification. The difference between rules useful for classification and useful rules useful for description can be identified as follow. Consider the zoo dataset (discussed in 3). If we want to find a rule that classifies its elements in the most effective way, we can consider the single rule $Mammal \sqsubseteq Milk$, that allow to exhaustively classify all the mammals. However it might be interesting to derive other common properties of mammals, as for instance $Mammal \sqsubseteq \neg Poisonous$. These other rules would not be extracted by the CN2 algorithm that would stop at the first rule sufficient for its purpose of classification.

In the preceding section, we introduced the first significant divergence in our approach: rather than employing the conventional CN2 algorithm, we are in the process of developing a modified version wherein a target class is deliberately selected, as it can be seen in Algorithm 1, line 1 and 6. This choice stems from our primary motivation, which is the necessity for descriptive rules. In this context, our objective is to generate a greater number of rules related to a specific class, rather than striving for a minimal set of classification rules. It is important to note that by selecting a "target class", we effectively reduce the problem to a two-class classification scenario, distinguishing between the chosen target class and all other classes. Given that computational efficiency is not a primary objective in our goal, it is worth noting that the CN2 algorithm can be executed without incurring significant extra work. Consequently, we have adopted a novel

---

Algorithm 1. Modified CN2 Algorithm

1: **procedure** MODIFIEDCN2(*all_examples*,*verbose*,*target_class*)
2:      *examples* ← Copy of *all_examples*
3:      *conflicts* ← Empty set
4:      *rule_list* ← Empty List
5:      **while** the number of rules in *rule_list* < maximum number of rules **do**
6:          *all_rules* ← Generate all rules for *target_class* using *examples* and *conflicts*
7:          *best_rule* ← FINDBESTRULE(*all_rules*, *conflicts*)
8:          **if** there is no *best_rule* **then**
9:              Break the loop
10:         **else**
11:             Add *best_rule* to *rule_list*
12:             **if** verbose is 'total' **then**
13:                 Update *conflicts* with condition values from *rule_list*
14:             **else**
15:                 Remove examples covered by *best_rule* from the *examples* list
16:             **end if**
17:             **if** there are no examples belonging to the target class in *examples* **then**
18:                 **if** verbose is 'no' **then**
19:                     Break the loop
20:                 **else**
21:                     *examples* ← *all_examples*
22:                     Update *conflicts* with condition values from *rule_list*
23:                 **end if**
24:             **end if**
25:         **end if**
26:     **end while**
27:     Reset the generated rules to None
28:     **return** *rule_list*
29: **end procedure**

Algorithm 2. FindBestRule Function

1: **function** FINDBESTRULE(rules, examples)
2:      best_rule ← nil
3:      **for** rule in rules **do**
4:          significance ← Calculate significance of the rule using examples
5:          **if** best_rule is **nil or** significance > best_rule.significance **then**
6:              best_rule ← rule
7:          **end if**
8:      **end for**
9:      **return** best_rule
10: **end function**

---

strategy wherein we rerun the algorithm separately for each class within the dataset, with each class set as the target. This approach allows us to comprehensively describe and extract rules for all the classes present in the dataset, even if it necessitates multiple executions of the modified CN2 algorithm.

Moving on to the second distinction in our approach, we find it in the organization of rules, which differs notably from the conventional CN2 version. In the standard CN2 algorithm, rules are referred to as 'ordered' because they must be applied in the specific order they have been generated, an essential requirement in the realm of classification. However, in our modified version, where our primary focus is generating descriptive rules, the concept of order becomes less crucial. In our approach, we prioritize the creation of descriptive rules, and as such, they do not inherently require a predefined order for application. This adaptability in rule application order provides us with greater flexibility and allows us to address the dataset's characteristics more effectively. To achieve this flexibility, we introduce a novel feature: the ability to choose whether to remove examples from the dataset after each rule is generated. This choice enables us to fine-tune our rule generation process, either retaining the dataset's original structure or gradually reducing it as we proceed, depending on our specific objectives. To handle this new feature we use the variable "*verbose*".

As we delve deeper into the distinctions of our approach, the third significant difference lies in the flexibility to determine the style of rules we prefer; whether they are longer, incorporating more features in 'and' conditions, or shorter and more concise. This feature empowers us to make a conscious choice regarding the nature of rules we generate, aligning them with our specific objectives. In our example, we will favor the creation of shorter rules, emphasizing simplicity and conciseness in our rule generation process. Lastly, in our approach, we offer the unique capability to specify the precise number of rules we want as output. This level of control allows us to tailor our rule generation process according to our exact requirements. We have the option to request an exact number of rules or choose to generate rules until they effectively cover all the examples in the dataset, ensuring that our approach can be customized to match the specific objectives of our analysis.

In Algorithms 1 and 2 we present the pseudocode of our adapted CN2 algorithm. In examining this algorithm, a key difference lies within the "*verbose*" input parameter, which plays a crucial role in the rule generation process. This parameter offers three distinct settings, each influencing the behavior of the algorithm: (i) *total*: Signifies a continuous search for new rules without removing examples already covered. Each new rule must differ in conditions from the previous ones (updating conflicts); (ii) *yes*: Denotes that, at each iteration, the algorithm removes examples covered by the rules from the dataset. If these examples run out, the algorithm reverts the dataset to its original state and updates the conflicts; (iii) *no*: Indicates the removal of covered examples without repopulating or updating conflicts. Iterations conclude once the dataset contains no elements of the target class.

Tuning Parameters. In discussing the adapted CN2 Algorithm, it's vital to explore the additional parameters: the **Max star size** parameter dictates the maximum length permitted for the generated rules. Alongside this, the **Greedy rules** parameter influences whether the algorithm favors longer or shorter rules, always in accordance with the specified *max star size*. Moreover, the **Max rules** parameter determines the maximum number of rules allowed to be generated. The **Verbose** parameter is particularly noteworthy, and already explained in the previous subsection. Lastly, the **Evalution**

Function stands as a pivotal parameter governing the rule-generation process. Various evaluation functions have been proposed over time for the CN2 algorithm, including F1 score, the Gini index for entropy calculation, accuracy, and precision. In our implementation, we incorporated these functions and introduced additional ones, specifically focusing on Recall, and AUC-ROC. Other functions explored during our study include F1 score, Accuracy, Specificity, WRACC, Gini Index, and Information Gain, each offering unique insights into rule performance and dataset characteristics.

Evaluation Functions Example. Evaluation functions are pivotal in determining the performance of the algorithm. The choice among these functions can prioritize example coverage, rule precision, or balance between the two. Moving to the example itself, with fixed parameters: *max_star_size* = 2, *max_rules* = 2, *Intervals* = cut_3, *class* = virginica. At first we use the *accuracy* function to choose the best rules, and what we obtain is:

- **Rule 1**: IF petal width (cm) == (1.6, 2.5] THEN class == virginica : 0.94
- **Rule 2**: IF petal length (cm) == (4.9, 6.9] AND petal width (cm) == (0.87, 1.6] THEN class == virginica : 0.98

Now, with the *precision* function, we obtain the following rules:

- **Rule 1**: IF sepal width (cm) == (1.999, 2.9] AND petal width (cm) == (1.6, 2.5] THEN class == virginica : 0.99
- **Rule 2**: IF sepal width (cm) == (3.2, 4.4] AND petal length (cm) == (4.9, 6.9] THEN class == virginica : 0.99

The corresponding confusion matrices for these rules applied to the test set are as in Table 1. The poor performance of the precision evaluation function compared to perfect accuracy suggests that it fails to generalize sufficiently, covering only a few examples from the test set. Precision prioritizes avoiding false positives, while recall focuses on capturing all true examples. Therefore, experimenting with different evaluation functions is advisable, as results can vary significantly based on this parameter. Ultimately, the choice of evaluation functions should align with the intended objective of the rules.

Table 1. Confusion matrices of the rules for the "virginica" class relative to the test set.

	Real True	Real False
Predicted True	13	0
Predicted False	0	32

(a) Confusion Matrix using accuracy

	Real True	Real False
Predicted True	7	0
Predicted False	6	32

(b) Confusion matrix using precision

Ontology Creation and Rules Annotation. Once rules are extracted, assigning a truth value becomes essential to ensure compatibility with the logic T^{CL}. While evaluation functions might seem straightforward for this task, using them directly contradicts the logic's essence. Instead, the truth value should reflect how frequently the rule holds true, essentially aligning with recall (or frequency). However, relying solely on recall for evaluation might lead to selecting technically accurate but trivial rules, like $Mammal \sqsubseteq Alive$, lacking significant relevance.

To address this challenge, we differentiate between the evaluation function and the truth value determination. While any implemented evaluation function guides rule selection, truth values exclusively rely on recall.

Example. Now we can see a full working example on the zoo dataset [3]. This dataset is a widely recognized dataset utilized for machine learning and pattern recognition tasks. It was introduced by Richard Forsyth in 1990 and comprises a collection of attributes describing various animals. Containing a total of 101 instances and 18 attributes, the Zoo dataset encompasses valuable information regarding the characteristics and traits of diverse animals, including their habitats, number of legs, type of animal, and more. Each instance in the dataset represents a different animal, and the attributes are primarily binary and nominal. Therefore, the simplicity of this dataset makes it perfect for better understanding the functioning of the presented algorithm.

Suppose we want to extract the typical properties of mammals. The extracted rules will have the form $0.9 :: \mathbf{T}(Mammal) \sqsubseteq Attribute_1 \sqcap \cdots \sqcap Attribute_n$ which can be translated to words as "The typical mammal has $Attribute_1$ and ... $Attribute_n$ with frequency 0.9". The procedure will go through the following steps:

1. First, we choose the maximum length of the resulting rules, i.e. the number of conditions (or attributes) in logical conjunction. We choose to search for rules of length equal to one, in order to have logical assertions about individual attributes.
2. As the maximum number of rules, in order to avoid trivial or forced rules, we choose to create 10 for each target class, given that there are a total of 18 attributes.
3. We also opted for AUC-ROC as evaluation function to achieve a favorable trade-off between rules that are informative (hence, non-trivial) and rules that capture a substantial number of instances from the target class.
4. Finally, we choose to use verbose output as "total" in order to have the exact number of rules required, and at the same time without removing the examples covered by the rules. This will give us rules that are more probable on the entire set, as they will not be biased by the examples that have already been covered by other rules.

In practice, each rule will try to cover the same examples as best as possible, according to the evaluation function. This means that each rule will try to be as specific as possible, while still being general enough to cover a significant number of examples. In Table 2, we observe the contrast between the rules obtained using our algorithm with the aforementioned settings and the rule derived from a conventional CN2 algorithm. Specifically, it is noteworthy that by employing our algorithm, with the explicit selection of a target class, we generate a pool of rules that effectively characterizes the chosen class. In contrast, as already mentioned, if we had used the traditional algorithm, we would have found rules suitable for classification but lacks the depth that our approach provides. This distinction arises when the class is easily distinguishable from others with a single rule. Consequently, the traditional algorithm would not have found additional insights, which, in a descriptive task, still convey valuable information.

Table 2. Rules generated by the new algorithm with AUC-ROC as the evaluation function. In the lower section, the only rule generated by the classic CN2 algorithm.

Rule	Condition	T Value
1	$Mammal \sqsubseteq Milk$	-
2	$\mathbf{T}(Mammal) \sqsubseteq \neg Eggs$	0.976
3	$\mathbf{T}(Mammal) \sqsubseteq Hair$	0.951
4	$\mathbf{T}(Mammal) \sqsubseteq FourLegs$	0.756
5	$\mathbf{T}(Mammal) \sqsubseteq Toothed$	0.976

Rule	Condition	T Value
6	$\mathbf{T}(Mammal) \sqsubseteq Catsize$	0.78
7	$\mathbf{T}(Mammal) \sqsubseteq \neg Aquatic$	0.854
8	$Mammal \sqsubseteq Breathes$	-
9	$Mammal \sqsubseteq \neg Feathers$	-
10	$\mathbf{T}(Mammal) \sqsubseteq \neg Airborne$	0.951

Rule	Conclusion	Confidence
1	IF $Milk \rightarrow$ type=$Mammal$	-

4   Conclusion and Future Work

We have addressed the challenge of automatically extracting rules for typicality inclusions in the logic $\mathbf{T}^{CL}$. To tackle this issue, we have adapted the CN2 algorithm to extract descriptive rules, as opposed to classification rules, while ensuring adherence to the underlying logic. Finally, we have provided an illustrative example to enhance understanding of a potential application. This introductory paper aims to facilitate the comprehension of the new algorithm through a simple example. In the near future, our goal is to apply the algorithm to diverse and more complex datasets. Specifically, our initial focus will be on the domain of music. We plan to build upon and refine the concepts presented in [7], automating rule extraction from more intricate datasets, such as GTZAN. Subsequently, leveraging the extracted logical rules, we aim to combine musical concepts and create new subgenres.

References

1. Chiodino, E., Di Luccio, D., Lieto, A., Messina, A., Pozzato, G.L., Rubinetti, D.: A knowledge-based system for the dynamic generation and classification of novel contents in multimedia broadcasting. In: ECAI 2020. FAIA, vol. 325, pp. 680–687. IOS Press (2020)
2. Clark, P., Niblett, T.: The cn2 induction algorithm. Mach. Learn. **3**, 261–283 (1989)
3. Forsyth, R.: Zoo. UCI Machine Learning Repository (1990)
4. Lieto, A., Perrone, F., Pozzato, G.L., Chiodino, E.: Beyond subgoaling: a dynamic knowledge generation framework for creative problem solving in cognitive architectures. Cogn. Syst. Res. **58**, 305–316 (2019)
5. Lieto, A., Pozzato, G.L.: A description logic framework for commonsense conceptual combination integrating typicality, probabilities and cognitive heuristics. J. Exp. Theor. Artif. Intell. **32**(5), 769–804 (2020)
6. Lieto, A., Pozzato, G.L., Striani, M., Zoia, S., Damiano, R.: Degari 2.0: a diversity-seeking, explainable, and affective art recommender for social inclusion. Cogn. Syst. Res. **77**, 1–17 (2023)
7. Lieto, A., Pozzato, G.L., Valese, A., Zito, M.: A logic-based tool for dynamic generation and classification of musical content. In: Dovier, A., Montanari, A., Orlandini, A. (eds.) AIxIA 2023. LNCS, vol. 1396, pp. 313–326. Springer, Cham (2022). https://doi.org/10.1007/978-3-031-27181-6_22

8. Lieto, A., Pozzato, G.L., Zoia, S., Patti, V., Damiano, R.: A commonsense reasoning framework for explanatory emotion attribution, generation and re-classification. Knowl.-Based Syst. **227**, 107166 (2021)

9. Lieto, A., et al.: A sensemaking system for grouping and suggesting stories from multiple affective viewpoints in museums. Hum. Comput. Interact. **39**(1–2), 109–143 (2024)

10. Michalski, R.S., Mozetic, I., Hong, J., Lavrac, N.: The multi-purpose incremental learning system aq15 and its testing application to three medical domains. In: AAAI 1986, pp. 1041–1045

11. Osherson, D.N., Smith, E.E.: On the adequacy of prototype theory as a theory of concepts. Cognition **9**(1), 35–58 (1981)

12. Quinlan, J.R.: Induction of decision trees. Mach. Learn. **1**, 81–106 (1986)

13. Riguzzi, F., Bellodi, E., Lamma, E., Zese, R.: Probabilistic description logics under the distribution semantics. Semantic Web **6**(5), 477–501 (2015)

Neural Network and Natural Language Processing

LLMental: Classification of Mental Disorders with Large Language Models

Arkadiusz Nowacki$^{(\boxtimes)}$ [ID], Wojciech Sitek [ID], and Henryk Rybiński [ID]

Institute of Computer Science, Warsaw University of Technology, Warsaw, Poland
{arkadiusz.nowacki.stud,wojciech.sitek,henryk.rybinski}@pw.edu.pl

Abstract. The increasing number of mental disorders is a severe problem in the modern world and can even lead to suicide if left untreated. In the age of digitalization, we move part of our lives to social media, where we share both the good and the bad moments. This allows for the early detection of mental disorders (such as depression, excessive stress, or social phobia) of which the user may even be unaware. We propose to modify large language models, such as PHI-2, Mistral, Flan-T5, or LLaMA 2, to classify mental disorders and to add appropriate layers. This gives a better prediction performance than zero-shot/few-shot for LLMs and classification by BERT-based models. Using such an architecture makes it possible to return a label rather than text, thus allowing the output of the LLM model to be freely modified.

Keywords: mental disorders classification · mental health analysis · large language models · social media

1 Introduction

Mental disorders such as depression, suicidal thoughts, or social phobia are civilization diseases and constitute a global problem. According to the World Health Organization's Mental Health Action Plan for 2013–2020, 25% of the world's population suffers from mental disorders. Moreover, statistics show that three-quarters of people with severe mental disorders do not receive therapeutic intervention, worsening the problem (see, e.g., [21]). During times of societal stress, such as the COVID-19 pandemic, access to mental health care becomes even more critical. It is essential to address the disparity in access to treatment, especially for people contending with mental health challenges who may face obstacles in accessing mental health services.

Social media platforms have recently become popular channels for people to share information, express opinions, articulate thoughts, and convey emotions. However, they may also reveal symptoms of mental disorders. Research suggests a correlation between increased social media use and depressive symptoms in young adults [10]. Through this, the social media platforms themselves have mechanisms that detect content suggesting that the author of a post may be in

A. Appice et al. (Eds.): ISMIS 2024, LNAI 14670, pp. 35–44, 2024.
https://doi.org/10.1007/978-3-031-62700-2_4

a life-threatening situation and display messages suggesting that specialists be contacted.

Artificial intelligence is widely used to detect mental disorders. The most successful results have been achieved using transformers, especially Bidirectional Encoder Representations from Transformers (BERT) [8]. More recently, research attention has been paid to large language models (LLMs), most of which are based on the transformer architecture while being an extension of it. LLMs have demonstrated a unique ability to interpret textual context. The first attempts were made to apply LLMs to the detection of mental disorders, although the focus was primarily on explaining the cause of detection rather than on detection itself. As a result, detection performance was inferior compared to BERT-based solutions [22].

This paper presents the capabilities of LLMs for classifying mental disorders. Particular emphasis was placed on the possibilities of LLM classification. For this purpose, instead of using a zero-shot/few-shot approach, we have decided to modify the network architecture.

The zero-shot/few-shot method allows one to use the original knowledge of an LLM, which in many situations may be sufficient for tasks such as sentence translation or simple inference. Just give a command to the model (zero-shot) or give a command and a few examples of how to do it (few-shot) and get the answer [1]. However, for more complex problems, the zero-shot/few-shot methods may need to be more extensive due to gaps in model knowledge, or the instructions may need to be revised.

By adding appropriate layers and fine-tuning, we were able to create classifiers. We tested the capabilities of LLM on publicly available datasets and compared the results with existing solutions. Our tests have shown that the modified LLM architecture gives better results than LLM using instruction-tuning methods and methods based on BERT models.

LLM models that detect mental disorders could have applications in social networks to detect the first signs of mental disorders in users and inform them or emergency services. In addition, it could support therapies for mental disorders, with applications in chatbots where it could perform a preliminary classification of disorders in a patient.

2   Related Work

The detection of mental disorders has become a prominent topic in the field of deep learning due to its significant impact on people's lives. One of the first research directions was focused on using a natural language processing (NLP) based approach for text classification in medicine [2, 13].

The use of NLP in medicine has been the subject of many studies. One of the first was to use text processing to create computerized clinical decision support (CDS) [2]. It was aimed at supporting medical workers in decision-making. The scope of the activity was to be based on the possibility of analyzing medical knowledge or the patient's medical documentation. Such a system would allow medical professionals to ask questions and draw conclusions based on available data, translate documentation for patients, summarize treatments, or automatically select disease codes. The system proposed utilizing a dictionary based on the National Cancer Institute (NCI) Dictionary and the Named-Entity Recognition (NER) method to extract information from available documents. While NER performed well in extracting keywords, it cannot interpret the overall meaning of documents or the causal connections between subsequent paragraphs [2,9]. This is because the NER-based approach has problems with proper context identification, making it challenging to resolve word ambiguity.

In [13], a part of the research was to investigate the possibilities of using the NLP approach for analyzing medical texts in order to detect mental disorders, with a particular emphasis on identifying suicidal thoughts. It was one of the earliest attempts to identify mental disorders in texts, and it consisted in providing an algorithm for classifying suicide notes. The algorithm was designed to determine whether a suicide victim wrote a given note or not. The algorithm used various approaches, including decision trees, classification rules, and Bayesian classifiers, to search for keywords in the text and detect mood-defining words. The authors achieved comparable or even superior results to those of human annotators. However, the main challenge remained understanding the context. Some emotions were not presented directly but informed about the intentions of the note's author.

The solution to the problem of context analysis was to create an attention mechanism, thus presenting a transformer architecture. The attention mechanism mimics cognitive attention, allowing it to focus on selected text parts carrying a significant message. The attention mechanism in transformers works by calculating weights for each sequence element based on their relationship, allowing the model to focus on important parts of the data [19].

BERT was one of the first and most significant transformer architectures [3]. It was a breakthrough in understanding the context of words in sentences. Unlike previous models, BERT is bidirectional during analysis, allowing it to better interpret the meaning of words. A key element of BERT is massive unsupervised learning on large datasets, contributing to better word representations. In addition, BERT introduces a Masked Language Model (MLM) layer that allows the model to predict missing parts of a sequence, helping to learn more representative and context-sensitive word embeddings.

Due to the features of the BERT model, it has been successfully applied to text classification problems, modifying its structure accordingly. Classification modifications include adding classification layers at the network's end, which receive classify tokens ([CLS]). For example, these layers can be adapted to the specific classification task by adding fully connected layers and activation functions. The model is then trained in the classification stage, and these additional layers are adjusted in the fine-tuning process [16].

In [8], MentalBERT was successfully used to classify mental disorders, mainly because of its capability to comprehend the context of sentences, which is frequently not immediately apparent in the case of disorders. It has been possible to achieve better results in detecting disorders compared to LSTM or CNN. However, in more challenging cases where disorders are less apparent or when a specific category of mental disorder needs to be identified, more than this approach may be required.

One of the offshoots of NLP that is becoming increasingly distinct is the emergence of LLMs. These models represent the next evolution after transformers, with more parameters and training on larger datasets. As a result, they can better understand a sentence's context and generate textual responses to given inputs. They are now being used as automated chatbots, allowing users to translate sentences, search for answers to a question, including solving math tasks, or carry on a conversation like a human being [6,12,17]. Due to their versatility, LLMs have been used to detect mental disorders. Instruction tuning was used to feed a context into the analysis and obtain a prediction [20]. The LLaMA model was chosen for this purpose [22]. MentaLLaMA allowed the labels and explanations for the predictions to be obtained. However, this method could have improved the quality of the prediction. The metrics often showed lower values compared to BERT-based solutions.

3   Utilized Datasets and Applied Methods

Instruction tuning was the most successful method to classify mental disorders without modifying the LLM network [22]. In this paper, we created experimentally classifiers by extending LLM architectures. We have chosen four models for this purpose, namely PHI-2, Mistral, LLaMA, and Flan-T5. They were trained with the Low-Rank Adaptation (LoRA) tool [7]. The GPT-3.5 model (base gpt-3.5-turbo-1106) was also fine-tuned for comparison purposes. The model was trained to return only the necessary labels without additional text. No network elements were modified due to restricted access to the model. The performed experiments are presented below.

3.1   Datasets

Four open-source datasets were selected for the experiments based on their reproducibility in other scientific works, allowing the comparison of the developed methods with existing solutions. The datasets were already divided into training and test sets, so they were not further divided.

The **Dreaddit** [18] dataset is a valuable resource for binary classification in stress detection. It includes 3553 posts from Reddit, each labeled with a Yes/No indicator for stress detection. This makes it easy to compare models to assess the effectiveness of different approaches to identifying stress-indicating content. The binary labeling of the dataset simplifies evaluation, making it a standardized benchmark for training and fine-tuning models across different types of mental disorders.

Another dataset, **SAD** [11], is related to stress detection through multiclass classification. It consists of 6850 SMS-like messages labeled according to the type of stress they contain. The types of stress are assigned to nine categories: *Work, Other, Health, Fatigue, or Psychical Pain, School, Family Issues, Emotional Turmoil, Financial Problem, Social Relationships, Everyday Decision Making.*

Such as SAD, **CAMS** [4] is a publicly available multi-class dataset containing Reddit posts. It includes six labels that indicate the reason for the mental disorder: *No reason, Bias or abuse, Jobs/careers, Medication, Relationship, and Alienation.* The collection consists of 5051 records.

The **IRF** [5] dataset was the most recently used. It contains 3029 data on annotated risk factors for mental disorders related to interpersonal relationships. The dataset is based on tasks that involve detecting loneliness, which may be subjective, and identifying interpersonal risk factors.

3.2 PHI-2, LLaMA 2 and Mistral

Most LLMs are designed as causal language models, meaning they accept text as input and return text as output. Using them in classification or regression problems can be problematic. The model's number of outputs cannot be easily determined. The primary method used is instruction-tuning, which involves using appropriate prompts and output data to prompt the model to return the desired text-based classes [20].

In the case of open-source LLMs, it is possible to modify the structure of the models. Similar modifications to those used for the transformer architectures in BERT can be applied. In the case of BERT, predictions are based on a token with additional dropout and linear layers. This allows for flexibility in setting the model output [16].

The PHI-2, Mistral 7B, and LLaMA 2 7B models can use a twin method. As part of the experiment on the classification of mental disorders, a prediction method used the last token and a linear layer (Fig. 1). In this way, the neural network output can be freely modeled and thus applied to the classification problem.

Fig. 1. LLM network scheme (for LLaMA 7B, PHI-2, and Mistral 7B) for the classification problem.

Due to the size of LLMs, several additional solutions were used. One is using LoRA [7] for fine-tuning the models. This made it possible to train only a certain part of the model to prepare it for the classification problem, leaving much of its original *knowledge.*

The datasets may be unbalanced, which, given the small amount of data available, may be problematic in the LLM fine-tuning process. For this reason, we modified the Cross-Entropy loss function by adding weights for given classes. Weighted Cross-Entropy looks like this:

$$L_{wce} = -\frac{1}{N} \sum_{i=1}^{N} \sum_{j=1}^{C} w_j y_{ij} \log (p_{ij}) \tag{1}$$

where N is the number of samples, C is the number of classes, y_{ij} is the ground truth label of sample i for class j, p_{ij} is the probability of sample i for class j, and w_j is the weight assigned to class j.

3.3   Flan-T5

Unlike the other LLMs, the Text-To-Text Transfer Transformer, also known as Flan-T5 [15], required a different modification of the network layers. In this case, instead of a single linear layer based on the last token, a series of three layers was used: a linear layer, a dropout layer, and a target linear layer, thanks to which the network output can be modified (Fig. 2). Compared to other models, this model uses a more extensive architecture because the classification process does not involve the last token, unlike PHI-2.

The T5 model uses a text-to-text framework where every task, including classification, is cast as feeding the model text as input and training it to generate some target text. This means that the tasks are not technically limited to using the last token for classification, but instead, the framework is based on processing input text and producing output text. For this reason, to standardize the data, a linear layer is used at the beginning, which usually has the same number of features at the output as the input. Next, a dropout layer is added to prevent overfitting and improve model generalization. Finally, a final linear layer is added to reduce the features to the number of labels.

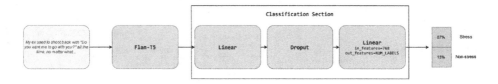

Fig. 2. Flan-T5 scheme for the classification problem.

Transformer-based language models, such as GPT-2, use the last token for sequence classification tasks. Each input token is processed through multiple layers of self-attention and feed-forward networks, generating a vector of embeddings that capture the contextual information of each token within the sequence. The last token's embedding is often considered the representation of the entire sequence, especially in models like GPT-2, which process sequences left-to-right and accumulate contextual information progressively [14]. This representation can be passed through additional layers to produce a probability distribution over possible classes, with the class with the highest probability being the model's prediction for the given sequence. The last token's unique ability to capture the complete context of the input sequence makes it useful for sequence-level predictions like classification. However, not all transformer models use this approach. Some models, such as BERT and its variants, use a [CLS] prepended to the sequence, which is then used for classification tasks [16].

3.4 GPT-3.5

Instruction tuning was used for classification, where it was impossible to modify the network structure. An example of such a model is OpenAI's solution, namely GPT-3.5. For this company's solutions, only an API is provided, using which and a certain coaching data structure, the available models can be fine-tuned. In the case of GPT-3.5, three elements comprise a single dataset: system, user, and assistant [12].

The system is a description of what the model is supposed to do. In this case, a single sentence was selected for the entire dataset, using which fine-tuning was performed. User is the data for the instructions contained in the system. Raw posts from social networks, which the model is supposed to classify, were used as a context for the user. Assistant contains what the model is supposed to return based on the previous two elements. In this case, the model returns classifications. The model returns *positive* or *negative* for binary classification and returns the class for multi-class classifications. An example of such a dataset is presented as follows:

```
{"messages": [
    {"role": "system", "content": "You assess whether the person
    writing  a particular post on a social network may have
    mental problems. Give the category of this problem."},
    {"role": "user", "content": "2 of my foster kittens died"},
    {"role": "assistant", "content": "Emotional Turmoil"}
]}
```

The experiment included a method to force specific classes for the dataset as the model's response. However, this was not the main focus of the experiment since it did not involve the appropriate layers for real classification. Therefore, it was treated as an add-on.

4 Results

For the previously presented datasets, fine-tuning was performed using the methods described. Weighted-averaged F1 score metrics were collected for each dataset to compare the prediction quality of the obtained models with existing models (Table 1). Although GPT-3.5 had a modified structure, it was placed together with classification models. This decision was influenced by the fact that it was possible to fine-tune the model to return only labels without additional text.

Table 1. The obtained results of the weighted-averaged F1 score parameter for the tested solutions: ZS—zero-shot, FS—few-shot, IT—instruction-tuning, C—classification, ITC—instruction-tuning for classification. The bold text represents the best performance. Data for the *Existing models* come from [22].

Model	Method	dreaddit	SAD	CAMS	IRF
Existing models					
GPT-3.5	ZS	71.79	54.05	33.85	41.33
GPT-3.5	FS	75.38	63.56	44.46	43.31
GPT-4	FS	78.18	55.68	42.37	51.75
MentaLLaMA-7B	IT	71.65	49.93	32.52	67.53
MentaLLaMA-chat-7B	IT	62.2	62.18	44.8	72.88
MentaLLaMA-chat-13B	IT	75.79	63.62	5.52	76.49
BERT-base	C	78.26	62.72	34.92	72.30
MentalBERT	C	80.04	67.34	39.73	76.73
RoBERTa-base	C	80.56	67.53	36.54	71.35
MentalRoBERTa	C	81.76	68.44	47.62	-
LLMental—fine-tuned models					
Flan-T5 (base)	C	81.34	75.92	60.35	**82.74**
GPT-3.5	ITC	83.79	**76.31**	55.95	79.54
LLaMA 2 7B	C	83.15	74.33	58.91	79.86
Mistral	C	84.54	74.93	57.07	80.84
PHI-2 2.7B	C	**84.80**	75.59	**62.37**	81.36

The modified LLM achieves the best results for all available datasets compared to all available methods. LLaMA 2 7B performs better than its pretrained counterpart, MentaLLaMA-7B. However, it should be emphasized that MentaLLaMA-7B is based on a previous-generation model and has a mechanism to explain why the model made such a classification in addition to class prediction [22].

A significant improvement can be seen in the case of the CAMS dataset. The text to be classified is much longer than in other datasets, which results in using a larger window to analyze the entire text. LLMs have this dependency in that they can better analyze a larger message context than solutions based on BERT, which may be a direct reason for the better value of the F1 score metric.

Mistral 7B and LLaMA 2 7B, with the same number of parameters, achieved similar F1 score metrics values for all sets, with an advantage for Mistral, which was better in three sets. Flan-T5 was the only one to achieve a lower result than the BERT-based solution for the Dreaddit set, but it also achieved the highest result for the IRF set. However, the most promising results were achieved by PHI-2, the second-smallest model in the experiment (Flan-T5 base has 248 million parameters, PHI-2 2.7 billion). PHI-2 achieved the highest F1 score result for two of the four datasets.

The GPT-3.5 model achieved the worst results for CAMS and IRF in the LLM category, but these results were higher than those of the previously created solutions. This is partly because the goal is not to classify based on structural changes but to force the return of classes through instruction tuning.

5   Conclusion

This paper presents the effectiveness of LLMs in detecting mental disorders. LLM models have achieved higher performance in the F1 score metric than previously used solutions. This provides an opportunity to develop these models further to improve detection.

It should be emphasized that after the structural modification, the models were not additionally tuned to the problem of mental disorder detection (as in the case of pre-trained BERT models) but were immediately tuned to the target datasets. This makes it possible to create pre-trained models that can achieve even better results than those presented here.

The PHI-2 model showed promising results, achieving the best F1 score parameter values for two of the four datasets, with values for the remaining datasets similar to the best. An additional advantage is the model's smaller size compared to the others. PHI-2 has 2.7 billion parameters, Mistral has 7 billion, and LLaMA 2 7B has 7 billion. This means that excellent results can be achieved with much less computing power.

References

1. Brown, T.B., et al.: Language models are few-shot learners (2020)
2. Demner-Fushman, D., Chapman, W.W., McDonald, C.J.: What can natural language processing do for clinical decision support? J. Biomed. Inform. **42**(5), 760–772 (2009). https://doi.org/10.1016/j.jbi.2009.08.007, https://www.sciencedirect.com/science/article/pii/S1532046409001087. biomedical Natural Language Processing
3. Devlin, J., Chang, M.W., Lee, K., Toutanova, K.: Bert: pre-training of deep bidirectional transformers for language understanding (2019)
4. Garg, M., et al.: Cams: an annotated corpus for causal analysis of mental health issues in social media posts. arXiv preprint arXiv:2207.04674 (2022)
5. Garg, M., Shahbandegan, A., Chadha, A., Mago, V.: An annotated dataset for explainable interpersonal risk factors of mental disturbance in social media posts (2023)
6. Gunasekar, S., et al.: Textbooks are all you need (2023)

7. Hu, E.J.,et al.: Lora: low-rank adaptation of large language models. CoRR **abs/2106.09685** (2021). https://arxiv.org/abs/2106.09685

8. Ji, S., Zhang, T., Ansari, L., Fu, J., Tiwari, P., Cambria, E.: Mentalbert: publicly available pretrained language models for mental healthcare (2021)

9. Li, J., Sun, A., Han, J., Li, C.: A survey on deep learning for named entity recognition. IEEE Trans. Knowl. Data Eng. **34**(1), 50–70 (2022). https://doi.org/10.1109/tkde.2020.2981314

10. Lin, L.Y., et al.: Association between social media use and depression among us young adults. Depress. Anxiety **33**(4), 323–331 (2016)

11. Mauriello, M.L., Lincoln, T., Hon, G., Simon, D., Jurafsky, D., Paredes, P.: Sad: a stress annotated dataset for recognizing everyday stressors in sms-like conversational systems. In: Extended Abstracts of the 2021 CHI Conference on Human Factors in Computing Systems. CHI EA 2021, Association for Computing Machinery, New York, NY, USA (2021). https://doi.org/10.1145/3411763.3451799

12. OpenAI: Gpt-4 technical report (2023)

13. Pestian, J.P., Nasrallah, H.A., Matykiewicz, P., Bennett, A.J., Leenaars, A.A.: Suicide note classification using natural language processing: A content analysis. Biomed. Inform. Insights **3** (2010). https://api.semanticscholar.org/CorpusID:7779743

14. Radford, A., Wu, J., Child, R., Luan, D., Amodei, D., Sutskever, I., et al.: Language models are unsupervised multitask learners. OpenAI blog **1**(8), 9 (2019)

15. Raffel, C., et al.: Exploring the limits of transfer learning with a unified text-to-text transformer (2023)

16. Sun, C., Qiu, X., Xu, Y., Huang, X.: How to fine-tune bert for text classification? (2020)

17. Touvron, H., et al.: Llama: open and efficient foundation language models (2023)

18. Turcan, E., McKeown, K.: Dreaddit: a reddit dataset for stress analysis in social media. arXiv preprint arXiv:1911.00133 (2019)

19. Vaswani, A., et al.: Attention is all you need (2023)

20. Wei, J., et al.: Finetuned language models are zero-shot learners (2022)

21. Windfuhr, K., Kapur, N.: Suicide and mental illness: a clinical review of 15 years findings from the UK national confidential inquiry into suicide. Br. Med. Bull. **100**(1), 101–121 (2011)

22. Yang, K., Zhang, T., Kuang, Z., Xie, Q., Ananiadou, S., Huang, J.: Mentallama: interpretable mental health analysis on social media with large language models (2023)

CSEPrompts: A Benchmark of Introductory Computer Science Prompts

Nishat Raihan[1(✉)], Dhiman Goswami[1], Sadiya Sayara Chowdhury Puspo[1],
Christian Newman[2], Tharindu Ranasinghe[3], and Marcos Zampieri[1]

[1] George Mason University, Fairfax, VA, USA
mraihan2@gmu.edu
[2] Rochester Institute of Technology, Rochester, NY, USA
[3] Aston University, Birmingham, UK

Abstract. Recent advances in AI, machine learning, and NLP have led to the development of a new generation of Large Language Models (LLMs) that are trained on massive amounts of data and often have trillions of parameters. Commercial applications (e.g., ChatGPT) have made this technology available to the general public, thus making it possible to use LLMs to produce high-quality texts for academic and professional purposes. Schools and universities are aware of the increasing use of AI-generated content by students and they have been researching the impact of this new technology and its potential misuse. Educational programs in Computer Science (CS) and related fields are particularly affected because LLMs are also capable of generating programming code in various programming languages. To help understand the potential impact of publicly available LLMs in CS education, we introduce CSEPrompts (https://github.com/mraihan-gmu/CSEPrompts), a framework with hundreds of programming exercise prompts and multiple-choice questions retrieved from introductory CS and programming courses. We also provide experimental results on CSEPrompts to evaluate the performance of several LLMs with respect to generating Python code and answering basic computer science and programming questions.

Keywords: Benchmark Dataset · Code LLM · Prompting

1 Introduction

In the last decade, NLP models have evolved from n-gram and word embedding models (e.g., word2vec [19], GloVe [20]) to advanced context-aware models like ELMo [21] and BERT [22], significantly improving performance in various tasks [14]. Recent developments in Large Language Models (LLMs), notably GPT-3 and GPT-4, have further revolutionized NLP by enabling human-like text generation and application in fields such as healthcare [8] and education [10], marking a new era in generative AI.

© The Author(s), under exclusive license to Springer Nature Switzerland AG 2024
A. Appice et al. (Eds.): ISMIS 2024, LNAI 14670, pp. 45–54, 2024.
https://doi.org/10.1007/978-3-031-62700-2_5

Several recent studies have addressed the impact of GPT models on education [16–18]. While these models bring several opportunities in educational technology, such as enhanced writing assistants, intelligent tutoring systems, and automatic assessment tools, concerns arise from the misuse of technology, particularly in coding tasks. The study conducted by Savelka et al. [27] shows that while GPT scores may not meet course completion criteria, the model shows notable capabilities, including correcting solutions based on auto-grader feedback. Students may take advantage of this technology to generate complete essays and programming assignments obtaining artificially high grades. Furthermore, it has been shown that ChatGPT excels in debugging, bug prediction, and explanation but has reasoning and integration limitations [29]. Recent studies investigate the use of GPT models on assessment in domains such law [9], mathematics and computer science [15], and medicine [11] evidencing the high quality of the models' output which would "pass the bar exam" [9].

In this paper, we investigate the impact of LLMs on CS education and assessment by carrying out an evaluation of the performance in introductory CS and programming course assignments. While all aforementioned studies are restricted to GPT [9,11,15], we present a more comprehensive evaluation of models that goes beyond GPT. With the goal of enabling reproducibility, we create CSE-Prompts, a framework comprising 219 programming prompts and 50 multiple-choice questions (MCQs) collected from coding websites and massive open online courses (MOOCs). We investigate the performance of eight models capable of generating both English text and programming code in Python. We address the following research questions:

- RQ1: How well do state-of-the-art LLMs perform on introductory CS assignments compared to existing Benchmarks?
- RQ2: Is there a significant difference in the performance of LLMs when completing assignments from coding websites compared to academic MOOCs?
- RQ3: Are state-of-the-art LLMs better at generating code or answering MCQs?
- RQ4: Are Code LLMs better at generating code and/or answering MCQs than raw LLMs?

2 Related Work

Prior to the revolution of generative pre-trained models, most coding tasks involved tasks such as code completion, code infilling, comment generation, and similar tasks that are often handled employing BERT-like [22] encoder-only models. Such models include CodeBERT [31], GraphCodeBERT [32] among others that are pre-trained on text-code pairs, often including ASTs (Abstract Syntax Trees) and CFGs (Control Flow Graphs) as well in the training corpus. However, encoder-only models are not primarily designed to be used as generative models and show subpar performance at Code Generation Tasks [33].

With the advent of generative models that are based on either encoder-decoder [34] or decoder-only [35] architecture, they start to show better code-generation capabilities - evidenced by the survey conducted by Zan et al. [28]. Hence, the need for a unified benchmark arises. Such datasets include HumanEval (introduced with GPT3.5 [1]) and MBPP [30] that contain a set of coding prompts, paired with human-generated solutions and three test cases for each task. Other similar datasets include CONCODE [36], DS-1000 [38] and the extensions of the two previously mentioned datasets as HumanEval+ [37] and MBPP+ [39]. Another similar task

However, these datasets do not focus on coding tasks that are often used in the educational domain - rather mostly common tasks that occur in software development or other aspects. The coding tasks in the education domain, focus more on the in-depth knowledge of the specific programming language that evaluates the core understanding of the syntax and semantics of the language, making them significantly different from the prompts that are included in the existing benchmarks. To bridge this gap, we introduce CSEPrompts, a framework with hundreds of programming exercise prompts and multiple-choice questions retrieved from introductory CS and programming courses. Each prompt is paired with five test cases, compared to three for most benchmarks.

3 CSEPrompts

We introduce CSEPrompts, a novel evaluation framework consisting of coding prompts from coding websites and academic MOOCs (Table 1). CSEprompts features a total of 269 exercise prompts as shown in Table 2. We include links to the questions and answers where the interested reader can download the data.

Table 1. List of Coding Websites & MOOCs

Name	Link
CodingBat	https://codingbat.com/python
Learn Python	https://www.learnpython.org
Edabit	https://edabit.com/challenges/python3
Python Principles	https://pythonprinciples.com/challenges/
Hacker Rank	https://www.hackerrank.com/domains/python
Edx	https://www.edx.org
Coursera	https://www.coursera.org
CS50 (Harvard)	https://learning.edx.org/course/course-v1:HarvardX+CS50S+Scratch/home
PforE (UMich)	https://www.coursera.org/learn/python/home
CS1301xI (GT)	https://learning.edx.org/course/course-v1:GTx+CS1301xI+1T2023/home
CS1301xII (GT)	https://learning.edx.org/course/course-v1:GTx+CS1301xII+1T2023/home
CS1301xIII (GT)	https://learning.edx.org/course/course-v1:GTx+CS1301xIII+1T2023/home
CS1301xIV (GT)	https://learning.edx.org/course/course-v1:GTx+CS1301xIV+1T2023/home

Coding Websites. We choose five leading online resources for introductory Python learning, as detailed in Table 2. These include *CodingBat*, offering a variety of coding challenges, *LearnPython* with its interactive tutorials, *Edabit* for a wide range of programming problems, *Python Principles* emphasizing practical exercises, and *HackerRank*, known for its diverse coding challenges and competitions. Our selection criteria for these platforms center on their interactivity, diversity of challenges, and structured learning approaches. Interactive methods, as provided by *LearnPython* and *Python Principles*, enable users to engage actively in learning. Meanwhile, *Edabit* and *HackerRank* cater to a broad spectrum of skill levels. These platforms are bolstered by strong communities and offer instant feedback, crucial for effective learning in programming.

MOOCs. Our study includes programming assignments, tasks, and multiple-choice questions from six MOOCs offered by Harvard University, the University of Michigan, and the Georgia Institute of Technology. These courses are available on online platforms like edx and Coursera, focusing on introductory Python programming for both beginners and those with some programming background. Details of these courses are presented in Table 2. *Harvard's CS50*, on edx, introduces basic programming concepts. *Michigan's Programming for Everybody (PforE)*, found on both edx and Coursera, is designed for programming beginners. Georgia Tech's courses, *CS1301xI, CS1301xII, CS1301xIII*, and *CS1301xIV*, cover a range of Python topics from beginner to advanced levels. These courses include practical programming exercises, but we exclude tasks involving File I/O or Command-Prompt/Terminal due to the limitations of Language Learning Models (LLMs) in such interactions. Finally, along with the coding tasks, we also gather several MCQs each containing a set of 5 to 10 options containing the correct answer and multiple distractors. We gather a total of fifty such MCQs from four courses from GT, mentioned in Section - CS1301xI, CS1301xII, CS1301xIII and CS1301xIV. A brief description is provided in the Table 2.

Table 2. Summary of Coding Prompts from Various Sources

Coding Websites		MOOCs - Coding Prompts			MOOCs - MCQs		
Platform	Prompts	University	Course	Prompts	University	Course	Prompts
CodingBat	24	Harvard	CS50	29	GT	CS1301xI	20
LearnPython	16	UMich	PforE	7	GT	CS1301xII	8
Edabit	29	GT	CS1301xI	11	GT	CS1301xIII	6
Python Principles	26	GT	CS1301xII	20	GT	CS1301xIV	16
HackerRank	23	GT	CS1301xIII	17			
Total	**118**	**Total**		**101**	**Total**		**50**

Dataset Statistics. The prompts are mostly shorter in the coding sites comparatively. A few key statistics are presented in Table 3, about the prompts. We gather at least 5 test cases for each prompt for our experiment. In most cases, the test cases are taken from the platforms from where the prompts themselves are taken. In cases where we have less, we generate more test cases using pynguin [12], an open-source unit test case generator tool for Python. For the MCQs, we gather the correct answer(s) from the platforms themselves. We further gather responses generated by the LLMs for each prompt. The responses are cleaned manually and only the code snippets are kept. We label each code snippet based on how many test cases they passed.

Table 3. Statistics for Prompts

Metric	CodingSites	Academic	MCQ
Total Prompts	118	101	50
Max. No. of Tokens	101	372	221
Min. No. of Tokens	5	17	15
Mean No. of Tokens	28	158	106
Standard Deviation	16	72	51

Data Collection Strategy. Unlike HumanEval [1] or MBPP [30] benchmarks that are generated by human, specifically for the purpose of evaluating code generation capabilities of Large Language Models; we take a different approach. In order to obtain the academic nuances and intricacies of coding tasks, we collect them from real-world academic courses and coding websites. They are collected manually without any web scrapper or extractor. It is also ensured manually that no prompts and/or tasks are duplicated.

4   Experiments

LLMs There are several proprietary LLMs like GPT-3.5 [1], PaLM-2 [2] etc., and Open-Source LLMs like Llama-2 [3], Falcon [4], StableLM [6], Pythia [7] etc. that are used for a wide variety of tasks, including coding problems as well. The eight LLMs used for our experiments are briefly mentioned in Table 4. For the MCQs, the prompt has a minor modification as shown in Fig. 2.

Table 4. LLMs Used on CSEPrompts

LLM	Parameter	Model Type	Reference
GPT3.5	175B	Base	[1]
Llama2	7B	Base	[3]
Falcon	7B	Base	[4]
MPT	7B	Base	[5]
Code-Llama	7B	Fine-tuned	[35]
StarCoder	7B	Fine-tuned	[25]
WizardCoder	7B	Fine-tuned	[26]
Mistral	7B	Base	[24]

Code Generation. We first prepare our prompts and then test each model for all the tasks and questions from CSEPrompt. The prompts have a simple format, as shown in Fig. 1 and 2. Each model is prompted with the same prompt. The models generate code, but they also generate more texts, including pseudo-codes, explanations, etc. The responses are then cleaned manually to exclude everything other than the code. The codes are then tested for all the test cases using the pytest [13] framework, which facilitates the easy creation of small and readable unit tests for Python codes.

'You are a helpful AI assistant. You are given the following problem: '

Write a function named capital_indexes. The function takes a single parameter, which is a string. Your function should return a list of all the indexes in the string that have capital letters.

'Please write a Python code snippet to solve the problem. Thanks.'

Fig. 1. Sample Prompt for Coding Tasks.

'You are a helpful AI assistant. You are given a Multiple Choice Question. '

(False and True) or (False or True)
                 Is this statement resolved to True or False?

– True
– False
– Statement will not compile

'You need to pick one or multiple correct answers from the given ones. Thanks.'

Fig. 2. Sample Prompt for MCQs.

5 Results

For the Code Generation task, there are a few widely used benchmark datasets like HumanEval [1] and MBPP [30]. HumanEval has 164 coding tasks paired with three test cases each and MBPP compiles a larger datset with 974 tasks, also with 3 test cases each. Code LLMs often present their results on these datasets based on the *Pass@K* metric - which means how many attempts the model take to pass all the test cases. We compare CSEPrompts with these two benchmark datasets based on Pass@1. Since these are Code Generation tasks, we exclude the MCQ tasks from CSEPrompts during the comparison (shown in Fig. 3).

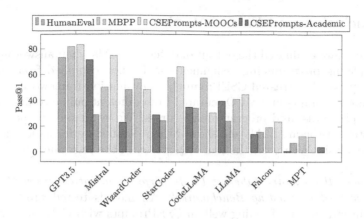

Fig. 3. Comparing CSEPrompts with HumanEval and MBPP based on **Pass@1**.

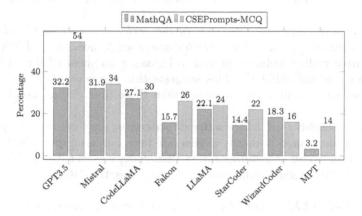

Fig. 4. Comparing CSEPrompts-MCQ with MathQA based on Zero Shot Prompting (in percentage).

For the MCQ task, however, there are no similar benchmarks to compare with. The most similar existing dataset is the MathQA-Python benchmark (introduced with MBPP [30]). This contains coding-related question-answer pairs in Python. However, no work explores the Code LLMs' expertise in the aspect of Multiple Choice Question Answering. In contrast, the Academic-MCQ subset of CSEPrompts is the first dataset Code-MCQ benchmark in the domain. We compare the results for several LLMs on it and MathQA-Python. As shown in Fig. 4, the MCQ task is comparatively easier for the models compared to QA tasks - most likely because more context and a set of possible answers guide the model to generate a better answer.

6   Conclusion and Future Work

In this paper, we evaluated the output of different LLMs when answering MCQs and completing programming assignments in Python compiled from introductory CS courses. We created CSEPrompts, an evaluation framework containing programming prompts and MCQs curated from different online coding websites, academic platforms, and programming courses. We evaluated the performance of eight state-of-the-art LLMs and showed their detailed performance along with an error analysis. We revisit the four RQs posed in the introduction.

RQ1: How well do state-of-the-art LLMs perform on introductory CS assignments compared to existing Benchmarks? All models tested produced high-quality output, thus performing well on CSEPrompts with GPT outperforming the other seven models. Compared to existing benchmarks, the LLMs perform better at *CSEPrompts [MOOCs]* and worse on *CSEPrompts [Academic]*.

RQ2: Is there a significant difference in the performance of LLMs when completing assignments from coding websites compared to academic MOOCs? The prompts from coding websites proved to be easier for most of the LLMs than those from academic MOOCs. This suggests that the prompts from academic MOOCs are more challenging/advanced than those from coding websites.

RQ3: Are state-of-the-art LLMs better at generating code or answering MCQs? As LLMs are developed primarily to generate text, our assumption is that LLMs would perform better at generation text than code, however, the LLMs we evaluated, are better at generating code than answering the MCQs.

RQ4: Are Code LLMs better at generating code and/or answering MCQs than raw LLMs? GPT3.5 being the bigger model among the ones we tested - outperformed others on all tasks - while being a general-purpose LLM. But for most cases, general-purpose LLMs perform better at MCQs and Code LLMs perform better on coding tasks.

In future work, we would like to carry out an even more comprehensive study by increasing the number of coding prompts and MCQs in the framework. Furthermore, we would like to incorporate other LLMs not included in this study.

References

1. OpenAI, GPT-4 Technical Report. ArXiv arxiv:2303.08774 (2023)
2. Anil, R., et al.: PaLM 2 Technical Report (2023)
3. Touvron, H., Martin, L., et al.: Llama 2: open foundation and fine-tuned chat models (2023)
4. Penedo, G., Malartic, Q., Hesslow, D., et al.: The RefinedWeb dataset for falcon LLM: outperforming curated corpora with web data, and web data only (2023)
5. The MosaicML NLP Team, MPT-30B: raising the bar for open-source foundation models (2023)

6. Islamovic, A.: Stability AI launches the first of its StableLM suite of language models (2023)
7. Biderman, S.: Pythia: a suite for analyzing large language models across training and scaling. In: EleutherAI (2023)
8. Nori, H., King, N., McKinney, S.M., Carignan, D., Horvitz, E.: Capabilities of gpt-4 on medical challenge problems. arXiv preprint arXiv:2303.13375 (2023)
9. Katz, D.M., Bommarito, M.J., Gao, S., Arredondo, P.: Gpt-4 passes the bar exam. SSRN (2023)
10. Tack, A.: The AI teacher test: measuring the pedagogical ability of blender and GPT-3 in educational dialogues (2022)
11. Haruna-Cooper, L., Rashid, M.A.: GPT-4: the future of artificial intelligence in medical school assessments. J. Roy. Soc. Med. 01410768231181251 (2023)
12. Lukasczyk, S., Fraser, G.: Pynguin: automated unit test generation for python, pp. 168–172 (2022)
13. Krekel, H., Pytest-dev team.: Pytest: helps you write better programs (2023)
14. Rogers, A., Kovaleva, O., Rumshisky, A.: A primer in BERTology: what we know about how BERT works. Trans. Assoc. Comput. Linguist. 8, 842–866 (2020)
15. Zhang, S.J., Florin, S., Lee, A.N., et al.: Exploring the MIT mathematics and EECS curriculum using large language models. arXiv preprint arXiv:2306.08997 (2023)
16. Lo, C.K.: What is the impact of ChatGPT on education? a rapid review of the literature. Educ. Sci. 13(4), 410 (2023)
17. Sok, S., Heng, K.: ChatGPT for education and research: a review of benefits and risks. SSRN 4378735 (2023)
18. Halaweh, M.: ChatGPT in education: strategies for responsible implementation. Contemp. Educ. Technol. 15 (2) (2023)
19. Mikolov, T., Sutskever, I., Chen, K., Corrado, G.S., Dean, J.: Distributed representations of words and phrases and their compositionality (2013)
20. Pennington, J., Socher, R., Manning, C.D.: Glove: global vectors for word representation (2014)
21. Peters, M.E., et al.: Deep contextualized word representations (2018)
22. Devlin, J., Chang, M.W., Lee, K., Toutanova, K.: Bert: pre-training of deep bidirectional transformers for language understanding (2018)
23. Roziere, B., Gehring, J., Gloeckle, F., Sootla, S., et al.: Code llama: open foundation models for code. arXiv preprint arXiv:2308.12950 (2023)
24. Jiang, A.Q., Sablayrolles, A., Mensch, A., et al.: Mistral 7B. arXiv preprint arXiv:2310.06825 (2023)
25. Li, R., Allal, L.B., Zi, Y., Muennighoff, N., Kocetkov, D., et al.: StarCoder: may the source be with you!. arXiv preprint arXiv:2305.06161 (2023)
26. Luo, Z., et al.: WizardCoder: empowering code large language models with evol-instruct. arXiv preprint arXiv:2306.08568 (2023)
27. Savelka, J., Agarwal, A., Bogart, C., Song, Y., Sakr, M.: Can generative pre-trained transformers (GPT) pass assessments in higher education programming courses?. arXiv preprint arXiv:2303.09325 (2023)
28. Zan, D., et al.: Large language models meet NL2Code: a survey. In: Proceedings of the 61st Annual Meeting of the Association for Computational Linguistics, vol. 1: Long Papers (2023)
29. Surameery, N.M.S., Shakor, M.Y.: Use chat gpt to solve programming bugs. Int. J. Inf. Technol. Comput. Eng. (IJITC) 3(01) , 17–22 (2023). ISSN: 2455-5290
30. Austin, J., Odena, A., Nye, M., Bosma, M., et al.: Program synthesis with large language models. arXiv preprint arXiv:2108.07732 (2021)

31. Feng, Z., Guo, D., Tang, D., Duan, N., et al.: CodeBERT: a pre-trained model for programming and natural languages (2020)
32. Guo, D., Ren, S., Lu, S., Feng, Z., Tang, D., et al.: Graphcodebert: pre-training code representations with data flow. arXiv preprint arXiv:2009.08366 (2020)
33. Wang, X., et al.: Syncobert: syntax-guided multi-modal contrastive pre-training for code representation. arXiv preprint arXiv:2108.04556 (2021)
34. Wang, Y., Wang, W., Joty, S., Hoi, S.C.H.: CodeT5: identifier-aware unified pre-trained encoder-decoder models for code understanding and generation (2021)
35. Roziere, B., Gehring, J., Gloeckle, F., Sootla, S., et al., Code llama: open foundation models for code. arXiv preprint arXiv:2308.12950 (2023)
36. Iyer, S., Konstas, I., Cheung, A., Zettlemoyer, L.: Mapping language to code in programmatic context (2018)
37. Liu, J., Xia, C.S., Wang, Y., Zhang, L.: Is your code generated by chatgpt really correct? rigorous evaluation of large language models for code generation. arXiv preprint arXiv:2305.01210 (2023)
38. Lai, Y., Li, C., Wang, Y., Zhang, T., Zhong, R.: DS-1000: a natural and reliable benchmark for data science code generation (2023)
39. Guo, W., Yang, J., Yang, K., Li, X., et al.: Instruction fusion: advancing prompt evolution through hybridization. arXiv preprint arXiv:2312.15692 (2023)
40. Babe, H.M., et al.: StudentEval: a benchmark of student-written prompts for large language models of code. arXiv preprint arXiv:2306.04556 (2023)

Semantically-Informed Domain Adaptation for Named Entity Recognition

Mariya Borovikova[1,3](✉) [iD], Arnaud Ferré[1] [iD], Robert Bossy[1] [iD],
Mathieu Roche[2,3] [iD], and Claire Nédellec[1] [iD]

[1] MaIAGE, Université Paris-Saclay, INRAE, Domaine de Vilvert,
78352 Jouy-en-Josas, France
mariya.borovikova@universite-paris-saclay.fr
[2] CIRAD, 34398 Montpellier, France
[3] TETIS, Univ. Montpellier, AgroParisTech, CIRAD, CNRS, INRAE,
34090 Montpellier, France

Abstract. Named Entity Recognition (NER) is an important task in Natural Language Processing that involves identifying entities in unstructured text. State-of-the-art NER methods often require extensive manual labeling for training. To bridge this gap, this paper introduces a domain adaptation technique that leverages semantic information about entity types using Sentence-BERT embeddings of their textual descriptions. We conduct experiments across various datasets from both general and biological domains, evaluating our approach in standard and zero-shot settings. Our experiences demonstrate the effectiveness of our method, which outperforms existing zero-shot techniques on certain datasets. Our findings underscore the importance of accurate semantic representations for entity types. This paper contributes to the advancement of zero-shot domain adaptation for NER and opens avenues for future research in improving NER systems' adaptability and performance across diverse domains.

Keywords: Named Entity Recognition · Domain Adaptation

1 Introduction

Named Entity Recognition (NER) plays an important role in Natural Language Processing, focusing on the task of identifying named entities within unstructured text (see Fig. 1). State-of-the-art NER methods require extensive manual labeling, which is inherently resource-intensive. Unsupervised and semi-supervised approaches, although capable of overcoming this limitation, are often less efficient within domain-specific contexts. Consequently, we have developed a method that leverages textual descriptions of entity types relevant to specific domains or datasets, offering a resource-efficient solution.

This paper introduces a domain adaptation technique, i.e. a strategy that involves adjusting a model trained on a source domain to perform effectively on

A. Appice et al. (Eds.): ISMIS 2024, LNAI 14670, pp. 55–64, 2024.
https://doi.org/10.1007/978-3-031-62700-2_6

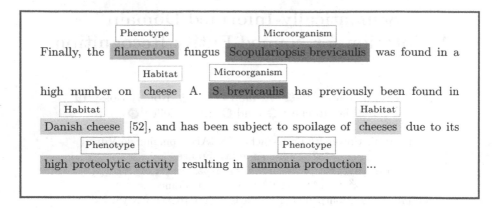

Fig. 1. NER task. In this text passage, the identified named entities are highlighted.

a different target domain. Our adaptation strategy leverages the semantics of entity types for NER. More precisely, we employ textual descriptions of entity types, derived from their names, processed by a Sentence-BERT [15] model to process. The resulting semantic representations are seamlessly integrated into our NER model as input features. To validate our approach, we conduct experiments on several datasets from both general and specific domains (i.e. biological). Our approach is tested in two settings: (1) the traditional training mode, where the model is trained and tested on distinct dataset partitions, hence on the same entity types, and (2) a mode where the model is trained and tested on different datasets, enabling testing on previously unseen entity types.

2 Related Work

The state-of-the-art NER models rely on supervised Machine Learning algorithms such as BiLSTM+CRF [22], BERT [3], generative models [17] or prompt engineering [20].

Unsupervised approaches to NER take different forms, including lexicon-based methods [14], clustering-based strategies [19] and generative models [6]. Additionally, prompting engineering strategies are explored in [21,23]. Although not dependent on annotated data, unsupervised approaches, often show lower efficiency compared to supervised methods. Alternatively, they may depend on very large language models, necessitating substantial computational resources, even during the prediction phase.

While unsupervised approaches address challenges associated with annotated data and computational resources, few-shot learning strategies represent a middle ground between unsupervised and supervised methods. These strategies require less data while still maintaining an acceptable level of performance. Few-shot strategies predominantly rely on transfer learning, aiming to leverage knowledge gained during training for application to related tasks. Domain adaptation is a particular case of transfer learning.

Traditional approaches to domain adaptation in NER commonly rely on bootstrapping techniques as seen in [2]. Modern methodologies frequently use difference between domains, as demonstrated in [7]. However, the approach that captured our attention is described in [12]. In this method, the authors leverage two BERT models: one to encode input text and another to encode labels, namely, entity types. The model combines these embeddings using a dot product, and the prediction is based on the maximum value. This approach has proven successful, especially in few-shot learning adaptation. However, it requires at least a small dataset from the target domain and is not fully zero-shot.

Our approach draws inspiration from the concepts outlined in [12]. Similar to the strategy described in that work, we use label semantics, transforming them through a BERT-based model. However, our approach does not entail model training for labels representation.

3   Methodology

The domain adaptation methodology proposed in this paper is based on the following hypothesis: Using the semantic representations of entity types can significantly improve the algorithm's overall performance and compensate for the lack of training data. To assess this hypothesis, we used the Sentence-BERT model [15] to semantically represent entity types. This section provides a description of our methodology and the datasets employed. For a visual overview of our approach, please refer to Fig. 2.

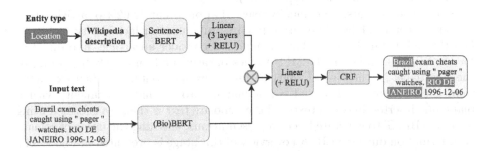

Fig. 2. Overview of the Semantic NER approach. Entity types are initially described in textual form (see Fig. 3) and subsequently processed through Sentence-BERT before integration into the NER pipeline.

3.1   NER Model Description

Recent NER approach employs a neural network that takes tokenized text as input and predicts each token's probability of belonging to predefined classes. However, this model cannot be directly applied to classes unseen during the training without modification. To address this limitation, we incorporate the representations of the entity types to be predicted as cues, as described in Sect. 3.2.

We use the BERT [3] and BioBERT [10] models as the base of the algorithm. Specifically, we use BERT for general domain datasets and BioBERT for those of the biological domain. We combine the output from BERT (or BioBERT) with the output from Sentence-BERT [15] and add three Linear Layers and a Conditional Random Field (CRF) layer. Before combining the embeddings from BERT (or BioBERT) and Sentence-BERT, a linear layer is applied to perform a linear transformation on the input data. This step is necessary because the (Bio)BERT embedding and Sentence-BERT embedding cannot be directly multiplied due to incompatible dimensions. Subsequently, the combined embeddings undergo three additional linear transformations, which gradually reduce the dimensionality. Finally, the output is passed through a CRF layer to consider the dependencies between adjacent labels, thereby enhancing the model's ability to identify entity boundaries. Since texts often contain multiple entity types, we generate one input sample per entity type for each text. We freeze the weights of BERT (or BioBERT) and Sentence-BERT and adjusted only the last Linear and CRF layers. This approach enables transitions between language models based on the domain of the dataset. We provide an overview of our approach in Fig. 2.

3.2 Semantic Entity Type Representation

The process to represent each entity type is the following. We start by taking an entity type name and attempt to retrieve a corresponding Wikipedia article with a matching title (e.g., 'location' for the entity type 'location'). In cases where no such article exists, we select the top result from a Wikipedia search (e.g., 'calendar date' for the entity type 'date').

In our experiments, we manually selected approximately 62% of articles, particularly when the labels of entity types are ambiguous or lack clarity, as identified through manual review. For example, we consider an article 'film director' for the 'director' label in the MIT Movies dataset, and an article 'location' for the 'loc' label in the CoNLL-2003 dataset. Textual content is then extracted from each selected article and concatenated into a single document, with a blank line inserted between texts. This composite text is then transformed using Sentence-BERT to generate latent representations, enabling the usage of semantic information during NER. An overview of this stage is presented in Fig. 3. The resulting latent representation is subsequently integrated into the NER pipeline.

3.3 Datasets

For our experiments we use datasets from diverse domains to enhance the algorithm's robustness and generalizability, validating its effectiveness beyond specific niches and confirming its adaptability in addressing a wide range of real-world applications. Specifically, we use six datasets, consisting of four from the general domain: CoNLL-2003 [16] (subcorpus in English), MIT movie review semantic corpus [11], MIT restaurant review [11], GeoVirus [5]; along with two from the biological domain: Bacteria Biotope [1] (BB) and NCBI diseases [4]. An overview of datasets is presented in Table 1.

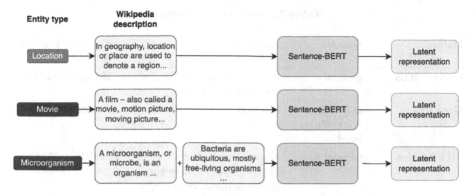

Fig. 3. Semantic Information Extraction Process. This figure illustrates the step-by-step process of extracting semantic information for entity types. Beginning with the entity type name (e.g., *Location*, *Movie*, *Microorganism*, etc.), it details the retrieval of related Wikipedia articles and the subsequent generation of latent semantic representations using Sentence-BERT.

In the pursuit of a zero-shot domain adaptation approach, we execute the following procedure for each dataset (denoted as n): creating a training set from all training partitions, excluding the dataset n, and subsequently assessing its performance on the test set of dataset n.

4 Experiments and Results

4.1 Experimental Setup

As explained in Sect. 3.1, we freeze the weights of BERT (or BioBERT) and Sentence-BERT while adjusting only the classifier weights. Since our model predicts one entity type per iteration, for a thorough evaluation of our model, we generate predictions for each entity for each input text from a corresponding test set. Subsequently, we merge our predictions into a comprehensive output encompassing all entity types, as illustrated in Fig. 4. This approach enables a fair comparison of our model with existing models. In addition to our primary classifier-based model, we incorporate two baselines. In our first baseline, we use cosine similarity instead of a classifier. Specifically, we calculate the cosine similarity between the entity type embedding and each token embedding, comparing it to a threshold. If the value is below, we predict the class with the lowest value; otherwise, we classify a token as not belonging to any class. The threshold was empirically selected through fine-tuning the model for one epoch using various thresholds, with 0.4 ultimately producing the optimal results. In the second baseline, we adopt a conventional training model without integrating entity type semantics.

Table 1. Datasets description

Dataset	Domain	Entity types	Size (in tokens)	Entities
CoNLL-2003	general	miscellaneous, location, organisation, person	287553	34788
GeoVirus	general, epidemiology, legal	location	258182	7258
MIT Movies	general	actor, award, character, character name, director, genre, opinion, origin, plot, quote, rating, ratings average, relationship, review, song, soundtrack, title, trailer, year	125736	55347
MIT Restaurants	general	dish, price, restaurant name, cuisine, amenity, location, hours, rating	81569	18514
BB	biological	phenotype, microorganism, habitat	50003	3435
NCBI diseases	biological	specific disease, modifier, disease class, composite mention	251443	6910

We evaluated our approaches in both standard and zero-shot modes. In the standard mode, we (1) trained a model and tested it on respective partitions of the same dataset, and (2) trained our model using all datasets and tested it on a single dataset. In the zero-shot mode, our model was trained on all datasets except one, which was reserved for testing purposes.

4.2   Results

Tables 2 and 3 present the F-measure scores of our algorithm, two baselines outlined in Sect. 4.1, and state-of-the-art (SOTA) models. The "Dataset" column identifies the evaluation dataset. The "S" column represents scores obtained using SOTA algorithms. It is important to note that the scores we present are sourced directly from the associated papers. Some datasets lack scores because, to our knowledge, there is no zero-shot algorithm evaluated on these datasets.

In Table 2, the "Same Dataset" column denotes scores achieved by training and testing the model on corresponding partitions of the same dataset, while the "Cross-Datasets" column reflects scores attained by training the model on all training partitions across datasets and testing on the test partition of the respective dataset. In Table 3, the "Cosine Similarity" column shows scores acquired with the baseline outlined in Sect. 4.1, while the "Classifier" column

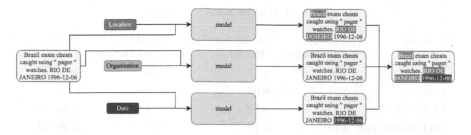

Fig. 4. Integration of predictions. This diagram visually represents the process of consolidating entity predictions for multiple entity types into a unified output. This method enables a holistic evaluation of our model's performance across various entity types.

showcases scores obtained with our algorithm. The "Standard Cross-Dataset" column reflects scores obtained when training our model conventionally, without incorporating entity type semantics. In other words, we used only the BERT (or BioBERT) model with a CRF layer on top.

Table 2. F-measure of standard training

Dataset	Same Dataset (Intra-Dataset)	Cross-Datasets (Inter-Dataset)	SOTA
BB	0.56	0.58	0.89 [9]
GeoVirus	0.46	0.78	0.91 [18]
CoNLL-2003	0.88	0.91	0.95 [22]
MIT restaurants	0.62	0.75	0.72 [13]
MIT movies	0.58	0.62	0.81 [13]
NCBI diseases	0.54	0.42	0.90 [8]

Our experiments yield several observations: (1) Models trained on all datasets demonstrate superior quality compared to models trained exclusively on individual datasets (with the exception of NCBI Disease dataset); (2) The classifier exhibits higher quality than both baselines (with the exception of GeoVirus dataset); (3) Our model performs better in a standard mode than in an zero-shot mode (with the exception of GeoVirus datset); (4) Notably, our zero-shot model outperforms the SOTA algorithm on the MIT Restaurants dataset and achieves performance comparable to the zero-shot SOTA score on the MIT Movies dataset.

5 Discussion

Based on our findings, we can affirm that incorporating entity type semantics significantly enhances NER results. Upon manual examination of predictions, we

Table 3. F-measures of zero-shot

Dataset	Classifier (Our Model)	Cosine Similarity (Baseline)	Standard Cross-Dataset	SOTA
BB	0.34	0.17	0.00	–
GeoVirus	0.54	0.07	0.79	–
CoNLL-2003	0.79	0.12	0.24	0.93 [19]
MIT restaurants	0.46	0.12	0.09	0.36 [23]
MIT movies	0.62	0.02	0.12	0.63 [21]
NCBI diseases	0.24	0.15	0.00	–

observed substantial variations in results among entity types. For instance, in the BB dataset, the F-measure for "Microorganism" is 0.75, contrasting with 0.15 for "Habitat" and 0.12 for "Phenotype." Our interpretation is as follows: Our approach relies on Wikipedia descriptions of entity types, and some entities are inherently more challenging to elucidate. In the BB dataset, the concept of "Microorganism" is more evident than that of "Habitat" or "Phenotype," for which even human annotators can make errors. Moreover, entities labeled as "Microorganism" are commonly denoted in text through specific Latin terms, thereby aiding in their identification. This intuition is further supported by the fact that we also conducted experiments by automatically selecting Wikipedia articles with an entity type as the article title, and the results were similar to a baseline. Thus, the quality of the texts chosen for describing entity types is crucial.

Another notable point concerns the automated acquisition of semantic representations for entity types. This method faces challenges as the type labels may not precisely capture the semantics of the entity type. For example, in the NCBI diseases dataset, if a disease description lacks a representation by a noun (e.g., *VHL* in *the VHL gene*), it is annotated with a "Modifier" entity type. Regrettably, this does not faithfully convey the semantics of the entity type. However, it is worth noting that despite these challenges, obtaining relevant textual descriptions for each entity type remains a significantly less time and resource-consuming task compared to annotating a corpus for those entity types.

A surprising observation arises from the GeoVirus corpus, where the zero-shot mode outperforms the standard mode. We attribute this phenomenon to an extensive overlap of "Location" entities with the Conll corpus, which is larger compared to the GeoVirus corpus.

Additionally, it is worth noting that we conducted comparison between our zero-shot algorithm and the best-performing zero-shot models for each dataset. However, the same algorithms do not necessarily outperform our approach on other datasets. For instance, [21] obtained a score of 0.21 on MIT Movies, which is significantly lower than other results. Similarly, [23] reported an F-measure of 0.49 for MIT Restaurants, which is comparable to our score. Additionally, all SOTA systems were trained on multiple datasets, with a substantial overlap in common labels, particularly evident in the CoNLL-2003 dataset. Consequently, the task becomes simpler for these entities, making a direct comparison with our approach less meaningful and honest. These findings raise questions about the robustness of existing systems.

6   Conclusion

In this work we presented an original approach to zero-shot domain adaptation for NER that requires minimal annotated data. We tested our approach on several datasets from general and biological domains. Our results demonstrate the superior performance of our approach compared to existing zero-shot techniques on some datasets. Looking ahead, future research will focus on examining the robustness and practical applications of this model within existing systems. Additionally, we aim to explore the automation of semantic representations for entity types, further advancing the versatility and effectiveness of our proposed approach.

Acknowledgements. The authors would like to express their sincere gratitude to the ANR-20-PCPA-0002, BEYOND for providing the funding that made this research possible. Special thanks to the "Mésocentre" computing center of Univ. Paris-Saclay, CentraleSupélec, École Normale Supérieure Paris-Saclay supported by CNRS and Région Île-de-France for granting access to a GPU cluster.

References

1. Bossy, R., Deléger, L., Chaix, E., Ba, M., Nédellec, C.: Bacteria biotope at bionlp open shared tasks 2019. In: Proceedings of the 5th Workshop on BioNLP Open Shared Tasks, pp. 121–131 (2019)
2. Chaudhary, A., Xie, J., Sheikh, Z., Neubig, G., Carbonell, J.: A little annotation does a lot of good: a study in bootstrapping low-resource named entity recognizers. In: Proceedings of the Conference on Empirical Methods in Natural Language Processing and the 9th International Joint Conference on Natural Language Processing (EMNLP-IJCNLP), pp. 5164–5174 (2019)
3. Devlin, J., Chang, M.W., Lee, K., Toutanova, K.: BERT: pre-training of deep bidirectional transformers for language understanding. In: Proceedings of the Conference of the North American Chapter of the Association for Computational Linguistics.: Human Language Technologies, vol. 1, pp. 4171–4186 (2019)
4. Doğan, R.I., Leaman, R., Lu, Z.: Ncbi disease corpus: a resource for disease name recognition and concept normalization. J. Biomed. Inf. **47**, 1–10 (2014)
5. Gritta, M., Pilehvar, M.T., Collier, N.: Which Melbourne? augmenting geocoding with maps. In: Proceedings of the 56th Annual Meeting of the Association for Computational Linguistics, vol. 1: Long Papers, pp. 1285–1296 (2018)
6. Iovine, A., Fang, A., Fetahu, B., Rokhlenko, O., Malmasi, S.: Cyclener: an unsupervised training approach for named entity recognition. In: Proceedings of the ACM Web Conference 2022, pp. 2916–2924 (2022)
7. Jia, C., Liang, X., Zhang, Y.: Cross-domain ner using cross-domain language modeling. In: Proceedings of the 57th Annual Meeting of the Association for Computational Linguistics, pp. 2464–2474 (2019)

8. Kocaman, V., Talby, D.: Biomedical named entity recognition at scale. In: Pattern Recognition. ICPR International Workshops and Challenges: Virtual Event, pp. 635–646 (2021)

9. Le Guillarme, N., Thuiller, W.: Taxonerd: deep neural models for the recognition of taxonomic entities in the ecological and evolutionary literature. Methods Ecol. Evol. **13**(3), 625–641 (2022)

10. Lee, J., et al.: Biobert: a pre-trained biomedical language representation model for biomedical text mining. Bioinformatics **36**(4), 1234–1240 (2020)

11. Liu, J., Pasupat, P., Cyphers, S., Glass, J.: ASGARD: a portable architecture for multilingual dialogue systems. In: 2013 IEEE International Conference on Acoustics, Speech and Signal Processing, pp. 8386–8390 (2013)

12. Ma, J., et al.: Label semantics for few shot named entity recognition. In: Findings of the Association for Computational Linguistics: ACL 2022, pp. 1956–1971 (2022)

13. Palm, R.B., Hovy, D., Laws, F., Winther, O.: End-to-end information extraction without token-level supervision. In: Proceedings of the Workshop on Speech-Centric Natural Language Processing, pp. 48–52 (2017)

14. Popovski, G., Kochev, S., Korousic-Seljak, B., Eftimov, T.: Foodie: a rule-based named-entity recognition method for food information extraction. ICPRAM **12**, 915 (2019)

15. Reimers, N., Gurevych, I.: Sentence-bert: sentence embeddings using siamese bert-networks. In: Proceedings of the Conference on Empirical Methods in Natural Language Processing and the 9th International Joint Conference on Natural Language Processing (EMNLP-IJCNLP), pp. 3982–3992 (2019)

16. Sang, E.T.K., De Meulder, F.: Introduction to the conll-2003 shared task: language-independent named entity recognition. In: Proceedings of the Conference on Natural Language Learning at HLT-NAACL, pp. 142–147 (2003)

17. Shen, Y., Song, K., Tan, X., Li, D., Lu, W., Zhuang, Y.: Diffusionner: boundary diffusion for named entity recognition. arXiv preprint arXiv:2305.13298 (2023)

18. Wang, J., Hu, Y.: Are we there yet? evaluating state-of-the-art neural network based geoparsers using eupeg as a benchmarking platform. In: Proceedings of the 3rd ACM SIGSPATIAL International Workshop on Geospatial Humanities, pp. 1–6 (2019)

19. Wang, S., et al.: k nn-ner: named entity recognition with nearest neighbor search. arXiv preprint arXiv:2203.17103 (2022)

20. Wang, S., et al.: Gpt-ner: named entity recognition via large language models. arXiv preprint arXiv:2304.10428 (2023)

21. Wang, X., et al.: Instructuie: multi-task instruction tuning for unified information extraction. arXiv e-prints pp. arXiv–2304 (2023)

22. Wang, X., et al.: Automated concatenation of embeddings for structured prediction. In: Proceedings of the 59th Annual Meeting of the Association for Computational Lingistiucs and the 11th International Joint Conference on Natural Language Processing, pp. 2643–2660 (2021)

23. Zhou, W., Zhang, S., Gu, Y., Chen, M., Poon, H.: Universalner: targeted distillation from large language models for open named entity recognition. arXiv preprint arXiv:2308.03279 (2023)

Token Pruning by Dimensionality Reduction Methods on TCT-ColBERT for Reranking

Nazish Hina[(⊠)], Mohand Boughanem, and Taoufiq Dkaki

Institut de Recherche en Informatique de Toulouse (IRIT), Toulouse, France
{hina.nazish,Mohand.Boughanem,Taoufiq.Dkaki}@irit.fr

Abstract. High-dimensional embeddings from dense retrieval models pose challenges in information retrieval and reranking because they need a lot of resources and computing power. The TCT-ColBERT model, introduced recently, has made retrieval systems more efficient, especially for indexing documents offline. However, it makes the index bigger because it needs to store each token for a particular document. Dimensionality reduction can help by reducing storage needs, noise reduction, and simplifying computations. Dimensionality reduction methods are used in many fields, especially when the dataset grows. So, it is worth exploring these methods for dense retrieval models. Using specific dimensionality reduction techniques, we can prune the number of dimensions and focus on the most important parts of the text embeddings that affect similarity calculations. This research explores two linear dimensionality reduction methods: PCA (Principal Component Analysis) and truncated-SVD (truncated Singular Value Decomposition). The main objective was to find out the most important parts of the embeddings for determining the similarity in documents and queries. The MS MARCO passage collection (test-2019) was used for experiments. The results showed that it could reduce token embeddings up to 95% without appreciable loss in performance in terms of recall and precision.

Keywords: PCA · TCT-ColBERT · SVD · Information retrival

1 Introduction

In search systems, Retrieving useful information on a large scale is crucial, as the amount of data is increasing continuously. Various search systems [19, 29] use a common strategy: a multi-stage ranking that balances search efficiency and effectiveness. In this process, a proficient first-stage retriever efficiently acquires a condensed set of documents from the complete corpus. Subsequently, a powerful but slower reranking stage employs a more intricate and effective ranking model to refine and enhance the list of ranked documents generated by the preceding stage. This "retrieval and reranking" pipeline has gained widespread adoption in

A. Appice et al. (Eds.): ISMIS 2024, LNAI 14670, pp. 65–74, 2024.
https://doi.org/10.1007/978-3-031-62700-2_7

academia [5,17], and industry [14]. In the first stage of information retrieval, lexical retrieval models, such as BM25 [23], have been employed. The goal of first stage retrieval is to maximize recall. For the rerank, it requires maximizing precision. To rerank, a second stage neural ranking model is used that captures the complex semantics of the data. Two main families of neural models have been developed:

– Representation-based [22]: In this approach, queries and documents undergo separate encoding processes, resulting in a single representation for each. The scoring is performed based on the distance between these representations.
– Interaction-based [18]: In the interaction-based paradigm, a query-document pair is jointly processed by a neural network to generate the score. This method treats the query and document as a unified entity during the scoring process.

The representation-based approach proves to be highly efficient, allowing the indexing of document representations. During inference, only the query needs to be computed. On the other hand, the interaction-based approach exhibits superior performance by conducting a more comprehensive scoring process between queries and documents.

The ColBERT [10] model introduced a unique strategy to bridge the gap between these two families. It indexes one representation per token, enabling the pre-computation of document representations and incorporating some of the capabilities of an interaction model. Specifically, each contextualized token of the query interacts with each pre-computed contextualized token of the document. The primary drawback of ColBERT is its large index size. This is because every token needs to be indexed, rather than a pooled version of a document. Despite this trade-off, ColBERT presents an innovative approach that combines the efficiency of representation-based methods with the enhanced performance of interaction-based methods. Utilizing token-level indexing, the approach achieves balance by enabling the pre-computation of document representations. Simultaneously, it maintains interaction capabilities between query and document tokens.

To address the challenge posed by the large index collection, the study [12,13] suggested a method that uses knowledge from a multi-representation model (ColBERT) and combines it into a single representation model (TCT-ColBERT). The authors incorporated a pooling layer to merge embeddings from diverse input tokens, specifically opting for AvgPool. In TREC (2021) [6], many researchers used this approach. For this reason, we consider this approach to be used in the study. This study explores methods to extract the reduced number of embedding representations that pose highly relatable knowledge instead of using a pooling layer to average the multiple representations into a single representation. This can be achieved by reducing the dimension of dense vectors. The fundamental concept of feature dimensionality reduction involves transforming a data sample from a high-dimensional space to a comparatively lower-dimensional space. The main goal is to find the mapping and uncover an efficient low-dimensional space hidden within the visible high-dimensional space [20].

In this paper, dimension reduction algorithms take center stage as they demonstrate high capabilities in tackling challenges found in real-world datasets. Dimension reduction algorithms, categorized into feature selection and feature extraction, play an important role. Feature selection involves identifying the most informative features while discarding less informative ones. On the other hand, feature extraction combines features through algebraic transformations, resulting in a smaller set of new features [2]. We employed two feature extraction methods: PCA and truncated-SVD.

In the process of PCA, the high-dimensional embedding space is projected into a new embedding subspace, which symbolizes the directions of maximum data variance [7]. PCA requires examining the cumulative explained variance ratio to decide how many components to keep [1]. Several benefits are offered by PCA [9]: it is non-iterative and, as a result, is less time-consuming.

Rank reduction is used to approximate a matrix by selecting only a subset of its singular values and associated vectors. It works by decomposing a matrix into three matrices: U, Σ, and V, where U represents the left singular vectors, Σ is a diagonal matrix containing the singular values, and V represents the right singular vectors. The reduced matrix is derived from truncated matrices, that have a reduced dimension [3]. Scikit-learn uses the 'TruncatedSVD' estimator to perform a truncated-SVD, which linearly extracts features. In this process, instead of using all singular values and vectors, it selects a fixed number of them, typically the top k, where k is smaller than the original rank of the matrix.

The study is focused on how the dimensionality reduction methods impact the performance of the dense retrieval model in reranking. Would it be worthwhile to utilize PCA or truncated SVD to reduce the representation of queries and documents? The study shows that these two methods improved efficiency in terms of computation with minimal loss of precision and recall.

2   Related Work

In information retrieval (IR), significant efforts have been dedicated to enhancing the effectiveness of pre-trained language models. Research has extensively explored methods such as quantization and distillation for efficiency improvements [24]. Power-BERT [8] and length-adaptive transformers [11] are designed to reduce the number of floating point operations per second (FLOPS). They address token elimination within the Pretrained Language Model (PLM) layers.

In the era of information retrieval, the conventional practice of pruning indexes has long been used to improve latency and memory demands. Numerous studies have extensively investigated this technique, as described in [4, 30].

One approach to address the challenge of ColBERT's large index size involves decreasing dimensionality and employing quantization on the document tokens. It is important to mention that this is already built into the system, because the output of the Pretrained Language Model, which is usually 768 dimensions, gets transformed into a smaller space of 128 dimensions. The original research [10] demonstrates promising results through dimensionality reduction and token quantization. Similarly, a binarization technique [28] has been proposed for information retrieval, and recent experiments [25] suggest the feasibility of binarizing ColBERT tokens. Our current study diverges from the established research direction by aiming to reduce the representations through feature extraction rather than size reduction. We perceive the combination of these research directions as essential for enhancing ColBERT models, but leave the integration as a potential avenue for future exploration. Furthermore, ColBERT has been expanded to incorporate pseudo-relevance feedback in [27], while query pruning has been explored to enhance latency in [26].

3   Feature Extraction

This study aims to reduce the dimensionality of document embeddings and query embeddings to a practical set of representations. We propose the application of dimensionality reduction methods, specifically PCA and SVD, to the embedding vectors obtained from the fine-tuned model, each having dimensions of 512×768. The implementation of `scikit-learn` [21] is employed for PCA and truncated SVD with default parameters. The objective is to extract dimensions or components contributing more significantly to the query and document similarity scores.

When PCA or truncated SVD is applied to the dense vector of 512×768 dimensions, it performs mathematical computations to extract useful components. Initially, we set the number of components to 40. Subsequently, we use the cumulative explained variance ratio attribute to address size concerns and enhance performance. This attribute, utilized in previous studies to determine the optimal number of components [1] to be retained, represents the percentage of variance explained by each selected component. We experiment with three different thresholds—0.85, 0.95, and 0.99—and identify 0.95 as the optimal threshold. For the 0.85 threshold, mostly only the first component of documents is selected, leading to a decline in performance. As for the 0.99 threshold, there is a noticeable decrease in performance. The 0.95 threshold yields the best performance, retaining up to 4 components for embeddings. In this case, the 0.95 threshold limits the number of components contributing to a cumulative sum of explained variance ratio equal to or above 0.95. After determining the optimal number of components for query and document embeddings, the dot product for each query component is computed with all components of document embeddings using the `numpy` library. This results in a vector of dimensions equal to the length of query components multiplied by the length of document components. Subsequently, we apply similarity metrics, as explained in Sect. 4.3. In the `state-of-the-art` comparison, the embedding vectors of 512*768 dimensions are obtained from the fine-tuned model, and the same metrics are applied to compute the similarity score. This acts as a baseline for the study.

4   Experimental Settings

4.1   Fine-Tuning

The fine-tuning of TctColBERT is done using a `Pyterrier` dense retrieval library [16]. The authors in [31] reported that student initialization from the teacher model is not required; It yielded comparable results while training on a smaller dataset compared to the amount of data reported in the original work [24]. The model is instantiated from the Huggingface Transformers checkpoint bert-base-uncased. The batch size is 8 and the in-batch negative is set to true. The learning rate is set to `lr = 3e-6`, and the optimizer used is `AdamW` with `eps = 1e-8`. The MS MARCO train judged dataset consisting of 270 million docPairs is used as training data, and 2500,000 iterations are performed, every iteration consists of 4 docPairs. Therefore, training takes 10 million examples in total.

4.2 Dataset

The experiments utilize the TREC 2019 Deep Learning passage ranking dataset [6] and the PyTerrier Information Retrieval experimentation platform [15]. The test dataset consists of 200 queries. The top 100 relative passages are retrieved against each query using BM25 with default parameters implemented in the Pyterrier toolkit. Then, for the retrieved data, different experiments are performed as described in Sect. 3. The embeddings of query and passage from fine-tuned TctColBERT are used for experiments.

4.3 Similarity Metrics

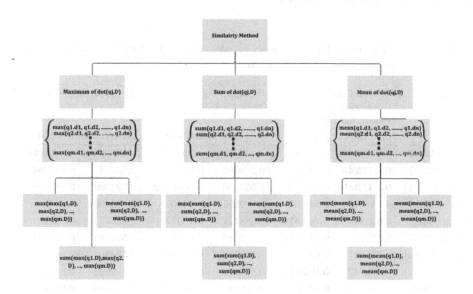

Fig. 1. Similarity metrics used in the study

For the similarity score between the query and document, the dot product of each query embedding, one by one with all document embeddings was calculated. Let

Q= [q1, q2, ..., qm] be the vector of query embeddings with the total number of 1,2,....,m embeddings, and D= [d1, d2, ..., dn] be the matrix containing document embeddings with the total number of 1,2,...,n embeddings. The dot product of the jth query embedding with all document embeddings is given by:

$$\text{DotProduct}(q_j, D) = [q_j \cdot d_1, q_j \cdot d_2, \ldots, q_j \cdot d_n]$$

Then, three different ways to calculate the final similarity are applied: maximum similarity, summation, and mean pooling of the computed dot product. It is described in Fig. 1.

4.4 Evaluation Metrics

Our modular approach is assessed following the TREC protocol, employing the official metric specific to in-domain collections: nDCG@10 for TREC DL passage reranking. Additionally, we provide results for recall@10 and P@10. The official evaluation measure for MS MARCO dataset, RR@10, is also observed. The evaluation metrics are calculated utilizing the `pytrec_eval` library, integrated within the `pyterrier` toolkit. This toolkit is a Python wrapper for the extensively employed `trec_eval` evaluation tool.

5 Results and Discussion

5.1 Efficiency and Accuracy

Table 1. nDCG@10 and RR@10 results on MS MARCO -2019. Bold values are the best for the baseline and Dimension pruning

Similarity Method	Without Any Dimensionality Reduction Method		By Applying TruncatedSVD		By Applying PCA	
	nDCG@10	RR@10	nDCG@10	RR@10	nDCG@10	RR@10
Max (Max)	0.550086	0.785880	**0.634925**	0.903876	**0.634888**	**0.903876**
Sum (Max)	**0.641245**	0.891473	0.559682	0.811240	0.559682	0.834884
Mean (Max)	**0.641245**	0.891473	0.523590	**0.914729**	0.523590	0.882429
Max (Sum)	0.554987	0.848283	0.580096	0.849059	0.580096	0.849059
Sum (Sum)	0.635742	**0.903876**	0.488121	0.775221	0.488573	0.775221
Mean (Sum)	0.635742	**0.903876**	0.522089	0.883721	0.522089	0.883721
Max (Mean)	0.554987	0.848283	0.523562	0.882171	0.523562	0.882171
Sum (Mean)	0.635742	**0.903876**	0.522089	0.883721	0.522089	0.883721
Mean (Sum)	0.635742	0.903876	0.522089	0.883721	0.522089	0.883721

In terms of computation cost, for all 512 embeddings of query and document, it took 130 min to complete, while for PCA and truncatedSVD, it took 18 min. Therefore, applying dimensionality reduction provides 86% cost-effectiveness.

In terms of efficiency, as in Table 1, in all 9 ways to measure similarity, RR@10 is highest for truncatedSVD, but for nDCG@10 it loses by 0.99%. The RR@10 for PCA and the baseline are the same for different measuring mechanisms, but the nDCG@10 is less for PCA. The SVD loses the p@10 by 2.20% and the recall@10 by 2.72% as reported in Table 2. The paired t-test is done to measure the p-value, which comes to 0.03 for the nDCg@10, 0.0085 for the p@10, and 0.0034 for recall@10. Therefore it did not make a significant difference in performance reduction.

Table 2. p@10 and Recall@10 results on MS MARCO -2019. Bold values are the best for the baseline and Dimension pruning

Similarity Method	Without Any Dimensionality Reduction Method		By Applying TruncatedSVD		By Applying PCA	
	P@10	Recall@10	P@10	Recall@10	P@10	Recall@10
Max (Max)	0.683721	0.141575	**0.730233**	**0.146220**	**0.730233**	**0.146220**
Sum (Max)	**0.746512**	**0.150253**	0.658140	0.128165	0.658140	0.128165
Mean (Max)	**0.746512**	**0.150253**	0.583721	0.114830	0.583721	0.114830
Max (Sum)	0.648837	0.132021	0.676744	0.136021	0.676744	0.136021
Sum (Sum)	0.734884	0.147360	0.579070	0.112229	0.579070	0.112229
Mean (Sum)	0.734884	0.147360	0.581395	0.114383	0.581395	0.114383
Max (Mean)	0.648837	0.132021	0.583721	0.114830	0.583721	0.114830
Sum (Mean)	0.734884	0.147360	0.581395	0.114383	0.581395	0.114383
Mean (Sum)	0.734884	0.147360	0.581395	0.114383	0.581395	0.114383

We compare our approach against two baselines, including lexical retrieval methods (BM25) and dense retrieval models (bi-encoders) (TCT-ColBERT) without any dimensionality reduction method. Table 3 shows the performance of these methods in terms of official task evaluation metrics.

Table 3. Reranking effectiveness on the TREC DL Passage ranking tasks. The best performance is highlighted in bold.

Model	nDCG@10	RR@10	P@10	Recall@10
BM25	0.479540	0.79438	0.597674	0.12234
TCT-ColBERT (All 512 Embeddings)	**0.641245**	0.903876	**0.746512**	**0.150253**
TCT-ColBERT (With truncated-SVD)	0.634925	**0.914729**	0.730233	0.146220
TCT-ColBERT (With PCA)	0.634888	0.903876	0.730233	0.146220

5.2 Analysis of Principle Components

We analyzed the components selected by the reduction mechanism on the passage set. It selects one to three components for different passages. For instance, Fig. 2 shows the selected components for two passages to demonstrate the covariance parameter. It means that for some passages, only the first component captures the maximum explained_ variance_ ratio_ to the threshold, i.e. 0.95% or above.

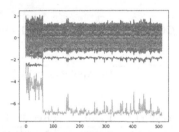

(a) Before TruncatedSVD, all 512 embeddings of passage 1

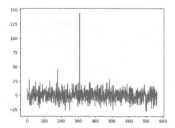

(b) After TruncatedSVD, 1 component is selected

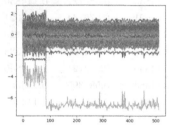

(c) Before TruncatedSVD all 512 embeddings of passage 2

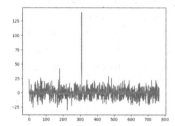

(d) After TruncatedSVD, 3 components are selected

Fig. 2. Dimensionality Reduction Analysis

6 Conclusion

In this research, the TCT-ColBERT model has been explored and assessed using dimensionality reduction techniques to decrease its tokens for reranking purposes. The resource requirements are initially examined for reranking using 512 token embeddings generated from TCT-ColBERT. Certain uncomplicated dimensionality reduction approaches enable the removal of 95% of embeddings from passage components with minimal impact on performance. Integrating these dimensionality reduction techniques with previously investigated compression methods have significant potential for increasing TCT-ColBERT-based information retrieval systems performance.

Acknowledgments. The authors are thankful to the Higher Education Commission (HEC) of Pakistan for financial support of the thesis, which is the source of this work.

Disclosure of Interests. The authors have no competing interests to declare that are relevant to the content of this article.

References

1. Abdi, H.: The Eigen-decomposition: eigenvalues and Eigenvectors. In: Encyclopedia of Measurement and Statistics, pp. 304–308 (2007)
2. Abe, S.: Support Vector Machines for Pattern Slassification, vol. 2. Springer, Heidelberg (2005). https://doi.org/10.1007/978-1-84996-098-4
3. Anowar, F., Sadaoui, S., Selim, B.: Conceptual and empirical comparison of dimensionality reduction algorithms (PCA, KPCA, LDA, MDS, SVD, LLE, ISOMAP, LE, ICA, t-SNE). Comput. Sci. Rev. **40**, 100378 (2021)
4. Carmel, D., et al.: Static index pruning for information retrieval systems. In: Proceedings of the 24th Annual International ACM SIGIR Conference on Research and Development in Information Retrieval, pp. 43–50 (2001)
5. Chen, R.C., Gallagher, L., Blanco, R., Culpepper, J.S.: Efficient cost-aware cascade ranking in multi-stage retrieval. In: Proceedings of the 40th International ACM SIGIR Conference on Research and Development in Information Retrieval, pp. 445–454 (2017)
6. Craswell, N., Mitra, B., Yilmaz, E., Campos, D., Voorhees, E.M.: Overview of the TREC 2019 deep learning track. arXiv preprint arXiv:2003.07820 (2020)
7. Ghojogh, B., et al.: Feature selection and feature extraction in pattern analysis: a literature review. arXiv preprint arXiv:1905.02845 (2019)
8. Goyal, S., Choudhury, A.R., Raje, S., Chakaravarthy, V., Sabharwal, Y., Verma, A.: PoWER-BERT: accelerating BERT inference via progressive word-vector elimination. In: International Conference on Machine Learning, pp. 3690–3699. PMLR (2020)
9. Karamizadeh, S., Abdullah, S.M., Manaf, A.A., Zamani, M., Hooman, A.: An overview of principal component analysis. J. Signal Inf. Process. 4(3B), 173 (2013)
10. Khattab, O., Zaharia, M.: ColBERT: efficient and effective passage search via contextualized late interaction over BERT. In: Proceedings of the 43rd International ACM SIGIR Conference on Research and Development in Information Retrieval, pp. 39–48 (2020)
11. Kim, G., Cho, K.: Length-adaptive transformer: train once with length drop, use anytime with search. arXiv preprint arXiv:2010.07003 (2020)
12. Lin, S.C., Yang, J.H., Lin, J.: Distilling dense representations for ranking using tightly-coupled teachers. arXiv preprint arXiv:2010.11386 (2020)
13. Lin, S.C., Yang, J.H., Lin, J.: In-batch negatives for knowledge distillation with tightly-coupled teachers for dense retrieval. In: Proceedings of the 6th Workshop on Representation Learning for NLP (RepL4NLP-2021), pp. 163–173 (2021)
14. Liu, S., Xiao, F., Ou, W., Si, L.: Cascade ranking for operational e-commerce search. In: Proceedings of the 23rd ACM SIGKDD International Conference on Knowledge Discovery and Data Mining, pp. 1557–1565 (2017)
15. Macdonald, C., Tonellotto, N.: Declarative experimentation in information retrieval using PyTerrier. In: Proceedings of ICTIR 2020 (2020)

16. Macdonald, C., Tonellotto, N., MacAvaney, S., Ounis, I.: PyTerrier: Declarative experimentation in Python from BM25 to dense retrieval. In: Proceedings of the 30th ACM International Conference on Information & Knowledge Management, pp. 4526–4533 (2021)

17. Matveeva, I., Burges, C., Burkard, T., Laucius, A., Wong, L.: High accuracy retrieval with multiple nested ranker. In: Proceedings of the 29th Annual International ACM SIGIR Conference on Research and Development in Information Retrieval, pp. 437–444 (2006)

18. Nogueira, R., Cho, K.: Passage re-ranking with BERT. arXiv preprint arXiv:1901.04085 (2019). https://doi.org/10.48550/arXiv.1901.04085

19. Nogueira, R., Yang, W., Cho, K., Lin, J.: Multi-stage document ranking with BERT. arXiv preprint arXiv:1910.14424 (2019)

20. Örnek, C., Vural, E.: Nonlinear supervised dimensionality reduction via smooth regular embeddings. Pattern Recogn. **87**, 55–66 (2019)

21. Pedregosa, F., et al.: Scikit-learn: machine learning in python. J. Mach. Learn. Res. **12**, 2825–2830 (2011)

22. Reimers, N., Gurevych, I.: Sentence-bert: sentence embeddings using siamese bert-networks. arXiv preprint arXiv:1908.10084 (2019)

23. Robertson, S.E., Walker, S.: Some simple effective approximations to the 2-poisson model for probabilistic weighted retrieval. In: SIGIR 1994: Proceedings of the Seventeenth Annual International ACM-SIGIR Conference on Research and Development in Information Retrieval, organised by Dublin City University, pp. 232–241. Springer, Heidelberg (1994). https://doi.org/10.1007/978-1-4471-2099-5_24

24. Sanh, V., Debut, L., Chaumond, J., Wolf, T.: DistilBERT, a distilled version of BERT: smaller, faster, cheaper and lighter. arXiv preprint arXiv:1910.01108 (2019)

25. Santhanam, K., Khattab, O., Saad-Falcon, J., Potts, C., Zaharia, M.: Colbertv2: effective and efficient retrieval via lightweight late interaction. arXiv preprint arXiv:2112.01488 (2021)

26. Tonellotto, N., Macdonald, C.: Query embedding pruning for dense retrieval. In: Proceedings of the 30th ACM International Conference on Information & Knowledge Management, CIKM 2021, pp. 3453–3457. Association for Computing Machinery, New York (2021). https://doi.org/10.1145/3459637.3482162

27. Wang, X., Macdonald, C., Tonellotto, N., Ounis, I.: Pseudo-relevance feedback for multiple representation dense retrieval, ICTIR 2021, pp. 297–306. Association for Computing Machinery, New York (2021).https://doi.org/10.1145/3471158.3472250

28. Yamada, I., Asai, A., Hajishirzi, H.: Efficient passage retrieval with hashing for open-domain question answering. arXiv preprint arXiv:2106.00882 (2021)

29. Yan, M., et al.: IDST at TREC 2019 deep learning track: deep cascade ranking with generation-based document expansion and pre-trained language modeling. In: TREC (2019)

30. Zobel, J., Moffat, A.: Inverted files for text search engines. ACM Comput. Surv. (CSUR) **38**(2), 6–es (2006)

31. Wang, X., MacAvaney, S., Macdonald, C., Ounis, I.: An inspection of the reproducibility and replicability of TCT-ColBERT. In: Proceedings of the 45th International ACM SIGIR Conference on Research and Development in Information Retrieval, pp. 2790–2800 (2022)

AI Tools and Models

Exploiting microRNA Expression Data for the Diagnosis of Disease Conditions and the Discovery of Novel Biomarkers

Daniele Rosa[1]📧, Antonio Pellicani[1,2]📧, Gianvito Pio[1,2](✉)📧,
Domenica D'Elia[3]📧, and Michelangelo Ceci[1,2,4]📧

[1] Department of Computer Science, University of Bari, Bari, Italy
{daniele.rosa,antonio.pellicani,gianvito.pio,michelangelo.ceci}@uniba.it
[2] Big Data Laboratory, National Interuniversity Consortium for Informatics,
Rome, Italy
[3] Institute for Biomedical Technologies, National Research Council, Bari, Italy
domenica.delia@cnr.it
[4] Department of Knowledge Technologies, Jožef Stefan Institute, Ljubljana, Slovenia

Abstract. MicroRNAs (miRNAs) play key roles in diseases and their detection in circulating biofluids makes them optimal candidate as disease biomarkers for improved diagnosis, prognosis, and therapy. Therefore, a thorough understanding of miRNAs in diseases is crucial for realizing their clinical potential.

This paper introduces a pipeline of analysis aimed at predicting disease conditions of patients by leveraging the large amount of data available in ExomiRHub, a database that stores extracellular and intracellular miRNAs data from multiple clinical studies. The proposed pipeline solves the inconsistencies raised by the integration of data collected and stored following different protocols, and enables learning predictive models from a proper amount of training data. We also show how the learned models from such data can be exploited to identify novel disease biomarkers by means of explainability techniques. Our experiments show the effectiveness of predicting disease conditions using microRNA expression values, as well as the potential of such models as a tool to discover novel non-invasive disease biomarkers.

Keywords: Bioinformatics · Prediction of disease conditions · Biomarker Identification

1 Introduction

In recent years, the exploration of *microRNAs* (*miRNAs*) as potential disease biomarkers raised new opportunities to improve diagnosis, prognosis, and therapeutic strategies. MiRNAs are small non-coding RNAs consisting of approximately 22 nucleotides. Their main function is to influence (mostly inhibit) gene expression by binding to specific *messenger RNA* (*mRNA*) targets and preventing protein synthesis [8]. The studies conducted so far showed that miRNA

A. Appice et al. (Eds.): ISMIS 2024, LNAI 14670, pp. 77–86, 2024.
https://doi.org/10.1007/978-3-031-62700-2_8

expression levels are often altered in many diseases, such as cancer, neurodegenerative disorders, cardiovascular diseases, and infections [5]. Therefore, they can serve as powerful diagnostic tools for the identification of diverse pathological conditions. However, elucidating the role of miRNAs in specific diseases remains a significant challenge, worsened by the resource-intensive nature of experimental investigations. In this context, many computational methods have been proposed [1,3,7], providing cost-effective approaches to formulate biological hypotheses that can be subsequently validated through in-vitro experiments.

One common problem faced by computational approaches is the limited amount of data available about known miRNA-disease associations, that restricts their ability to build reliable models. As a result, constructing association networks from heterogeneous biological data has emerged as a primary goal in overcoming this obstacle [3]. *LP-HCLUS* [1] is an example of an approach that aims to reach this goal by solving a link prediction task on heterogeneous graphs to unveil previously unknown RNA-disease associations. It exploits validated relationships to predict novel associations between non-coding RNAs and diseases, demonstrating its potential in elucidating the functional role of miRNAs in disease onset or progression.

Several other computational models have been proposed to predict miRNA-disease associations, including *NTSMDA* [9] and *EDTMDA* [4]. While these models have shown promising results, they have some limitations. For example, *NTSMDA* discovers new miRNA-disease associations starting from a set of known miRNA-disease associations, but cannot be applied to diseases for which there is no miRNA related to them. On the other hand, *EDTMDA* exploits ensemble learning, matrix factorization, and dimensionality reduction to generate predictions, but its calculation of miRNA-disease similarity is imperfect, since it discards some relevant biological information [4].

In general, the adoption of computational approaches and, more specifically, of machine learning techniques in the biomedical field must address several challenges: complex and heterogeneous input data, massive datasets with many examples and features (often with few labeled data), multiple data sources that may be subject to different data processing, and the need to balance the quality of the result with the invasiveness in terms of collection of personal data.

In this paper, we propose a novel pipeline to identify disease biomarkers by predicting the healthy or disease condition of patients from miRNA expression data, and by identifying the most relevant miRNAs that contribute to such predictions. In this way, such miRNAs can be considered possible indicators (thus, potentially novel biomarkers) of disease/healthy conditions.

The primary data source for our study is ExomiRHub [6], a comprehensive platform designed for integrating and analyzing the human extracellular miRNA transcriptome to facilitate the discovery of non-invasive biomarkers. ExomiRHub aggregates extracellular miRNA expression data across various diseases, mostly related to different types of cancer, by merging data from different studies reported in Gene Expression Omnibus (GEO)[1]. While ExomiRHub is a

---

[1] https://www.ncbi.nlm.nih.gov/gds.

valuable resource, it exhibits some limitations: the data collected from different studies may not be suitable for an integrated analysis due to different isolation methods and detection platforms adopted, that make the reported measurements hardly comparable [6]. Analyzing data from ExomiRHub can also be challenging due to a high variability in terms of quality and quantity of data related to the different studies considered: each study has unique characteristics, such as sample size, sample type, and experimental conditions. Moreover, even within studies sharing certain characteristics, significant differences in terms of miRNA expression values can be found. To the best of the authors' knowledge, no previous work leveraged data related to multiple studies, such as those of ExomiRHub, to predict disease conditions related to miRNAs and/or to discover new biormarkers.

The approach we propose addresses these limitations through a sophisticated pre-processing pipeline to enhance and align the data related to multiple studies, enabling their simultaneous exploitation. Subsequently, we adopt the pre-processed data to train two machine learning models through Random Forest (RF) and Multi-Layer Perceptron (MLP), with the goal of predicting healthy/disease conditions of patients from miRNA expression data and sample metadata. By exploiting some techniques based on explainability, we also identify some potentially novel disease biomarkers.

2 Materials and Methods

In the field of biological research, numerous digital resources have been developed to assist scientists in collecting and analyzing data. One such resource is ExomiRHub [6], a web platform that aggregates extracellular miRNA expression data for various diseases. Specifically, ExomiRHub contains data of 29,198 samples coming from a total of 191 studies obtained by querying GEO for specific keywords such as *exosome, extracellular vesicle, plasma,* or *serum miRNA,* and then selecting studies related to human miRNAs from the obtained candidates. Data about the studies have been annotated with biomedical information about the sample, including the disease condition and sample type. As a result, each study in ExomiRHub consists of two types of data: the first containing metadata about the samples, and the second containing miRNA expression values.

As mentioned in the introduction, data about different studies in ExomiRHub are not easily integrable for several reasons, including *i)* the type of technology used for the expression profile analysis, *ii)* the different subsets of miRNAs considered in each study, and *iii)* the diverse names used for identifying the same features across studies. Furthermore, the number of samples for each study fluctuates dramatically, ranging from 1 to approximately 5000, with the larger study accounting for almost 20% of the total number of samples. These inconsistencies within the data may negatively affect the training of a machine learning model, potentially compromising the accuracy and reliability of the results. Therefore, in this specific scenario, performing an accurate data pre-processing is essential for extracting valuable knowledge from the available data.

2.1   The Proposed Pre-processing Pipeline

The proposed pipeline includes several data preparation steps for both sample metadata and miRNA expression values, with the ultimate goal to integrate data about multiple studies reported in ExomiRHub. Figure 1 shows the entire pipeline, from the raw data to the prediction of the disease through machine learning models and the identification of novel biomarkers.

The data preparation begins with the **Study Selection** phase. In this phase, we first identify and exclude studies that are not specifically related to a given disease. Subsequently, redundant or mutually inconsistent studies are removed. For instance, the studies *EMIR00000091*, *EMIR00000092*, *EMIR00000097*, and *EMIR00000101* are discarded because they are subsets of *EMIR00000102*. Similarly, *EMIR00000177* and *EMIR00000180*, contain common samples with inconsistent miRNA expression levels. Therefore, only the larger dataset, i.e., *EMIR00000177*, is preserved. Finally, the miRNA expression profiling method used in each study is retrieved from the GEO website, since not directly available in ExomiRHub, and studies with unknown (i.e., reported as "other" in GEO) miRNA profiling are excluded from the analysis, to avoid the use of data coming from unclear or not-standard profiling strategies. After this initial study selection phase, only 48 studies out of the initial 191 studies are kept.

Subsequently, a **Sample Selection** phase is performed. Specifically, 18 samples not related to Homo sapiens are removed. Furthermore, we exclude 189 samples based on two-channel microarrays as profiling technique. This step was made to preserve samples that are as homogeneous as possible in terms of miRNA expression profiling techniques.

Finally, we performed a **Feature Selection** phase, consisting in preprocessing the miRNA expression data and the sample metadata, separately, related to the remaining studies. Specifically, in the context of miRNA expression data,

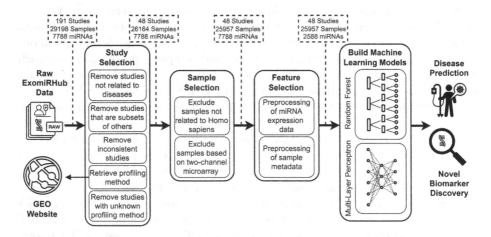

Fig. 1. The proposed pipeline for the prediction of disease conditions and the discovery of novel biomarkers from sample metadata and miRNA expression data.

the preprocessing aims to standardize the feature names representing the miR-NAs. Indeed, the selected studies lack a consistent representation for miRNA names, resulting in a high number of distinct identifiers (7,788), with multiple different identifiers possibly representing the same miRNA (673 duplicate identifiers). Next, miRBaseConverter [11] is used to trace back all the identifiers to the last version of miRBase (v22). Unfortunately, various studies contain identifiers which do not comply with the standards used in the literature. In particular, 4,527 identifiers are not present in any version of miRBase. Since these identifiers cannot be considered reliable for discovering novel biomarkers, they are excluded from the analysis. After this phase, a total of 2,588 miRNAs are kept.

On the other hand, as regards sample metadata, it is noteworthy that descriptive information are represented as flat features for some studies and as nested/hierarchical structures (in a JSON-like format) for other studies. Therefore, we first flatten features containing nested structures. Moreover, to finally make the representation of the same feature coherent along all the studies, all the features undergo a standardization process for both their names and values. This step is necessary because different studies use different feature names (e.g., *sex* and *gender*) to represent the same characteristic or different strings (e.g., *female, f, F*, etc.) to represent the same feature value. Finally, features with > 95% of missing values are discarded from the analysis, because their contribution towards the construction of a predictive model would be irrelevant. The final set of features preserved from sample metadata is *sample_id, technology* (profiling technique), *country, city, sample_type* (type of the sample, like *serum circulating miRNA, serum derived exosomes*, etc.), *gender, age, disease* (which is the target attribute of the predictive model to build).

2.2 Construction of Predictive Models and Biomarkers Identification

After data preparation, the proposed pipeline includes training machine learning models to predict the healthy/disease condition for each sample. Note that the considered classification task is not binary, namely, we are not interested in solely detecting the fact that a sample has any disease or not, but also in detecting which specific disease. Therefore, the considered task is a multi-class classification problem. To solve this task, we opted for two different categories of models: Random Forests and Multi Layer Perceptron (MLP). Specifically, we chose to use a tree-based classifier, as it already exhibited state-of-the-art performance on multi-class classification tasks in the biological domain [4,10].

Decision trees consist of nodes and branches, and are learned using a top-down approach that recursively divides the data set. Each node in the tree considers a specific feature and a value or threshold, and uses this information to split the data into smaller groups. The process of splitting the data is repeated until the leaf nodes are reached, where the predicted labels or numerical values are found. The splits are chosen greedily, with the goal of maximizing a particular heuristic (the reduction of the Gini Index [2], in our case). The adopted Random Forests method consists in learning multiple decision trees from different subsets of samples and/or features, which tends to reduce variance, namely huge changes in the output due to small changes in the training instances.

The second method employed is the Multi-Layer Perceptron (MLP), a type of artificial neural network (ANN) commonly used for supervised learning tasks. An MLP consists of multiple layers of neurons, with each layer processing the input data in a different way. The first layer, known as the *input layer*, takes in the raw data on which the network will be trained on. Subsequent layers, referred to as *hidden layers*, perform nonlinear transformations on the data, enabling the network to learn more complex patterns and relationships. Finally, the *output layer* produces the ultimate output of the network, i.e., the predicted class of disease. The choice of MLP is based on its inherent ability to capture complex relationships within the data, making it a suitable candidate for exploring the distinctive contributions of sample metadata and miRNA expression values.

Finally, we apply an explainability technique to identify those features that provide the highest contribution to the prediction of a given disease condition. MicroRNAs highly contributing to the prediction can be considered potentially novel biomarkers, that would deserve to be better studied through in-vitro experiments. Methodologically, we adopt the permutation feature importance post hoc interpretation method, which measures the relevance of a given feature through the decrease in the model performance after the feature's values are shuffled in the test set: the highest the decrease in the performance after the shuffling, the highest the contribution provided by the feature.

3   Results and Discussion

We conducted a comparative analysis to evaluate if the available data can fruitfully be exploited to predict disease conditions and, therefore, to identify candidate biomarkers. We also evaluated the predictive performances when using only sample metadata or only miRNA expression values.

All the experiments were run in a stratified 5-fold cross validation setting, using the implementation of Random Forests and MLP available in scikit-learn (https://scikit-learn.org/1.4/), with their default parameter configurations. As evaluation measures, we considered the *accuracy*, the *sensitivity*, the *specificity*, and the *informedness*. The accuracy assesses the proportion of correctly classified instances out of all instances in the test set. However, it is not able to properly evaluate the model performance when classes are imbalanced, like in our case (see the column #*ex* in Tables 1 and 2). Therefore, we also report two metrics that are commonly adopted in biomedicine: *sensitivity* and *specificity*, that correspond to the true positive rate (TPR) and to the true negative rate (TNR), respectively, in the machine learning field. Finally, we collect the results in terms of informedness, that is computed as *sensitivity* + *specificity* $- 1$.

We collect specificity, sensitivity and informedness for each class, separately, and also compute an arithmetic average as well as a weighted average, according to the number of instances for each class.

Table 1. Sensitivity, specificity, informedness, and overall accuracy obtained by RF.

Class	#ex	Metadata			miRNA			Combined		
		Sens.	Spec.	Infor.	Sens.	Spec.	Infor.	Sens.	Spec.	Infor.
Acquired immune deficiency synd.	5	1.00	0.00	0.00	0.80	0.50	0.30	0.80	0.50	0.30
Benign bone and soft tissue dis.	88	0.02	0.99	0.01	0.71	0.89	0.60	0.71	0.92	0.62
Benign breast dis.	13	0.00	1.00	0.00	0.08	1.00	0.08	0.08	1.00	0.08
Benign ovarian dis.	8	0.00	1.00	0.00	0.38	1.00	0.38	0.00	1.00	0.00
Benign pancreatic dis.	4	0.00	1.00	0.00	0.00	1.00	0.00	0.25	1.00	0.25
Benign prostate dis.	29	0.00	1.00	0.00	0.45	1.00	0.45	0.41	1.00	0.41
Benign thyroid nodule	2	0.00	1.00	0.00	0.00	1.00	0.00	0.00	1.00	0.00
Biliary tract cancer	38	0.00	1.00	0.00	0.45	0.97	0.42	0.40	0.98	0.37
Bladder cancer	89	0.00	1.00	0.00	0.89	0.96	0.85	0.49	0.96	0.46
Borderline ovarian tum.	13	0.00	1.00	0.00	0.00	1.00	0.00	0.00	1.00	0.00
Breast cancer	339	0.01	1.00	0.01	0.85	0.98	0.83	0.88	0.98	0.86
Cholelithiasis	2	0.00	1.00	0.00	0.00	1.00	0.00	0.00	1.00	0.00
Chronic hepatitis C	25	0.44	0.94	0.38	0.84	0.75	0.59	0.76	0.81	0.57
Chronic pancreatitis	2	0.50	0.75	0.25	1.00	1.00	1.00	0.50	1.00	0.50
Colon cancer	10	0.00	1.00	0.00	0.20	0.99	0.19	0.20	1.00	0.20
Colon carc.	18	0.94	0.14	0.09	1.00	0.86	0.86	1.00	0.43	0.43
Colorectal cancer	53	0.00	1.00	0.00	0.15	1.00	0.15	0.17	1.00	0.17
Colorectal carc.	3	1.00	0.00	0.00	0.00	1.00	0.00	0.00	1.00	0.00
Esophageal cancer	52	0.00	1.00	0.00	0.21	1.00	0.21	0.48	1.00	0.48
Esophageal squamous cell carc.	123	0.02	0.99	0.02	0.67	1.00	0.66	0.92	1.00	0.91
Fulminant myocarditis	1	0.00	1.00	0.00	0.00	1.00	0.00	0.00	1.00	0.00
Gestational diabetes mellitus	3	0.00	1.00	0.00	0.00	1.00	0.00	0.00	1.00	0.00
Glioblastoma	8	0.00	1.00	0.00	0.75	0.60	0.35	0.75	0.60	0.35
Glioma	58	0.03	1.00	0.03	0.76	0.99	0.75	0.53	0.99	0.52
Head and neck cancer	6	0.00	1.00	0.00	0.33	1.00	0.33	0.33	1.00	0.33
Hepatocellular carc.	54	0.24	0.99	0.23	0.41	0.99	0.40	0.28	1.00	0.27
Int. bone and soft tissue tum.	29	0.00	1.00	0.00	0.00	0.99	-0.01	0.00	1.00	0.00
Intrahepatic cholangiocarcinoma	20	0.40	0.62	0.02	0.95	0.73	0.68	0.95	0.73	0.68
Liver cancer	10	0.00	1.00	0.00	0.60	0.99	0.59	0.60	0.99	0.59
Liver cirrhosis	1	0.00	0.00	-1.00	0.00	1.00	0.00	0.00	1.00	0.00
Lung adenocarcinoma	7	0.86	1.00	0.86	0.00	0.91	-0.09	0.14	0.82	-0.04
Lung cancer	360	0.64	0.86	0.50	0.96	0.97	0.92	0.94	0.94	0.88
Lung cancer (post-operation)	36	0.00	1.00	0.00	1.00	1.00	1.00	1.00	1.00	1.00
Lung carc.	41	1.00	0.00	0.00	1.00	0.00	0.00	1.00	0.00	0.00
Malignant bone and soft tissue tum.	83	0.12	0.94	0.06	0.83	0.83	0.66	0.83	0.90	0.73
Meningioma	3	0.00	1.00	0.00	0.00	1.00	0.00	0.00	1.00	0.00
Metastatic brain tum.s	6	0.00	1.00	0.00	0.00	1.00	0.00	0.00	1.00	0.00
Multiple myeloma	6	0.50	0.50	0.00	0.33	1.00	0.33	0.33	1.00	0.33
Nasopharyngeal carc.	24	1.00	0.00	0.00	0.96	0.42	0.38	0.96	0.42	0.38
Non-alcoholic steatohepatitis	1	0.00	1.00	0.00	0.00	1.00	0.00	0.00	1.00	0.00
Non-small cell lung cancer	4	0.50	1.00	0.50	1.00	0.00	0.00	1.00	0.00	0.00
Obesity-related dis.	2	1.00	0.50	0.50	1.00	0.50	0.50	1.00	0.50	0.50
Oral squamous cell cancer	5	0.40	1.00	0.40	1.00	1.00	1.00	1.00	1.00	1.00
Ovarian cancer	81	0.07	1.00	0.07	0.64	0.99	0.63	0.89	0.96	0.85
Pancreatic cancer	85	0.12	1.00	0.11	0.58	0.99	0.57	0.60	0.99	0.59
Perioperative myocardial injury	1	0.00	1.00	0.00	0.00	1.00	0.00	0.00	1.00	0.00
Polycystic ovary synd.	3	0.00	1.00	0.00	0.33	0.67	0.00	0.33	0.67	0.00
Preeclampsia	8	0.00	1.00	0.00	0.13	1.00	0.13	0.00	1.00	0.00
Pre-term birth	8	0.00	1.00	0.00	0.00	1.00	0.00	0.00	1.00	0.00
Central nervous system lymph	8	0.00	1.00	0.00	0.00	1.00	0.00	0.00	1.00	0.00
Prostate cancer	261	0.35	0.86	0.20	0.90	0.92	0.83	0.95	0.87	0.82
Retinoblastoma	5	0.00	1.00	0.00	1.00	0.75	0.75	0.80	0.75	0.55
Sarcoma	41	0.00	1.00	0.00	0.61	1.00	0.61	0.46	1.00	0.46
Small cell lung carc.	1	1.00	1.00	1.00	0.00	1.00	0.00	0.00	1.00	0.00
Squamous cell	1	1.00	1.00	1.00	0.00	1.00	0.00	0.00	1.00	0.00
Stomach cancer	65	0.08	1.00	0.08	0.51	0.99	0.50	0.35	0.99	0.35
Thyroid cancer	3	1.00	0.00	0.00	1.00	0.00	0.00	1.00	0.00	0.00
Type 1 autoimmune pancreatitis	2	0.00	1.00	0.00	0.00	1.00	0.00	0.00	1.00	0.00
Ventricular septal defect in the fetus	2	0.00	1.00	0.00	1.00	1.00	1.00	1.00	1.00	1.00
Wilms tum. (before chemotherapy)	6	0.00	1.00	0.00	0.67	1.00	0.67	0.67	0.93	0.60
Wilms tum. (post chemotherapy)	6	0.00	1.00	0.00	0.67	0.86	0.52	0.83	0.86	0.69
Avg.		0.23	0.85	0.09	0.47	0.88	0.35	0.45	0.88	0.33
Weighted Avg.		0.22	0.91	0.13	0.73	0.93	0.66	0.73	0.92	0.65
Accuracy		0.62			0.86			0.87		

84   D. Rosa et al.

Table 2. Sensitivity, specificity, informedness, and overall accuracy obtained by MLP.

Class	#ex	Metadata			miRNA			Combined		
		Sens.	Spec.	Infor.	Sens.	Spec.	Infor.	Sens.	Spec.	Infor.
Acquired immune deficiency synd.	5	1.00	0.00	0.00	1.00	0.00	0.00	1.00	0.00	0.00
Benign bone and soft tissue dis.	88	0.03	0.99	0.03	0.75	0.84	0.59	0.61	0.90	0.51
Benign breast dis.	13	0.00	1.00	0.00	0.00	1.00	0.00	0.00	1.00	0.00
Benign ovarian dis.	8	0.00	1.00	0.00	0.38	1.00	0.37	0.25	1.00	0.25
Benign pancreatic dis.	4	0.00	1.00	0.00	0.00	1.00	0.00	0.00	1.00	0.00
Benign prostate dis.	29	0.00	1.00	0.00	0.62	1.00	0.62	0.69	1.00	0.68
Benign thyroid nodule	2	0.00	1.00	0.00	1.00	0.33	0.33	0.00	1.00	0.00
Biliary tract cancer	38	0.00	1.00	0.00	0.24	1.00	0.23	0.21	0.99	0.20
Bladder cancer	89	0.00	1.00	0.00	0.75	0.97	0.72	0.90	0.94	0.84
Borderline ovarian tumor	13	0.00	1.00	0.00	0.08	0.99	0.07	0.00	1.00	−0.01
Breast cancer	339	0.00	1.00	0.00	0.84	0.98	0.83	0.88	0.98	0.86
Cholelithiasis	2	0.00	1.00	0.00	0.00	1.00	0.00	0.00	1.00	0.00
Chronic hepatitis C	25	0.52	0.88	0.40	0.88	0.81	0.69	0.96	0.75	0.71
Chronic pancreatitis	2	0.00	1.00	0.00	0.00	0.75	−0.25	0.00	1.00	0.00
Colon cancer	10	0.00	1.00	0.00	0.70	0.86	0.56	0.80	0.81	0.61
Colon carc.	18	1.00	0.14	0.14	0.94	0.29	0.23	0.94	0.43	0.37
Colorectal cancer	53	0.00	1.00	0.00	0.15	1.00	0.15	0.17	1.00	0.17
Colorectal carc.	3	1.00	0.00	0.00	1.00	0.00	0.00	0.67	1.00	0.67
Esophageal cancer	52	0.00	1.00	0.00	0.04	1.00	0.04	0.12	0.99	0.10
Esophageal squamous cell carc.	123	0.03	1.00	0.03	0.52	1.00	0.52	0.61	1.00	0.61
Fulminant myocarditis	1	0.00	1.00	0.00	1.00	1.00	1.00	1.00	1.00	1.00
Gestational diabetes mellitus	3	0.00	1.00	0.00	0.33	1.00	0.33	0.33	1.00	0.33
Glioblastoma	8	0.00	1.00	0.00	0.75	0.60	0.35	1.00	0.40	0.40
Glioma	58	0.00	1.00	0.00	0.74	0.98	0.73	0.67	0.98	0.65
Head and neck cancer	6	0.00	1.00	0.00	1.00	0.50	0.50	0.00	1.00	0.00
Hepatocellular carc.	54	0.24	0.99	0.23	0.33	0.99	0.32	0.37	0.99	0.37
Int. bone and soft tissue tumor	29	0.00	1.00	0.00	0.00	1.00	0.00	0.00	1.00	0.00
Intrahepatic cholangiocarcinoma	20	0.95	0.08	0.03	0.90	0.69	0.59	0.90	0.85	0.75
Liver cancer	10	0.00	1.00	0.00	0.40	0.95	0.35	0.60	0.93	0.53
Liver cirrhosis	1	0.00	1.00	0.00	0.00	1.00	0.00	0.00	1.00	0.00
Lung adenocarcinoma	7	0.00	1.00	0.00	0.00	1.00	0.00	0.29	0.73	0.01
Lung cancer	360	0.49	0.90	0.39	0.89	0.97	0.86	0.83	0.97	0.81
Lung cancer (post-operation)	36	0.00	1.00	0.00	0.89	1.00	0.89	0.89	1.00	0.89
Lung carc.	41	1.00	0.00	0.00	1.00	0.00	0.00	1.00	0.00	0.00
Malignant bone and soft tissue tum.	83	0.17	0.93	0.10	0.75	0.92	0.66	0.77	0.92	0.69
Meningioma	3	0.00	1.00	0.00	0.33	0.98	0.32	0.00	0.99	−0.01
Metastatic brain tumors	6	0.00	1.00	0.00	0.17	0.99	0.16	0.00	0.99	−0.01
Multiple myeloma	6	0.83	0.00	−0.17	0.83	1.00	0.83	0.83	0.75	0.58
Nasopharyngeal carcinoma	24	1.00	0.00	0.00	0.63	1.00	0.63	1.00	0.17	0.17
Non-alcoholic steatohepatitis	1	0.00	1.00	0.00	1.00	1.00	1.00	0.00	1.00	0.00
Non-small cell lung cancer	4	0.00	1.00	0.00	1.00	0.00	0.00	0.50	1.00	0.50
Obesity-related dis.	2	1.00	0.50	0.50	0.50	0.00	−0.50	0.00	1.00	0.00
Oral squamous cell cancer	5	0.80	0.33	0.13	1.00	1.00	1.00	0.80	1.00	0.80
Ovarian cancer	81	0.07	1.00	0.07	0.43	0.99	0.42	0.54	0.99	0.53
Pancreatic cancer	85	0.00	1.00	0.00	0.49	0.98	0.48	0.66	0.96	0.62
Perioperative myocardial injury	1	0.00	1.00	0.00	0.00	1.00	0.00	1.00	1.00	1.00
Polycystic ovary synd.	3	0.00	1.00	0.00	0.67	0.33	0.00	0.67	0.33	0.00
Preeclampsia	8	0.00	1.00	0.00	0.75	0.77	0.52	0.50	0.54	0.04
Pre-term birth	8	0.00	1.00	0.00	0.38	0.93	0.30	0.75	0.50	0.25
Central nervous system lymph	8	0.00	1.00	0.00	0.00	0.99	−0.01	0.00	0.98	−0.02
Prostate cancer	261	0.35	0.86	0.20	0.88	0.81	0.70	0.91	0.83	0.74
Retinoblastoma	5	1.00	0.00	0.00	0.80	0.25	0.05	0.80	1.00	0.80
Sarcoma	41	0.00	1.00	0.00	0.73	1.00	0.73	0.49	1.00	0.48
Small cell lung carc.	1	0.00	1.00	0.00	0.00	1.00	0.00	0.00	1.00	0.00
Squamous cell carc.	1	1.00	1.00	1.00	0.00	1.00	0.00	0.00	0.80	−0.20
Stomach cancer	65	0.11	1.00	0.11	0.39	0.99	0.37	0.35	0.98	0.34
Thyroid cancer	3	1.00	0.00	0.00	0.33	1.00	0.33	1.00	0.00	0.00
Type 1 autoimmune pancreatitis	2	0.00	1.00	0.00	0.00	0.75	−0.25	0.00	1.00	0.00
Ventricular septal defect in the fetus	2	0.00	1.00	0.00	1.00	1.00	1.00	1.00	1.00	1.00
Wilms tumor (before chemotherapy)	6	1.00	0.00	0.00	0.50	0.86	0.36	0.50	1.00	0.50
Wilms tumor (post chemotherapy)	6	0.00	1.00	0.00	0.00	1.00	0.00	0.83	0.79	0.62
Avg.		0.24	0.81	0.05	0.52	0.82	0.34	0.50	0.85	0.36
Weighted Avg.		0.20	0.91	0.11	0.68	0.91	0.60	0.70	0.91	0.61
Accuracy		0.61			0.83			0.84		

Table 3. Top 10 miRNAs contributing to the prediction for each considered method.

	Top 10 miRNAs (prefix *hsa-miR-* omitted for space constraints)
RF	1228-5p, 5100, 1343-3p, 1199-5p, 8059, 1207-5p, 1225-5p, 192-5p, 548x-3p, 3201
MLP	1290, 10a-3p, 221-3p, 3162-5p, 663a, 939-5p, 197-5p, 30c-1-3p, 210-3p, 25-3p

In Tables 1 and 2, we show the results obtained using only the sample metadata, only the miRNA expression values, and both (indicated as *Combined* in the tables), obtained by RF and MLP, respectively.

By observing the results, it is clear that the use of only sample metadata is not enough to learn effective models. Indeed, RF and MLP obtained an accuracy of 0.62 and 0.61, respectively, with a very low informedness, emphasizing the incapability of modeling under-represented classes.

Looking at the models trained from miRNA expression values, we can see how the results significantly improve for each metric, with an accuracy of 0.86 for RF and 0.83 for the MLP. Notably, the sensitivity almost doubled, which corresponds a better ability to recognize a specific disease, while the specificity remained constant. This also reflects on the results in terms of informedness.

Finally, the models trained on the combination of sample matadata and miRNA expression values achieved results comparable to those trained only from miRNA expression values, with a slight improvement achieved in the case of the MLP. In general, RF achieved better results than MLP, possibly due to its capability to exploit the large amount of available features. As expected in all the cases, diseases with a larger number of samples achieved better results. This highlights the importance of having a significant number of samples per disease.

Finally, in Table 3 we report the top 10 features that mostly contributed to the predictions, according to the strategy introduced in Sect. 2. Such miRNAs represent possible novel candidate biomarkers that would deserve to be experimentally validated in a laboratory. It is noteworthy that the miRNAs suggested by the models learned through RF and MLP are not the same. This is possibly due to the fact that they are based on totally different principles, that also reflect on the way each feature influences the output. In general, this is not a negative aspect, since different approaches may lead to the discovery of possible novel biomarkers exploiting different viewpoints.

4   Conclusions

In this paper, we introduced a novel pipeline for the effective integration of data related to multiple diseases from different studies. We used ExomiRHub as data source and trained two different models to predict disease conditions of patients. The extensive preprocessing phase of the proposed pipeline allowed us to overcome multiple criticisms naturally present in ExomiRHub, which would have compromised the simultaneous exploitation of data related to all the available studies. As expected, our experimental evaluation emphasized that miRNA expression values are much more helpful in supporting the identification of disease conditions than sample metadata. Moreover, by relying on an explainability approach, we can also highlight which miRNAs can be considered as potential biomarker, that deserves to be confirmed by in-lab experiments.

For future work, we will explore the adoption of multi-view learning approaches, rather than feature concatenation, to properly capture the information conveyed by sample metadata and miRNA expression values. Moreover, we will work on designing an ensemble method to exploit the output of several models and aggregate the miRNAs they suggest as potential biomarkers.

Acknowledgments. This work has been partially supported by the European Union - NextGenerationEU through the Italian Ministry of University and Research, Projects PRIN 2022"BA-PHERD: Big Data Analytics Pipeline for the Identification of Heterogeneous Extracellular non-coding RNAs as Disease Biomarkers", grant n. 2022XABBMA, CUP: H53D23003690006, CUP: B53D23013260006, and CN3 RNA - "National Center for Gene Therapy and Drugs based on RNA Technology", CUP: H93C22000430007.

References

1. Barracchia, E.P., Pio, G., D'Elia, D., Ceci, M.: Prediction of new associations between ncRNAs and diseases exploiting multi-type hierarchical clustering. BMC Bioinf. **21**(1), 1–24 (2020)
2. Breiman, L., Friedman, J., Stone, C.J., Olshen, R.A.: Classification and Regression Trees. CRC Press, Boca Raton (1984)
3. Chen, X., Xie, D., et al.: MicroRNAs and complex diseases: from experimental results to computational models. Brief. Bioinf. **20**(2), 515–539 (2019)
4. Chen, X., Zhu, C.C., Yin, J.: Ensemble of decision tree reveals potential miRNA-disease associations. PLoS Comput. Biol. **15**(7), e1007209 (2019)
5. Condrat, C.E., et al.: miRNAs as biomarkers in disease: latest findings regarding their role in diagnosis and prognosis. Cells **9**(2), 276 (2020)
6. Liu, Y., Min, Z., Mo, J., Ju, Z., Chen, J., et al.: ExomiRHub: a comprehensive database platform to integrate and analyze human extracellular miRNA transcriptome for discovering non-invasive biomarkers (2023)
7. Pio, G., Ceci, M., Loglisci, C., D'Elia, D., Malerba, D.: Hierarchical and overlapping co-clustering of mRNA: miRNA Interactions. In: ECAI 2012. Frontiers in Artificial Intelligence and Applications, vol. 242, pp. 654–659. IOS Press (2012)
8. Ranganathan, K., Sivasankar, V.: Micrornas-biology and clinical applications. J. Oral Maxillofacial Pathol. JOMFP **18**(2), 229 (2014)
9. Sun, D., Li, A., Feng, H., Wang, M.: NTSMDA: prediction of miRNA-disease associations by integrating network topological similarity. Mol. BioSyst. **12**(7), 2224–2232 (2016)
10. Williams-DeVane, C.R., Reif, D.M., Cohen Hubal, E., et al.: Decision tree-based method for integrating gene expression, demographic, and clinical data to determine disease endotypes. BMC Syst. Biol. **7**, 1–19 (2013)
11. Xu, T., et al.: miRBaseConverter: an R/Bioconductor package for converting and retrieving miRNA name, accession, sequence and family information in different versions of miRBase. BMC Bioinf. **19**(19), 179–188 (2018)

HERSE: Handling and Enhancing RDF Summarization Through Blank Node Elimination

Amal Beldi[1,2(✉)], Salma Sassi[2], Richard Chbeir[2], and Abderrazek Jemai[1,3]

[1] Faculty of Mathematical Physical and Natural Sciences of Tunis, SERCOM Laboratory, Tunis El Manar University, 1068 Tunis, Tunisia
`abderrazekjemai@yahoo.co.uk`
[2] University Pau & Pays Adour, LIUPPA, Anglet 64600, France
`{amal.beldi,salma.tissaoui,richard.chbeir}@univ-pau.fr`
[3] Polytechnic School of Tunisia, SERCOM Laboratory, INSAT, Carthage University, 1080 Tunis, Tunisia

Abstract. The rapid expansion of digital data and its inherent heterogeneity present challenges in extracting meaningful information, particularly when dealing with substantial volumes of unstructured data distributed across diverse sources. Blank nodes, also known as anonymous nodes or bnodes, play a pivotal role in the Resource Description Framework (RDF) utilized in the semantic web. These nodes, which lack unique identifiers, are frequently employed in RDF/S knowledge bases to represent complex attributes or resources with known properties but unknown identities. This paper introduces an innovative methodology for the detection and conversion of blank nodes within RDF graph summaries. Our approach leverages centralized and decentralized Skolemization to transform these nodes into Skolem Uniform Resource Identifiers (URIs), there by enhancing the clarity and interoperability of RDF data. By improving the accuracy and relevance of RDF summaries in relation to original datasets, we enhance the efficiency and effectiveness of subsequent queries against these summaries.

Keywords: RDF graph · RDF summary · blank nodes · detection · Skolemization

1 Introduction

At the core of the Semantic Web lies the Resource Description Framework (RDF), which acts as a standard for distributing graph-structured data. RDF utilizes Internationalized Resource Identifiers (IRIs) as universal identifiers, enabling graphs from diverse locations on the Web to interact and share information about identical resources. The adoption of RDF on the Web has been steadily increasing, as evidenced by the numerous datasets published in RDF format, adhering to Linked Data and RDF summarization principles [6]. These

A. Appice et al. (Eds.): ISMIS 2024, LNAI 14670, pp. 87–101, 2024.
https://doi.org/10.1007/978-3-031-62700-2_9

datasets cover a broad spectrum of domains, including collections from governmental organizations, the medical and scientific communities, social platforms, media sources, online encyclopedias, and more [5]. Additionally, a significant number of websites, amounting to hundreds of thousands, along with hundreds of millions of web pages, now incorporate embedded RDFa [17]. In the current context, RDF includes blank nodes, which represent resources without specific identifiers. The presence of these blank nodes in RDF graphs introduces an additional layer of complexity to the operations involved. In the initial W3C recommendation for RDF, released in 1999 [3], the concept of anonymous nodes was introduced as a means to represent resources without specified identifiers. This feature was justified by various use cases, including the depiction of resource collections in RDF, the application of reification to treat RDF statements as if they were standalone resources, and the description of resources without a native Uniform Resource Identifier (URI) or Internationalized Resource Identifier (IRI) association. In the context of graph summarization, previous studies on detecting blank nodes in RDF graphs have not extensively explored this issue. Today, blank nodes present different challenges, such as making summarized RDF graphs more difficult to comprehend and analyze since blank nodes may obscure critical information about the underlying data. Additionally, blank nodes can affect the performance of applications utilizing summarized RDF graphs, as they might require more resources to process than non-blank nodes. Moreover, blank nodes can be exploited to manipulate summarized RDF graphs, potentially concealing sensitive information or fabricating false connections between elements. This paper primarily addresses these challenges:

- **Challenge 1:** How can we effectively identify and locate blank nodes in our summarized RDF graphs?
- **Challenge 2:** What strategies can be employed to handle blank nodes and enhance the visualization of information in summarized RDF graphs?

The paper's structure is outlined as follows: Sect. 2 elaborates on the significance of handling and enhancing RDF summarization through blank node elimination. Section 3 provides an overview of RDF summarization techniques and discusses their limitations, particularly regarding blank node handling. Section 4 delves into the proposed HERES framework for detecting blank nodes, with a focus on Skolemization, highlighting the Blank Node Resolution phase. Section 5 covers the experimental evaluation, detailing the conducted experiments, their results, and ensuing discussions. Finally, Sect. 6 concludes the study and outlines future research directions.

2 Why is it Important to Detect Blank Nodes in RDF Summary Graphs?

Despite the significant increase in RDF usage on the Web in recent years [7], blank nodes, a crucial aspect of RDF, often encounter challenges in their perception and utilization. Blank nodes, which lack identifiable properties, can emerge due to a variety of reasons, including data errors, data compression, or the use of summarizing techniques. These nodes have frequently been misunderstood, misinterpreted, or ignored by implementer, other standards, and the wider Semantic Web community. In our research [14], we primarily focus on developing an RDF schema approach to summarize heterogeneous data sources, explicitly aiming to minimize the reliance on representing data through blank nodes. However, the significance of addressing blank nodes becomes evident when we examine specific use cases within our application domain, such as the medical field. Consider a scenario within the medical domain where blank nodes might manifest in an RDF graph, particularly when documenting patient data. Imagine utilizing an RDF database to compile a comprehensive medical history for a patient. The absence of certain details, like family medical history, lifestyle habits, or specific medical data due to it being unprovided or unavailable, could result in the generation of blank nodes in the patient's RDF graph summary. For example, a scenario where a patient's family medical history is not documented in the database or specific health details are omitted during medical consultations would lead to the creation of blank nodes for this missing information. Such occurrences complicate data analysis and highlight the critical need for both thorough data collection and the integrity of medical data. This situation underscores two key points: the necessity for diligent data gathering processes and how the existence of blank nodes in RDF graphs can signify information voids, potentially affecting the depth and precision of medical analyses. Consequently, initiatives aimed at improving data completeness and quality are vital for the trustworthiness of medical RDF graphs and the analyses derived from them. The detection of blank nodes in an RDF graph summary plays a vital role in various aspects as Data Completeness, Data Quality, Data Integrity and Decision-Making. Several alternative methods have been proposed for handling blank nodes in RDF graphs, particularly within the RDF 1.1 Working Group [4]. These approaches are crucial for improving graph comprehension and analysis, enhancing application performance, and maintaining data integrity. They encompass various techniques [1] such as RDF syntax analysis, graph processing algorithms, semantic analysis methods, and statistical analysis, along with the transformation of anonymous nodes into URI references. In RDF graphs, blank nodes represent resources without specific identifiers and are indispensable for preserving data integrity and accurately representing the underlying data. However, when it comes to summarized RDF graphs, which are often created to simplify complex RDF data or summarize large datasets, the treatment of blank nodes may vary. Some approaches may choose to omit blank nodes altogether to streamline the graph and highlight essential information relevant to the summarization objective. Addressing blank nodes in RDF summaries not only ensures data integrity and facilitates analysis but also contributes to the overall effectiveness and relevance of the summarized data for various applications and use cases.

3   Overview of RDF Summarization Techniques and Blank Node Handling

Our objective was to investigate different RDF summarization techniques to assess their handling of blank nodes in RDF summaries. This section presents an overview of relevant systems dedicated to condensing RDF graph summaries and identifying blank nodes. These systems often utilize predefined rules or patterns to extract key information and fundamental concepts from RDF graphs. Considerable research endeavors have resulted in the development of various strategies for RDF graph summarization, categorized broadly into four main groups: structural techniques, statistical methods, pattern mining strategies, and hybrid approaches. Structural Methods: capture the inherent structural characteristics of RDF graphs by identifying key elements, such as nodes and relationships, that represent the overall structure. Two types are structural quotient methods [8], which use quotient graphs to assign representatives to equivalent nodes, and structural non-quotient methods [9,10], which connect important nodes using centrality measures. Statistical Methods approaches provide quantitative summaries of RDF graph contents. Researchers, such as [13], explore statistical approaches involving counting occurrences, constructing histograms, and analyzing quantitative measures. The goal is to capture and summarize the statistical properties and distributions of data within the RDF graph. Pattern Mining approaches leverages patterns present in the RDF graph to generate the summary. For example, Zneika [11] proposes an approach that involves mining approximate graph patterns to construct the summary, represented as an RDF graph. This allows more efficient querying and analysis of data. Hybrid Methods approaches combine structural, statistical, and pattern mining techniques [12]. By integrating multiple techniques, they aim to provide more comprehensive and accurate summaries of RDF graphs, leveraging the strengths of different methods to offer a holistic view of the graph's structure, content, and patterns. Our analysis underscores that each method initiates with an RDF graph to construct a corresponding summary. While specific investigations, as referenced in [13] and [11] address distinct facets of user requisites, there persists a conspicuous lacuna in comprehensively fulfilling user preferences and demands. Significantly, existent summarization techniques exhibit deficiencies in detecting and managing blank nodes, revealing a substantial domain for advancement and refinement within RDF graph summarization. This inadequacy accentuates the imperative for more sophisticated methodologies adept at surmounting this challenge effectively. For these reasons, we propose in this paper a novel approach to managing blank nodes based on centralized and decentralized Skolemization [1].

4   The HERSE Framework: A Novel Approach to Blank Node Management in RDF Summary

We introduce the "HERSE" Framework (Handling and Enhancing RDF Summarization through blank node Elimination), presenting a novel approach for managing blank nodes in RDF summaries. HERSE is designed to achieve two primary objectives: (a) Effective Identification and Localization of Blank Nodes and (b) Proposing Strategies for Enhanced Information Visualization. HERSE is based on Skolemization techniques [4]. Our choice is motivated by the fact that Skolemization ensures the unique identification of blank nodes by replacing them with unique identifiers (Skolem URIs), thereby facilitating their use across different contexts and enhancing interoperability among RDF datasets. Additionally, Skolemization simplifies queries and reasoning on RDF graphs by reducing ambiguity and preserving the original semantics of the data. We have chosen to integrate both centralized and decentralized Skolemization [1, 4] approaches within HERSE to provide users with versatile solutions. Centralized Skolemization offers a centralized service for generating unique URIs, while decentralized Skolemization empowers publishers to generate constants locally. Both approaches aim to replace blank nodes with well-known URIs, promoting data consistency and avoiding naming conflicts among documents. By leveraging Skolemization techniques, HERSE endeavors to enhance the quality and usability of RDF summaries while upholding data integrity and interoperability. Recognizing the importance of user-centric design, HERSE offers users the flexibility to choose between centralized and decentralized Skolemization approaches, empowering them to tailor the framework according to their specific requirements and preferences.

4.1 HERSE Architecture

In this paper, we build upon our prior work presented in [14] where we introduced a novel approach to RDF graph summarization. Our current focus shifts towards the critical aspects of detection and resolution of blank nodes within RDF summaries. The HERSE (Handling and Enhancing RDF Summarization through Blank Node Elimination) framework presents an architecture tailored for the refinement of RDF graph summaries. By methodically managing blank nodes, HERSE stands out in its ability to generate summaries that are both lucid and actionable. Central to our architecture is the identification and resolution of blank nodes within heterogeneous datasets. HERSE accomplishes this by detecting these nodes and systematically assigning them unique Internationalised Resource Identifiers (IRIs) through decentralized Skolemization. This innovative approach not only fills data gaps but also brings about a more integrated and comprehensive representation of the dataset, elevating the summarization process to new heights of clarity and precision. The HERSE architecture is composed of two phases1 (Fig. 1):

- **Phase 1: RDF Summary Creation Phase:** This first phase aims to generate an RDF graph summary from heterogeneous databases. This phase consists of several modules:
    - **Preprocessing Module:** this module encompasses three key steps: data cleaning, data classification, and data identification

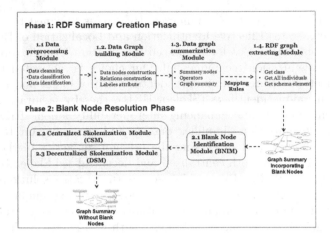

Fig. 1. The HERSE Framework Architecture for RDF Graph Summarization and Blank Node Resolution

- **Data Graph Building Module:** An aggregation process is applied to transformed input data to generate aggregate values. This module constructs a graph-based data model iteratively, consisting of aggregated items. [15]
- **Data Graph Summarization Module:** This module generates a succinct output by implementing a schema-driven approach considering both the structure and content of the data. [2]
- **RDF Graph Extracting Module:** In this module, a set of mapping rules is defined to construct an RDF graph according to the provided Graph Summary (GS).
- **Phase 2: Blank Node Resolution Phase:** This second phase is the core phase of the HERSE framework and consists of identifying and managing blank nodes. This phase is tailored to address the inherent challenges of blank nodes in RDF datasets.

It is composed by three modules:

- **Blank Node Identification Module (BNTM):** This module aims to detect blank nodes in the RDF summary graph resulted from phase 1.
- **Centralized Skolemization Module (CSM):** This module provides a centralized approach for managing blank nodes. When blank nodes are detected, they can be resolved centrally using this module. This involves assigning unique Skolem URIs to blank nodes through a centralized service, ensuring consistent identification and integration into the RDF graph.
- **Decentralized Skolemization Module (DSM):** In contrast to the centralized approach, the Decentralized Skolemization Module offers a decentralized method for managing blank nodes. Users have the option to handle blank nodes locally, generating unique Skolem URIs within their own environment.

This approach empowers users to manage blank nodes autonomously while maintaining control over the generation of Skolem URIs.

Our methodology leverages skolemization to transform blank nodes found in RDF summaries into Skolem URIs. This approach provides a standardized mechanism for node identification and facilitates the generation of an RDF summary free from the complications of blank nodes. The result is a more refined model that adeptly encapsulates diverse data sources' structure and content. The significance of this work lies in its commitment to improving the utility of RDF data by addressing the challenges associated with blank nodes. By incorporating Skolemization into our framework, we offer a practical solution that aligns with the foundational principles of the semantic web, optimizing the process of RDF data handling and analysis.

4.2   Decentralized Skolemization Module (DSM)

After identifying anonymous nodes in the RDF summary graph, this module utilizes decentralized Skolemization to handle blank nodes effectively. For instance, in the RDF graph shown in Fig. 2, nodes :b1 and :b2 are anonymous. Throughout our analysis, we reference this graph. Decentralized Skolemization involves generating Internationalized Resource Identifiers (IRIs) to replace blank nodes. This standardized process creates globally unique IRIs, termed Skolem IRIs, by combining a naming convention (e.g., "genid" or "bnode") with a unique identifier. This approach ensures robust identification and substitution of blank nodes, enhancing data consistency and interoperability within RDF summaries. By employing Skolem IRIs, it seamlessly integrates blank nodes into the RDF graph structure, improving overall coherence and comprehensibility.

Definition 1: Decentralized Skolemization of Blank Nodes in RDF Summary

We propose a mathematical formalism to the detection of empty nodes (or anonymous nodes) in an RDF graph using decentralized Skolemization in an RDF graph summary using decentralized Skolemization.

Let GS be an RDF graph summary containing blank nodes and defined as follows [14]:

$$GS = (C, O, M) \text{ be an RDF graph summary, where:}$$

- C is the set of classes,
- O is the set of objects,
- M is the set of metadata.

We also define:

- N as the set of blank nodes in RDF graph Gs,
- I as the set of unique identifiers assigned to blank nodes.

Attribution of Unique Identifiers Each blank node $n \in N$ is associated with a unique identifier $i \in I$:

$$\forall n \in N, \exists i \in I : \text{ID}(n) = i$$

After each occurrence of a blank node n in the RDF graph summary G has been replaced by its associated unique identifier I, the RDF graph without blank nodes is defined as follows:

$$Gs' = (C', O', M')$$

where

- $C' = C$ (unchanged classes),
- $O' = O \cup I$ (updated objects with unique identifiers of blank nodes),
- $M' = M \cup I$ (updated metadata with unique identifiers of blank nodes).

Thus, the RDF graph GS' obtained after decentralized Skolemization contains only unique identifiers for each blank node, ensuring a consistent and unambiguous representation of RDF data while preserving the elements of class, object, and metadata. The detection of empty nodes can be modeled using a decentralized Skolemization function: Skolem: S→S', where Sn is the set of unique identifiers generated to replace the blank nodes in S. For each blank node n ∈ N, the Skolemization function Skolem can be defined as follows: Skolem (n)= H(n) H(n) is the result of a decentralized hashing algorithm applied to the content or properties of a node0.

The identification of blank nodes in the RDF graph summary can be achieved through a decentralized Skolemization function $Skolem : Sn \rightarrow Sn'$, where Sn denotes the set of nodes in the graph. For each blank node $n \in N$, Skolem replaces it with a unique identifier $H(n)$ generated based on the node's content or properties. By applying a decentralized hashing algorithm to the characteristics of the blank node, $H(n)$ effectively serves as the replacement, facilitating its integration into the RDF graph summary.

Algorithm 1: Decentralized Skolemization using Hashing Algorithm

---

Function DecentralizedSkolemization(*node*):
    hash ← HashFunction(*node*) // Compute hash value of the node content
    skolemNode ← "skolem_" + *hash* // Generate skolem node identifier using hash value **return** *skolemNode*

This algorithm0, utilizing a Hashing Algorithm, outlines the procedure for Decentralized Skolemization, which is a technique used to assign unique identifiers to blank nodes in an RDF graph. The function named Decentralized Skolemization takes a node as input and performs the following steps:

1. **Compute Hash Value:** Firstly, the algorithm computes the hash value of the node's content using a hashing algorithm. Hashing algorithms, such as SHA-256 or MD5, transform the input data into a fixed-size string of characters, known as the hash value. This hash value uniquely represents the content of the node

2. **Generate Skolem Node Identifier:** Next, the algorithm generates a unique skolem node identifier by concatenating the string "skolem_" with the computed hash value. This concatenation ensures that the resulting identifier is distinct and unlikely to clash with other identifiers.

3. **Return Skolem Node Identifier:** Finally, the function returns the generated skolem node identifier, which serves as a replacement for the original blank node. This identifier enables the identification and integration of the blank node into the RDF graph.

Decentralized Skolemization leverages hashing algorithms to create globally unique identifiers for blank nodes. By using the content of the node to compute the hash value, this technique ensures that each blank node is assigned a distinct identifier based on its properties. These identifiers facilitate the seamless integration of blank nodes into RDF graphs while preserving their uniqueness and integrity. Additionally, the decentralized nature of this approach allows each node to generate its identifier independently, enhancing scalability and efficiency in handling large RDF datasets.

4.3   Centralized Skolemization Module (CSM)

The Centralized Skolemization Module (CSM) is integral to our framework for managing blank nodes within RDF summary graphs. Unlike decentralized skolemization, which operates independently, CSM employs a centralized approach for generating unique identifiers for blank nodes. Upon detection of blank nodes within the RDF summary graph, the CSM begins by initiating a process where unique identifiers are assigned to these nodes. This involves interaction with a centralized service responsible for generating Skolem URIs. The service ensures consistency and coherence in the generation of these identifiers across the entire RDF dataset. Functionally, the CSM acts as a central hub for handling blank nodes, ensuring that each blank node receives a unique identifier that is globally recognizable and consistent. This centralized approach streamlines the integration and interpretation of blank nodes, enhancing the overall coherence and interoperability of RDF datasets.

Definition 2: Centralized Skolemization-of Blank Nodes in RDF Summary

We propose a mathematical formalism to the detection of empty nodes (or anonymous nodes) in an RDF graph using decentralized skolemization in an RDF graph summary using centralized skolemization. Let Gs be an RDF graph summary containing blank nodes and defined as follows [14]:

Let $Gs = (C, O, M)$ be an RDF graph summary, where:

- C is the set of classes,
- O is the set of objects, and
- M is the set of metadata.

We also define:

- N as the set of blank nodes in RDF graph summary G,
- S as the distinguished subset of URIs, such that all Skolem constants belong to S,
- U as the set of all URIs.

After each occurrence of a blank node n in the RDF graph summary G has been replaced by the generated Skolem constant s:, the RDF graph without blank nodes is defined as follows:

$$Gs' = (C', O', M')$$

where

- $C' = C$ (unchanged classes),
- $O' = O \cup S$ (updated objects with the generated Skolem constant),
- $M' = M \cup S$ (updated metadata with the generated Skolem constant).

Thus, the RDF graph Gs' obtained after centralized skolemization ensures uniqueness of the generated constants on a global scale, enhancing the coherence and interoperability of RDF datasets.

The detection of empty nodes can be modeled using a centralized Skolemization function Skolem: S→S', where 'sn' is the set of unique identifiers generated to replace the blank nodes in s_n. For each blank node s_n Sn, the Skolemization function Skolem can be defined as follows, based on the centralized skolemization approach: **Unique Identifier Generation**: Skolem(s) = UID(s) where $U_{ID}(s)$ represents a centrally generated unique identifier assigned to node s. By using this centralized Skolemization function, the empty nodes in the RDF graph can be replaced with unique identifiers in a controlled and systematic manner.

This formalism describes the representation of an RDF summary Gs, where Sn is the set of nodes representing resources and O is the set of arcs representing RDF triples.

The detection of empty nodes in this graph is achieved using a centralized skolemization function Skolem : $Sn \rightarrow Sn'$, where Sn' is the set of unique identifiers generated to replace the empty nodes in Sn. For each blank node sn in Sn, the Skolem function assigns a centrally generated unique identifier, denoted as UID(sn). By employing this centralized skolemization function, the empty nodes in the RDF graph can be systematically replaced with unique identifiers in a controlled manner. The algorithm0 represents Centralized Skolemization function takes two parameters: node, which represents the node to be Skolemized, and counter, which keeps track of the current counter value for generating unique Skolem node identifiers. The function concatenates the string "*skolem*" with the value of the counter to generate a unique skolem node identifier, Skolem Node.

This algorithm is useful for centrally generating unique identifiers for Skolemization, ensuring that each blank node in the RDF graph receives a distinct identifier.

Algorithm 2: Centralized Skolemization using Unique Identifier Generation

---

Function CentralizedSkolemization(*node, counter*):

skolemNode ← "skolem_" + *counter* // Generate unique skolem node identifier

IncrementCounter(*counter*) // Increment counter for next skolem node **return** *skolemNode*

---

5 Experimental Evaluation

5.1 Experimental Protocol

The methodology for our experimental evaluation includes the following steps:

- Dataset Preparation: A collection of RDF datasets, featuring varied structures and blank node occurrences, is curated for a comprehensive assessment.
- RDF Summary Generation: This phase involves systematically generating RDF summaries for each prepared dataset. It lays the essential foundation for the subsequent Skolemization process, ensuring the summaries are precisely tailored for effective blank node detection and resolution.
- Skolemization Process: Each dataset is processed to detect and replace blank nodes with unique Skolem IRIs, utilizing both centralized and decentralized Skolemization techniques.
- Performance Metrics: Metrics such as detection accuracy, Skolemization time, and integrity of the RDF summaries are used for quantitative analysis.

Figure 2 illustrates the RDF graph before the process of skolemization. It is a visual representation of semantic data, where the nodes and edges denote the entities and their interrelationships, respectively. In this pre-skolemization state, several nodes are 'blank', indicated by labels such as ":b1" and ":b2", signifying the absence of Uniform Resource Identifiers (URIs). These blank nodes are typically utilized for data points where the identity is either unknown, unspecified, or not necessary for the given context. The predicates, represented by directed relations, establish the type of relationships between subjects and objects within the graph. For instance, predicates like "rdf:type" and "ex:conseille" define the nature of links between various entities such as "Ex:Cid", "Ex:Article", and "Ex:Doctorat". Blank nodes within this RDF graph complicate the semantic structure as they lack global identifiers, which hinders data interoperability and querying. The skolemization process aims to transform these blank nodes into Skolem URIs, thereby providing each previously anonymous node with a unique,

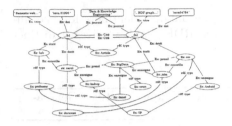

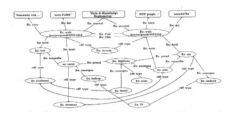

Fig. 2. RDF graph before skolemization

Fig. 3. RDF graph generated after skolemization

globally recognizable identifier. The transformation is expected to enhance the graph's utility by making it more accessible and manageable for querying, reasoning, and linking data across different datasets.

To address and rectify the presence of blank nodes within this RDF graph, we have applied the HERSE methodology. This process involves the systematic detection of blank nodes, followed by their conversion into Skolem URIs using both centralized and decentralized skolemization techniques as dictated by the framework.

5.2   Evaluation

Skolemization Time Analysis. For our evaluation, we intend to assess the impact of varying the number of anonymous nodes on the time required for their Skolemization. Figure 3 displays the RDF graph subsequent to the Skolemization procedure. The Skolemization process effectively substitutes the previously anonymous nodes, represented as _$b1$ and _$b2$, with explicit Skolem URIs. These nodes are now distinctly identified as :http://example.com/.well-known/genid/ 4855185542d68d0de95b5318261f and http://example.com/.well-known/genid/ 4555555a8537248cb28182c1313d. This is achieved through the application of an injective Skolemization function $SK : \beta \rightarrow I_{\text{Skolem}}$, where I_{skolem} is the set of IRIs reserved for Skolemization. These IRIs are unique to the Skolemized RDF graph, ensuring no overlap with the IRIs in the original graph G, i.e., $(I_{\text{Skolem}} \cap I)$.

The line graph in Fig. 5 depicts the Skolemization time, measured in seconds, as a function of the number of blank nodes in an RDF graph. It shows a clear linear progression, indicating that the time required for Skolemization increases in a predictable manner with the number of blank nodes. Starting from a low of 0.167 s for a graph with 20 blank nodes, there is a steady climb to 1.024 s for a graph with 120 blank nodes. This visualization aids in understanding the impact of blank node volume on the Skolemization process and suggests that the HERSE framework is capable of handling increasing quantities of blank nodes with a proportional growth in processing time (Fig. 4).

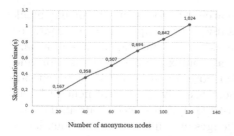

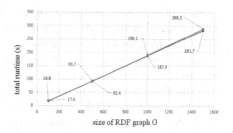

Fig. 4. Skolemization Time for Various Numbers of Blank Nodes

Fig. 5. Total Execution Time of G/w as a Function of RDF Graph Size

Impact of RDF Graph Complexity on Skolemization Performance.
In this exploratory analysis, our objective was to evaluate the influence of RDF graph dimensions on the cumulative execution duration required for our summarization procedures, drawing a comparative analysis with the framework established Čebirić et al. [17]. We systematically processed a series of RDF graphs, each exhibiting a distinct scale, and meticulously recorded the execution time for each. This time encapsulated the entire workflow: from the initial loading of the input graph, through the preprocessing phase addressing anonymous nodes, to the generation of both weak and strong summaries, culminating in the storage of the resultant graph summaries. Our findings revealed a direct, linear correlation between the input graph size and the total execution time, signifying that the summarization time is predominantly dictated by the data preprocessing phase. Figure 5 displays the results of our experimentation. To illustrate, the duration required for generating weak summaries ascended from 17.6 s for a graph containing 100 nodes to 281.7 s for a graph comprising 1500 nodes. A parallel increment was observed for strong summaries, with execution times escalating from 26.4 s to 406.1 s for graphs ranging from 100 to 1500 nodes in size.

The linear progression of execution time in relation to graph size underscores the scalability of our summarization approach within the HERSE framework, demonstrating its capability to efficiently manage RDF graphs of increasing complexity. This scalability is particularly vital for applications necessitating the processing of large-scale RDF datasets, affirming the practicality and effectiveness of our framework in extensive data environments.

6   Conclusion and Future Work

In conclusion, the HERSE framework presents a significant advancement in the realm of RDF graph summarization, offering a systematic approach to the challenge of blank nodes through Skolemization. Our experimental results affirm the framework's capability to handle extensive and complex RDF datasets efficiently, showcasing a promising linear increase in execution time in response to larger graph sizes.

Looking ahead, the next phase of research will focus on enhancing blank node management, particularly in scenarios involving complex data aggregations. We plan to explore sophisticated algorithms for the intelligent detection and advanced processing of blank nodes, such as context-aware Skolemization and adaptive summarization techniques. Furthermore, we are committed to extending the framework's functionality to support the advanced treatment of blank nodes in dynamic RDF graphs where real-time updates necessitate immediate re-Skolemization. By addressing these challenges, we aim to push the boundaries of semantic web technology and provide robust solutions for the next generation of RDF data summarization tools.

References

1. Mallea, A., Arenas, M., Hogan, A., Polleres, A.: On blank nodes. In: Aroyo, L., et al. (eds.) ISWC 2011. LNCS, vol. 7031, pp. 421–437. Springer, Heidelberg (2011). https://doi.org/10.1007/978-3-642-25073-6_27
2. Beldi, A., Sassi, S., Chbeir, R., Jemai, A.: DG summ: a schema-driven approach for personalized summarizing heterogeneous data graphs. Comput. Sci. Inf. Syst. **00**, 62–62 (2023)
3. Cyganiak, R., Wood, D., Lanthaler, M.: RDF 1.1 Concepts and Abstract Syntax. W3C Recommendation (2014). http://www.w3.org/TR/rdf11-concepts/
4. Hayes, P.J., Patel-Schneider, P.F.: RDF 1.1 Semantics. W3C Recommendation (2014). http://www.w3.org/TR/rdf11-mt/
5. Schmachtenberg, M., Bizer, C., Paulheim, H.: Adoption of the linked data best practices in different topical domains. In: Mika, P., et al. (eds.) ISWC 2014. LNCS, vol. 8796, pp. 245–260. Springer, Cham (2014). https://doi.org/10.1007/978-3-319-11964-9_16
6. Heath, T., Bizer, C.: Linked Data: Evolving the Web into a Global Data Space, vol. 1, no. 1, pp. 1–136. Morgan & Claypool (2011)
7. Klyne, G., Carroll, J.J.: Resource Description Framework (RDF): Concepts and Abstract Syntax. W3C Recommendation (2004)
8. Schätzle, A., Neu, A., Lausen, G., Przyjaciel-Zablocki, M.: Large-scale bisimulation of RDF graphs. In: Proceedings of the Fifth Workshop on Semantic Web Information Management, SWIM@SIGMOD Conference 2013, New York, NY, USA, 23 June 2013, pp. 1:1–1:8 (2013)
9. Chen, C., Yan, X., Zhu, F., Han, J., Yu, P.S.: Graph OLAP: towards online analytical processing on graphs. In: Proceedings of the 8th IEEE International Conference on Data Mining (ICDM 2008), 15–19 December 2008, Pisa, Italy (2008)
10. Rudolf, M., Paradies, M., Bornhövd, C., Lehner, W.: Synopsys: large graph analytics in the SAP HANA database through summarization. In: First International Workshop on Graph Data Management Experiences and Systems, GRADES 2013, Coloated with SIGMOD/PODS 2013, New York, NY, USA (2013)
11. Zneika, M., Lucchese, C., Vodislav, D., Kotzinos, D.: RDF graph summarization based on approximate patterns. In: Information Search, Integration, and Personalization-10th International Workshop, ISIP 2015, Grand Forks, ND, USA, 1–2 October 2015, Revised Selected Papers, pp. 69–87 (2015)
12. Spahiu, B., Porrini, R., Palmonari, M., Rula, A., Maurino, A.: ABSTAT: ontology-driven linked data summaries with pattern minimalization. In: SumPre (2016)

13. Hose, K., Schenkel, R.: Towards benefit-based RDF source selection for SPARQL queries. In: Proceedings of the 4th International Workshop on Semantic Web Information Management, SWIM 2012, Scottsdale, AZ, USA, 20 May 2012, p. 2 (2012)
14. Beldi, A., Richa, J.R., Sassi, S., Chbeir, R., Jemai, A.: A novel approach for extracting summarized RDF graph from heterogeneous corpus. In: 2023 International Conference on Innovations in Intelligent Systems and Applications (INISTA), pp. 1–7. IEEE (2023)
15. Beldi, A., Sassi, S., Chbeir, R., Jemai, A.: Schema formalism for semantic summary based on labeled graph from heterogeneous data. In: Szczerbicki, E., Wojtkiewicz, K., Nguyen, S.V., Pietranik, M., Krótkiewicz, M. (eds.) ACIIDS 2022. CCIS, vol. 1716, pp. 27–44. Springer, Singapore (2022). https://doi.org/10.1007/978-981-19-8234-7_3
16. Wang, X., et al.: PCSG: pattern-coverage snippet generation for RDF datasets. In: The Semantic Web-ISWC 2021: 20th International Semantic Web Conference, ISWC 2021, Virtual Event, 24–28 October 2021, Proceedings 20, pp. 3–20 (2021)
17. Cebiric, S., Goasdoué, F., Guzewicz, P., Manolescu, I.: Compact Summaries of Rich Heterogeneous Graphs. Research Report RR-8920, INRIA Saclay; Université Rennes 1 (2017)

Rough Sets for a Neuromorphic CMOS System

Rory Lewis[(✉)], Michael Bihn, and Katrina Nesterenko

University of Colorado at Colorado Springs, Colorado Springs, CO 80918, USA
rlewis5@uccs.edu

Abstract. The authors have recently produced a CMOS system that autonomously rewires it's circuitry to learn similarly to a biological brain. Biological brains learn very fast in that their synapses mysteriously ignore what is not relevant, and connect only the most important synapses to learn from an event. Conversely, man-made 'brains' as in, artificial intelligence systems, need to be trained to ignore all the unimportant issues - thus draining resources. Our hypothesis is to verify whether these variations of classical Rough Set algorithms can be implemented on our neuroCMOS-FPGA, and if so, under what circumstances of uncertainty, if at all, does one algorithm do better or worse than another. Herein, our next step is to 1) emulate how the neonate's brain grows, and 2) have a rough sets system 'allocate' learned synthetic synapses onto the CMOS system as it grows while discerning when one learning event has similarities to other learning events; how should a rough sets system blend the building of neonatal connectome, and at the same time meld variations of synapses that have similarities?

1 Introduction

When you were a child and first touched a hot plate on a stove, and burnt yourself, as trivial as this sounds, you did not say to yourself: "Well the stove was white, any other color stove will not burn me". Neither did you say: "Oh! I got burned by the stove in my mother's kitchen that has a neon light in her kitchen; when I go to another house that has an incandescent light in the kitchen it will be OK to touch the hot plate!" As absurd as the aforementioned sounds, the synapses in your brain ignored billions of moot attributes and connected only the exact, most critical, and precise synapses that ensured that you would never touch another hot plate again. Conversely, artificial intelligence (AI) researchers have to laboriously train their AI to i) learn to ignore thousands of inconsequential artifacts that have nothing to do with being burned, and teach the AI to recognize and associate thousands of variations of hot plates.

During neonate brain development there is an explosion of synapse formation between neurons called synaptogenesis. When IBM's TrueNorth engineers [3] and neuroscientists studied how to emulate synaptogenesis for their neuromorphic chip, they assumed that the voltage inside the orange arrow in Fig. 1a was a

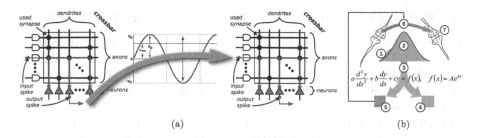

Fig. 1. TrueNorth hardware (a) Assumed signal onto IBMs spiking neuromorphic system [10]. **(b)** 2nd Order Diff Eq conversion.

typical sine wave. Herein, they developed a system where an incoming spike signal arriving from the red horizontal axons was collected at the end of the orange arrow by the red vertical dendrites. However, these *Axon-Hillock* neurons were misfiring [4,8]. Schmidt & Avitabile found that TrueNorth's sinusoidal wave was randomly connecting to any neuron [17] because, as shown in Fig. 1*b*, there is a short 'hidden state' [1] of hypopolarization, which precedes the depolarization, that forms a very small bump on the trailing edge of the sinusoidal wave as indicated by ① and was most likely overlooked by IBM's TrueNorth engineers. The realization of this hump ① shown in Fig. 1*b* formed the basis for the authors to take the 2nd order Differential Equation ③ of both ① and ② and split the combined area under the curve into gray area ④ that constitutes the power necessary to project the neuron signal ⑥ to its destination ⑦.

2   Experiment Setup for a Two-Fold Testing Scenario

The challenge presented in Phase III is that we expand Phase II's neuroCMOS-FPGA 32 states to 3D $250 \times 250 \times 250$ matrix, comprising 15,625,000 states. We are confronted with two rules: **Challenge 1**: We cannot organize its learned connected input linearly, because i) there will not be enough space between the FPGAs on the x, y, and z axis, and ii) the length of wiring connecting and overlapping each FPGA, located all over the CMOS landscape will slow the system down. Herein, we must study and emulate the mathematics and algorithms of neonatal brain growth, *see* Sect. 2.1. **Challenge 2**: We cannot feed the system binary, on or off logic. In order to emulate the biological brain we need to feed the system uncertain and imprecise data. Herein, rather than feeding the system on or off lighting to learn as we did in Phases I and II. In Phase III we will feed the system of 15,625,000 states, three colors of light; Red, Green, and White, and test Rough Sets, KNN, Fuzzy-Rough Nearest Neighbour (FRNN) without weights, FRNN with weights, and Fuzzy Rough One-Versus-One (FROVOCO) to see what system builds a synthesized connectome on our neuroCMOS-FPGA architecture that has the least amount of wiring between the FPGAs that the CMOS connects in order to simulate how a biological brain learns, *see* Sect. 2.2.

2.1   Challenge 1. Neonate Brain Growth on the neuroCMOS-FPGA

We collected data from the baby connectome project at [2] and downloaded from folder UNC-Infant-Cortical-Surface-Atlas both i) VTKFormat_V1.8.rar, and ii)Manual_V1.8.pdf. [15] and [19] developed the data. We unpacked files: "SparseAtlas_[01, 03, 06, 09, 12, 18, 24]_rh.InnerSurf.vtk". Next, we parsed the data fields with Matlab into the appropriate arrays of scalars and vectors. For each time point, there are three surfaces with 327,680 polygons each utilizing 3 of 163,842 points, where each polygon is a triangle. We chose the coordinates of the first vertex to apply our processing and collected the coordinates at 1, 3, 6, 9, 12, 18, and 24 months and stored them in our database. In neuroscience, it is generally accepted that an accurate curve of neonate explosions is essential to accurately emulate brain growth. Herein, we represent the curve curve fitting by utilizing Matlab's curve fitting tool on the data and then applied several methods of curve fitting methodologies including but not limited to *smoothingspline, pp* @0.9809 r^2, *pchip, pp* @1.0 r^2, *cubicinterp, pp* @1.0 r^2, and *spline, pp* @1.0 r^2. Where all of the methods marked with "*pp*", had piece-wise polynomial fitted curves, meaning that several polynomials are needed to describe the fit and the r^2 is an evaluation of fit. We then evaluated the polynomials and chose *Poly5* as it had the strongest R-squared, and yields a 5th degree polynomial to describe vertex motion. After applying the same *Poly5* fitting to the X, Y, and Z longitudinal values and applying these fitted curves for the coordinates over time, this yielded a definitive path as seen in Fig. 2a. Adding panning is shown in Fig. 2b and Fig. 2c shows combining both graphs together.

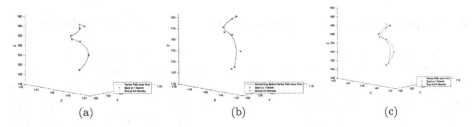

(a)   (b)   (c)

Fig. 2. Vertex 1 path over time: (a) *Poly5* fitting the X, Y, & Z longitudinal values over time yields a definitive path and panning in 3D. (b) *Smoothing spline* fit of non-linear path. (c) combining both the poly5 and the smoothing spline fits, where the smoothing spline in 'b' is in red. (Color figure online)

Continuing with the functions for the smoothed curves of the 5th polynomials, we take the following coefficients: p1 = 1.742e-05, p2 = −0.001163, p3 = 0.02857, p4 = −0.3194, p5 = 1.702, and p6 = 141.5 and produce a function for the x variable of the first vertex position as seen in Eq. 1.

$$X(t) = p1 * t^5 + p2 * t^4 + p3 * t^3 + p4 * t^2 + p5 * t + p6$$
$$X(t) = 1.742e - 05 * t^5 + -0.001163 * t^4 + 0.02857 * t^3 + \tag{1}$$
$$+ - 0.3194 * t^2 + 1.702 * t + 141.5$$

Similarly, for the functions of the smoothed curves of Y, from *Poly5* that are 5th polynomials, we take the following coefficients: p1 $= -9.346e-07$, p2 $= 7.466e-05$, p3 $= -0.0008573$, p4 $= -0.05254$, p5 $= 1.396$, and p6 $= 118.2$ that produces the function for the y variable of the first vertex position as shown in Eq. 2.

$$Y(t) = p1 * t^5 + p2 * t^4 + p3 * t^3 + p4 * t^2 + p5 * t + p6$$
$$Y(t) = -9.346e - 07 * t^5 + 7.466e - 05 * t^4 + \tag{2}$$
$$+ - 0.0008573 * t^3 + -0.05254 * t^2 + 1.396 * y + 118.2$$

Finally, the functions for the smoothed curves of Z from *Poly5* that are 5th polynomials, we take the following coefficients: p1 $= 6.965e-06$, p2 $= -0.0007443$, p3 $= 0.02737$, p4 $= -0.4555$, p5 $= 3.816$, and p6 $= 145.5$ that produce a function for the z variable of the first vertex position as shown in Eq. 3.

$$Z(t) = p1 * t^5 + p2 * t^4 + p3 * t^3 + p4 * t^2 + p5 * t + p6$$
$$Z(t) = 6.965e - 06 * t^5 + -0.0007443 * t^4 + 0.02737 * t^3 + \tag{3}$$
$$+ - 0.4555 * t^2 + 3.816 * t + 145.5$$

2.1.1   Building the Differential Equations For each of the variables X, Y, and Z, we build the differential equations describing their behavior over time by taking the first and second derivatives of each and adding them to the fitted function. For X we produce Eqs. 4, 5 & 6:

$$X(t) = 1.742e - 05 * t^5 + -0.001163 * t^4 + 0.02857 * t^3 + \tag{4}$$
$$+ - 0.3194 * t^2 + 1.702 * t + 141.5$$

$$X'(t) = (5 * 1.742e - 05) * t^4 + (4 * -0.001163) * t^3 + \tag{5}$$
$$+(2 * -0.3194) * t + 1.702$$

$$X''(t) = (4 * 5 * 1.742e - 05) * t^3 + (3 * 4 * -0.001163) * t^2 + \tag{6}$$
$$+(2 * 3 * 0.02857) * t + (2 * -0.3194)$$

Now adding them to the fitted function we produce Eq. 7.

$$X'' + X' + X = (4 * 5 * 1.742e - 05) * t^3 + \quad (7)$$
$$+(3 * 4 * -0.001163) * t^2 + (2 * 3 * 0.02857) * t + (2 * -0.3194)+$$
$$+(5 * 1.742e - 05) * t^4 + (4 * -0.001163) * t^3 + (3 * 0.02857) * t^2 +$$
$$+(2 * -0.3194) * t + 1.702 + +1.742e - 05 * t^5 + -0.001163 * t^4 +$$
$$+0.02857 * t^3 + -0.3194 * t^2 + 1.702 * t + 141.5$$

Where the X equation represents the x-coordinate of the position of the associated vertex at that time(t), the X' equation, the first derivative of the X equation, represents the velocity of the associated vertex in the x-coordinate, and the X'' equation, the second derivative of the X equation, represents the acceleration of the associated vertex in the x-coordinate. Here, the sum of these three equations provides a differential equation describing the exact behavior, position, velocity, and acceleration of the associated vertex in the x-coordinate. We repeat the same process for Y equations, Y, Y' and Y'' and the Z equations, Z, Z' and Z''.

2.2   Challenge 2. Handling Uncertain and Imprecise Data

We use combinations of red, green, and white lights as illustrated in Table 1 where 1 indicates a light of that attribute's color is "ON" and the 0 indicates it is "OFF". Looking at Tuple 2, attribute 2, a 100% "HIT" for RED, would be a sequence of red lights that go on, off, on, on, off, and on (101101). As a human, if we saw, some-where in the sequence of lights, red lights

Table 1. Perfect matches and imprecise data.

	RED	GREEN	WHITE
HIT	1 0 1 1 0 1	1 1 0 1 0 1	1 1 1 0 0 1
v1	1 0 1 1 0 α	1 1 0 1 0 α	1 1 1 0 0 α
v2	1 0 1 1 α 1	1 1 0 1 α 1	1 1 1 0 α 1
v3	1 0 1 α 0 1	1 1 0 α 0 1	1 1 1 α 0 1
v4	1 0 α 1 0 1	1 1 α 1 0 1	1 1 α 0 0 1
v5	1 α 1 1 0 1	1 α 0 1 0 1	1 α 1 0 0 1
v6	α 0 1 1 0 1	α 1 0 1 0 1	α 1 1 0 0 1

going 101101, we say that is a perfect match. Now, considering tuple v1 attribute RED (hereinafter $t^{v1} A^{RED}$); we see a 10110α where the α could mean one of three states: i) a zero, meaning "no light" or ii) a green light or iii)a white light. In essence, $t^{v1} A^{RED}$'s 10110α could represent i) 10110 [0], ii) 10110 [Green light "ON"] or iii) 10110 [white light "ON"]. Meaning as a human, any of these three aforementioned states of RED's 10110 α would have us say "Yes, that's pretty close", i.e., Rough Set Theory. Finally, each sequence of lights being received by the sensor of our neuroCMOS-FPGA will comprise, in 10 s, a series of 1,000 flashes of red, green, and white lights in on and off states. If in the series of flashes, it sees $t^{HIT} A^{RED}$ (101101), or $t^{HIT} A^{GREEN}$ (110101) or $t^{HIT} A^{WHITE}$ (111001) it will register a "HIT" meaning an exact match. Conversely, if it receives any combination of tuples v1, v2, v3, ... v6, it will register a roughly "yes". Note that for these experiments we are concerned about how to simulate neonate brain growth in the architecture when it has a plurality of rough sets of colors and HITs.

2.2.1   The Human Context. Given the aforementioned, how would you, the reader, classify a *"rough RED, a Green HIT, and two rough WHITES"*? In other words, when you were two weeks old, and there was an explosion of synaptogenesis in your neonate's brain, how would your brain organize itself to learn a *"rough RED, a Green HIT, and two rough WHITES"*? We have a partial answer as shown in Sect. 2.1.1 that has shown the authors how to physically organize 15,625,000 synthetic synapses (FPGAs). However, the second part of the answer is unknown in that it is not clear how our neuroCMOS-FPGA should optimally classify a plurality of uncertain and imprecise lights together with a few precise "HITS". Here, we turn to testing classical Rough Set Theory against various forms of offshoots such as K-Nearest Neighbor (KNN), Fuzzy Rough Set (FRS), Fuzzy-Rough Nearest Neighbor (FRNN), and a Fuzzy Rough One-Versus-One (OVO) combination(FROVOCO).

2.2.2   KNN is a simple and robust classifier that measures the distance between tested examples and training examples. KNN only has one hyperparameter k and in the Learning Phase, we use a training set $\mathcal{D} = \{(x_t, y_t)\}_{t=1}^N$ where the input vectors x_t and y_t correspond to the t−th instance of y and $\mathcal{D}$ holds the data set [12]. Once a value of k is chosen, we calculate the Euclidean of the training data points using $\sqrt{\sum_{i-1}^n (x_i - y_i)^2}$ that finds the k-nearest neighbors and assigns a class containing the maximum number of nearest neighbors. Lastly, we retrieve the KNN instances using $\mathcal{N}(x_t) = \{(x_t^{(i)}, y_t^{(i)})\}_{t=1}^k$ that are closest to x, and it yields $\hat{y}$ as the weighted combination of the labels $y^{(1)}, ..., y^{(k)}$ as follows:

$$\hat{y} = f(x; \mathcal{D}) = \frac{\sum_{t=1}^k w(d(x, x^{(i)})) \cdot y^{(i)}}{\sum_{t=1}^k w(d(x, x^{(i)}))} \tag{8}$$

Where k is the number of neighbors, d is the distance function, and as shown in this example and w is the weighting function. Note that when K is small, the noise will have a higher influence on the result [6] and when K is a large value it makes it computationally expensive [5].

2.2.3   FRNN Fuzzy Rough Sets (FRS) introduces a feature subset selection called reducts [13] by building fuzzy approximations using the nearest neighbors algorithm [9] based on the lower and upper approximations of a fuzzy set A in a universe $\mathbb{U}$ on which a fuzzy relation R is defined [7]. We define ℓ as a fuzzy implication and use $\ell(R(x,y), A(y))$ to express the extent that an element similar to x belongs to A [7]. We define the lower and upper approximations as:

$$(R \downarrow A)(x) = \inf_{y \in \mathbb{U}} \ell(R(x,y), A(y)) \tag{9}$$

$$(R \uparrow A)(x) = \inf_{y \in \mathbb{U}} \mathcal{T}(R(x,y), A(y)) \tag{10}$$

Where $\mathcal{T}(R(x,y), A(y))$ expresses the extent that an element similar to x belongs to $\mathcal{A}$. Now, segueing from FRS to the Fuzzy-Rough Nearest Neighbor (FRNN) algorithm; after finding the k nearest neighbors in the test object j. We define a class C such that if the sum of $R : (R \downarrow C)(y) + (R \uparrow C)y$ can reach the maximum, then the test object belongs to the class C [11]. In essence, if a value of a fuzzy lower approximation is high, it shows that neighbors of the new data belong to a particular class C. Conversely, a high value of a fuzzy upper approximation means that at least one neighbor belongs to that class.

2.2.4 FROVOCO. The Fuzzy Rough One-Versus-One (OVO) combination (FROVOCO) classification algorithm is designed for imbalanced data by balancing OVO decomposition with two global class affinity measures [14]. To classify a test instance, each binary classifier trained on a sub-task and computes confidence degrees for its classes. All values are grouped in a score-matrix $R(x)$ in the form of Eq. 11 and afterward aggregated to one class prediction for the target.

$$R(x) = \begin{pmatrix} - & r_{12} & \cdots & r_{1m} \\ r_{21} & - & \cdots & r_{2m} \\ \vdots & & & \vdots \\ r_{m1} & r_{m2} & \cdots & - \end{pmatrix}, \tag{11}$$

3 Experiments

To accommodate for our proposed 15.6 million FPGAs, representing 15.6 million synapses that are arranged in accordance to i) neonate synaptogenesis, and ii) Rough Set Theory (RST) analysis of uncertain data. In our case, with the Table 1 training data, one could envision three or more mountains of rocks, comprised of FPGAs that are arranged next to one another. Accordingly, *lithology* is the study of rocks and we are creating RST to form a neuromorphic mountain, not made of rock, but FPGAs synthetic synapses. Hence we coin the term ***Lithomorphology*** to analyze how RST and synaptogenesis algorithms work together to optimize connectivity length synapses inside large mountains of synthetic synapses.

OUR HYPOTHESIS is to verify whether these variations of classical Rough Set algorithms can be implemented on our `neuroCMOS-FPGA`, and if so, under what circumstances of uncertainty, if at all, does one algorithm do better or worse than another. Once this is achieved, *our ultimate goal is to then create an innovative Rough Set algorithm from scratch* that will combine, and set a road map for how we should design, our own, rough set algorithm - specifically designed to handle uncertain and certain data for a neuromorphic microprocessor that rewires itself, similar to how a biological brain connects synapses when it learns. Note that to test our hypothesis we bear in mind that this research is a step beyond our previous work [16] (ISMIS '22) where we ran similar tests on hardware and assessed the time and power consumption for training each model, averaged over 10 executions each as shown in Fig. 3b. For This research

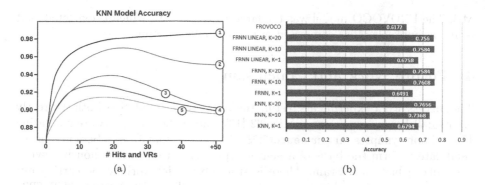

Fig. 3. (a) **Current Results** (1) Training on variations of Table 1, (2) KNN, (3) FRNN, (4) OVO and (5) FROVOCO., and (b) **Previous Results** on hardware KNN, FRNN, and FROVOCO model accuracy are compared for all tested values of k after training on an 80/20 split of the abalone dataset [16]

effort we are testing the viability of carrying out Rough Sets *using software to emulate a large hardware system comprised of 15.6 million synthetic synapses on a CMOS and FPGA system*. We use a two-pronged approach. i) emulate neonate synaptogenesis and then ii) use Rough Sets to cluster uncertain data and carry out straightforward and accurate tests. Here, it was evident that the way we set up the validation tests, completely random, KNN was outstanding and all the rest tapered off in accuracy as we increased the amount of hits and near misses.

3.1   Experiment Results

Looking at Fig. 3*a*: ① represents the training data comprised of the averaged mean of 12 runs in Table 1. ② represents the KNN results. The authors are not sure why KNN acted, almost too well. We understand that KNN is robust and performed the best in the previous experiments using the same algorithms on hardware. To test our skepticism we ran various instantiations of randomness of the validation data but to no avail; KNN was still incredibly powerful ③, ④, and ⑤ represent FRNN, OVO, and FROVOCO models respectively, and are grouped as such because they were all similar and far below the accuracy and robustness of KNN. Looking at Fig. 3*b* it is clear that KNN, again, is superior to the other models, even though in that case we were testing the power consumption across all 10 models. We did however note that both FRNN models, with and without weights, with k = 10 slightly outperformed standard KNN k = 10 on an embedded system with approximately 76% accuracy each, while standard KNN was 74% accurate. This warrants further investigation in certain research areas within resource-constrained environments where execution time is less important than accuracy [18]. It should be noted that the exported serialized models in Fig. 3*b* were the same size for all k values, with each of the KNN models being 503 kB and each of the weighted FRNN models being 2.07 MB. The non-weighted FRNN models were also 2.07 MB, except for the k = 1 model, which was 1.08

MB. The FROVOCO model was the largest at 2.75 MB, which is expected as it is an ensemble classifier comprised of many smaller classifiers.

4   Future Work and Conclusion

The success of our experiments testing the viability of Rough Sets both i) directly on a small hardware system ISMIS '22 [16], and ii) here on software emulating 15.6 million synapses on our `neuroCMOS-FPGA` prove that Rough Sets is a prime candidate to form the basis of a neuromorphic chip that autonomously rewires itself like a biological brain. There is still substantial work to be carried out before we can justify asking for funding a 15.6 million FPGA `neuroCMOS-FPGA` chip:

1. **Combining both sets of tests presented in this paper.** The inherent challenge is that the authors are forced to combine two disparate models. We first need one that constructs millions of synthetic synapses and organizes them to have the shortest amount of connectivity wiring between 15.6 million synapses. Accordingly, we use how biological brains optimize this exact issue when synaptogenesis occurs in the neonate brain. The next model needs to discern between direct hits and uncertain data. But how do we combine these two aforementioned models into one cohesive `neuroCMOS-FPGA` chip?
2. **Creating a novel *Lithomorphology* strain of Rough Sets for neuro-morphic chips.** Knowing that KNN outperformed the Rough Set variants, we need to create a novel Rough Set methodology from the ground up that retains the neuromorphic data's categorical and numerical features, discerns between a plethora of uncertain but loosely matched data and ignored missing values, and uses a different K for each query.

In conclusion, it is exciting to see how wonderfully explicit Rough Set Theory is at controlling an autonomous neuromorphic chip. In essence, a chip, without a clock, that evolves in its intelligence, learning like a biological brain and rewiring itself without a human touching it – except somewhere out there - Pawlak.

References

1. Action potential. https://www.kenhub.com/en/library/anatomy/action-potential. Accessed 24 Jan 2024
2. UNC infant surface atlases of cortical structures (2024). https://shorturl.at/jnsQ2
3. Akopyan, F., et al.: Truenorth: design and tool flow of a 65 mw 1 million neuron programmable neurosynaptic chip. IEEE Trans. Comput. Aided Des. Integr. Circuits Syst. **34**(10), 1537–1557 (2015)
4. Cover, K.K., Mathur, B.N.: Axo-axonic synapses: diversity in neural circuit function. J. Comp. Neurol. **529**(9), 2391–2401 (2021)
5. Dong, W., Moses, C., Li, K.: Efficient k-nearest neighbor graph construction for generic similarity measures. In: Proceedings of the 20th International Conference on World Wide Web, pp. 577–586 (2011)

6. García-Pedrajas, N., Ortiz-Boyer, D.: Boosting KNN classifier by means of input space projection. Expert Syst. Appl. **36**(7), 10570–10582 (2009)
7. He, R., Xu, C., Li, D., Hou, W., Yu, X., Zhang, H.: A fuzzy-rough-based approach for uncertainty classification on hybrid info sys. In: 2018 IEEE 3rd International Conference on Image, Vision and Computing (ICIVC), pp. 791–796. IEEE (2018)
8. Hosseini, M.J.M., et al.: Organic electronics axon-hillock neuromorphic circuit: towards biologically compatible, and physically flexible, integrate-and-fire spiking neural networks. J. Phys. D Appl. Phys. **54**(10), 104004 (2020)
9. Hussein, A.S., Khairy, R.S., Najeeb, S.M.M., ALRikabi, H.T., et al.: Credit card fraud detection using FRNN and sequential minimal optimization with logistic regression. Int. J. Interact. Mob. Tech. **15**(5) (2021)
10. Huttenlocher, P.R., Dabholkar, A.S.: Regional differences in synaptogenesis in human cerebral cortex. J. Comp. Neurol. **387**(2), 167–178 (1997)
11. Jensen, R., Cornelis, C.: Fuzzy-rough nearest neighbour classification and prediction. Theoret. Comput. Sci. **412**(42), 5871–5884 (2011)
12. Kang, S.: KNN learning with graph neural networks. Mathematics **9**(8), 830 (2021)
13. Kumar, A., Prasad, P.S.: Scalable fuzzy rough set reduct computation using fuzzy min-max neural network preprocessing. IEEE Trans. Fuzzy Syst. **28**(5), 953–964 (2020)
14. Lenz, O.U., Peralta, D., Cornelis, C.: *fuzzy-rough-learn* 0.1: a python library for machine learning with fuzzy rough sets. In: Bello, R., Miao, D., Falcon, R., Nakata, M., Rosete, A., Ciucci, D. (eds.) IJCRS 2020. LNCS (LNAI), vol. 12179, pp. 491–499. Springer, Cham (2020). https://doi.org/10.1007/978-3-030-52705-1_36
15. Li, G., Wang, L., Shi, F., Gilmore, J.H., Lin, W., Shen, D.: Construction of 4D high-definition cortical surface atlases of infants: methods and applications. Med. Image Anal. **25**(1), 22–36 (2015)
16. Nesterenko, K., Lewis, R.: Rough sets for intelligence on embedded systems. In: Ceci, M., Flesca, S., Masciari, E., Manco, G., Raś, Z.W. (eds.) ISMIS 2022. LNCS, vol. 13515, pp. 230–239. Springer, Cham (2022). https://doi.org/10.1007/978-3-031-16564-1_22
17. Schmidt, H., Avitabile, D.: Bumps and oscillons in networks of spiking neurons. Chaos Interdisc. J. Nonlinear Sci. **30**(3) (2020)
18. Taylor, B., Marco, V.S., Wolff, W., Elkhatib, Y., Wang, Z.: Adaptive deep learning model selection on embedded systems. ACM SIGPLAN Notices **53**(6), 31–43 (2018). https://doi.org/10.1145/3299710.3211336
19. Wang, F., et al.: Developmental topography of cortical thickness during infancy. Proc. Natl. Acad. Sci. **116**(32), 15855–15860 (2019)

Neural Network and Data Mining

Erasing the Shadow: Sanitization of Images with Malicious Payloads Using Deep Autoencoders

Angelica Liguori[3]([✉])[ID], Marco Zuppelli[1][ID], Daniela Gallo[2][ID],
Massimo Guarascio[3][ID], and Luca Caviglione[1][ID]

[1] Institute for Applied Mathematics and Information Technologies, Genova, Italy
{marco.zuppelli,luca.caviglione}@ge.imati.cnr.it
[2] University of Salento, Lecce, Italy
daniela.gallo@unisalento.it
[3] Institute for High Performance Computing and Networking, Rende, Italy
{angelica.liguori,massimo.guarascio}@icar.cnr.it

Abstract. Steganography is used by threat actors to avoid detection or bypass blockages. Among the various approaches, hiding data within digital images is now the preferred offensive technique. Alas, developing attack-agnostic mitigation mechanisms is difficult, especially due to the tight relation between the images and the steganographic approach. Therefore, this paper takes advantage of autoencoders for *sanitization*, i.e., to disrupt the malicious information hidden in images without altering the visual quality. To this aim, we used an enhanced U-Net-like neural architecture. Results obtained with realistic threats showcased that our approach can effectively disrupt cloaked data and prevent the recovery of the payload while preserving the original image quality.

Keywords: Deep Learning · Steganography · Sanitization

1 Introduction

The increasing effectiveness of security countermeasures imposes a rethink of the attack chain used by malicious actors. Modern malware deploys techniques for preventing its early detection or blockages such as firewalls [3]. To this aim, a popular approach relies upon some form of code obfuscation, e.g., adding junk code or diverting the normal execution flow to prevent signatures [6]. Another major technique is multi-stage loading where offensive routines are retrieved only when needed [20]. Indeed, a recent trend exploits information hiding and steganography [10].

The literature abounds in techniques for concealing data within network traffic, binary code, multimedia objects, and IoT nodes [4,17], but attacks observed

---

Angelica Liguori, Marco Zuppelli, and Daniela Gallo equally contributed to the paper and are considered the first authors.

© The Author(s), under exclusive license to Springer Nature Switzerland AG 2024
A. Appice et al. (Eds.): ISMIS 2024, LNAI 14670, pp. 115–125, 2024.
https://doi.org/10.1007/978-3-031-62700-2_11

"in the wild" primarily take advantage of digital pictures [5,10]. In more detail, the malicious information (e.g., a list of IP addresses to contact or configuration data) is hidden within a digital image acting as the carrier. The steganographic process should be kept as simple as possible to avoid signatures, lags, or transmission delays. Thus, malware mainly exploits Least Significant Bit (LSB) steganography, which hides content by overwriting the last bits of the color components of pixels composing the targeted image. For instance, several malicious actors used plain LSB approaches (e.g., 1 bit of data is hidden in each color channel), whereas others tried to smuggle payloads via the Invoke-PSImage mechanism, altering multiple bits for each color channel [16].

Even if threat actors taking advantage of steganographic techniques is a growing concern [10], only a few works address the issue of *sanitizing* digital images, i.e., the process of disrupting the hidden content while preserving the hosting media. For the specific case of using Artificial Intelligence (AI), [13] showcases a framework based on diffusion models to remove malicious contents hidden in an image while preventing its degradation. However, the considered threat model deals with an attacker hiding an image within another, which is seldom observed in realistic offensive campaigns [5,10]. A more realistic template has been considered in [21], where authors use AI to remove data cloaked via Invoke-PSImage. Moreover, [9] discusses how to use machine learning to detect malicious PowerShell scripts starting from signatures in the color histogram of altered pixels. Alas, the tight interdependence between the hiding technique and the targeted carrier makes the mitigation complex. For instance, some methods could be difficult to deploy in a more general setting, e.g., porting them from mobile ecosystems [18]. A workaround could take advantage of the ability of AI to "generalize" the sanitization process for a family of steganographic methods, e.g., plain LSB and deep approaches [14]. Yet, AI could be affected by "concept drifts" inducing degradation of the models and requiring further tweaks [8].

In this perspective, our work showcases the use of Autoencoders (AEs) [2] to sanitize a payload cloaked in a digital image via steganography. This requires disrupting the malicious information without altering the visual quality. The use of AEs has been driven by their properties and performances, especially for security-oriented applications. For instance, they demonstrated to be effective in identifying malicious patterns of system calls [7].

Compared to previous research [9,13,14,21], our work has the following improvements: *i*) it considers two realistic hiding templates observed in many attacks, i.e., plain LSB and OceanLotus, *ii*) it evaluates the use of an enhanced version of U-Net-like encoder-decoder architecture with further connections to improve the learning process stability, *iii*) it assesses the performance in terms of quality decays and computational constraints, which are crucial to understand whether a sanitization method could be deployed in realistic scenarios. We point out that, compared to our preliminary work [21], in this paper, we leveraged a more sophisticated neural architecture based on U-Net+.

The rest of the paper is structured as follows. Section 2 outlines the considered problem, whereas Sect. 3 introduces the neural architecture for sanitizing images. Section 4 showcases numerical results, and Sect. 5 concludes the paper.

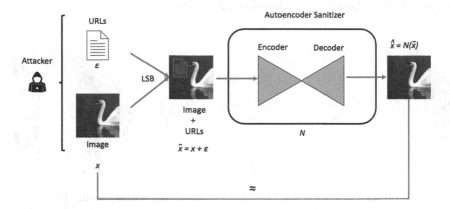

Fig. 1. Reference scenario of the considered sanitization problem. The malicious payload is disrupted by processing the altered image with an Autoencoder.

2 Problem Definition

As hinted, we address the problem of sanitizing images without altering the perceived quality. Figure 1 showcases the reference scenario. In more detail, we consider an attacker hiding arbitrary information denoted with ϵ in a legitimate image denoted with x. To model a realistic use case, we assume that ϵ is a list of URLs pointing at financial institutions, as observed in the ZeusVM trojan [11]. Concerning the steganographic techniques used to conceal the information, we consider two strategies observed in many malware samples and in the OceanLotus advanced persistent threat [5,10]. Specifically, for each pixel composing the image:

- LSB plain: the payload is hidden in the least significant bit of the red, green, and blue color channels;
- LSB variant: the payload is concealed in the 3 least significant bits of red and green color channels, and in the 2 least significant bits of the blue channel.

As a result, the original image x is modified, and the new object containing the malicious content is denoted with $\tilde{x} = x + \epsilon$. Referring again to Fig. 1, our goal is to find an estimate $\hat{x}$ that is as close as possible to the original, unmodified image x, such that $\hat{x} = N(\tilde{x})$. The function N maps the "compromised" image $\tilde{x}$ to its corresponding estimate $\hat{x}$. Hence, N is designed to minimize the dissimilarity between the estimated image $\hat{x}$ and the "clean" image x, measured using a suitable loss function $\mathcal{L}$. In our reference scenario, N acts as the sanitizer and takes the form of an AE, an unsupervised neural network model performing two main operations. First, it compresses the input data (i.e., the matrix representation of the image) within a latent space. Then, it reconstructs the original information provided as input. The pixel is represented by the RGB components and it is the smallest manageable element for the AE.

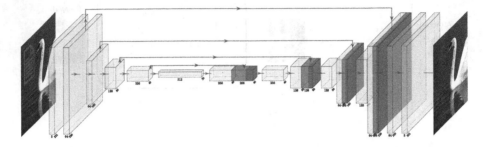

Fig. 2. U-Net-like architecture: grey blocks represent the corresponding blocks in the encoding phase, stacked with the outputs from deconvolutional blocks. The malicious input image is sanitized through deep convolutional layers.

3   Sanitization Through Deep Learning

The mapping function N implementing the sanitizer loosely resembles a U-Net [15]. Compared to a traditional AE, it includes skip connections that facilitate the multi-scale feature fusion, allowing the network to combine low- and high-level features from different stages of the encoder. Thus, the network can capture both local details and global context, leading to more accurate performances.

Figure 2 depicts the reference U-Net-like architecture: the input image is progressively halved in size and doubled in volume using convolutional blocks until a core compact representation of 512 layers of size 20×20 is obtained. The decoding phase is characterized by a series of deconvolutional blocks, each of them combined with the corresponding residual block from the encoding phase and transformed in volume through an additional convolutional block. The final image is then reconstructed by exploiting a sigmoid activation layer. Each block is composed of a convolution/transposed convolution layer, a rectified linear unit, a dropout, and a batch normalization layer.

The neural model is learned on a set $\mathcal{D} = \{(\tilde{x}_1, x_1), (\tilde{x}_2, x_2), \ldots, (\tilde{x}_n, x_n)\}$ of image pairs, where x_i represents the original "clean" image and $\tilde{x}_i = x_i + \epsilon_i$ the (possibly) "compromised" input. In more detail, the AE is fed with images containing a secret, while the legitimate ones are used to compute the loss values. Hence, the learning phase tries to optimize the network weights by minimizing the reconstruction loss. The idea is to teach the AE to reproduce altered images as clean ones. Specifically, we used the Mean Squared Error (MSE) since it measures divergences at a pixel level:

$$MSE(N, \mathcal{D}) = \frac{1}{n} \sum_{i=1}^{n} \|x_i - N(\tilde{x}_i)\|^2$$

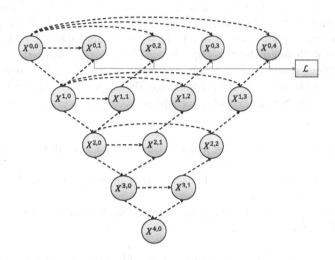

Fig. 3. Connection scheme for the U-Net+ like architecture.

As proposed in [19], the standard U-Net neural architecture has been extended with further hidden layers to yield different latent representations (i.e., representations with different sizes). This makes the learning process more stable and improves the effectiveness of the reconstruction. Figure 3 sketches the connection scheme of our architecture that represents a variant of the UNet+ presented in [19]. For the sake of simplicity, hereinafter, we denote our variant as U-Net+. Specifically, the input image is simultaneously reconstructed by four output layers denoted as $X^{0,1}, X^{0,2}, X^{0,3}$ and $X^{0,4}$, respectively. This allows to take advantage of the intermediate representations generated by the other layers. In particular, let $\hat{X}_i^{0,j}$ be the output of the j-th layer for the i-th instance. Hence, the loss function is defined as the sum of the errors of the output layers:

$$\mathcal{L} = \sum_{j=1}^{L} \frac{1}{n} \sum_{i=1}^{n} \left\| x_i - \hat{X}_i^{0,j} \right\|^2$$

4 Performance Evaluation

In this section, we introduce the setup used to assess the performance of our sanitization approach. Then, we present numerical and qualitative results.

4.1 Experimental Setup

As a first step, we created a suitable dataset of images to model the malware cloaking the URLs. To this aim, we used the Berkeley Segmentation Data Set [1] containing 500 legitimate images that have been cropped to 321×321 pixels to simplify the analysis. The URLs to be hidden have been borrowed from a public list[1] of real financial institutions. To model a ZeusVM-like attack template, each image contained a payload composed of 70 randomly-picked URLs, which has been separated through the *d* sequence. This allowed considering an attacker trying to retrieve single URLs (even if sanitized) by searching for the given escape sequence. To hide/recover the content, we developed suitable Python scripts implementing the two LSB techniques presented in Sect. 2 and the following hiding patterns:

- *sequential*: the payload is hidden starting from the first pixel (i.e., the one located in the upper left corner);
- *rows*: the payload is first divided into three equal parts and then hidden in parallel, interleaved areas;
- *squares*: the payload is hidden in equally-sized blocks of 107×107 pixels placed along the diagonal.

As a result, we obtained a dataset of $3,500$ images composed of 500 clean images, $1,500$ images with payloads hidden with the LSB plain method (500 images for each pattern), and $1,500$ images containing payloads hidden with the LSB variant method (500 images for each pattern). For each combination, we split the dataset into train, test, and validation sets, composed of 170, 165, and 165 images, respectively.

To implement our sanitization architecture, we used PyTorch [12]. Comparisons have been performed via a basic deep convolutional autoencoder, denoted in the following as DeepAE. In essence, it consists of two convolutional layers halving the size of the input image and producing a representation of 64 layers of size 160×160 pixels. The third convolutional layer is the latent space. The decoder doubled the size of the latent space by using two transposed convolutional layers. The other baseline architecture is the U-Net described in Sect. 3 but without the connection scheme depicted in Fig. 3. The models were trained with the Adam optimizer with a learning rate equal to 0.001.

To evaluate the performances, we used two metrics. The first is the MSE already introduced in Sect. 3. The second is the Peak Signal-to-Noise Ratio (PSNR), which measures the ratio between the maximum power of a signal and the power of the related noise. Specifically, the PSNR has been defined as follows:

$$PSNR(x, \hat{x}) = 20 \log_{10} \left(\frac{MAX(x)}{\sqrt{MSE(x, \hat{x})}} \right)$$

---

[1] https://github.com/cloudipsp/all_banks_ips.

where $MAX(x)$ is a function returning the maximum possible pixel value of the image x.

Lastly, experiments were executed on an NVidia DGX Station with 4 V100 GPUs with 32 GB of RAM.

Table 1. Comparative analysis among different neural models used for sanitization. Best results are reported in bold.

Neural Model	LSB Technique	Pattern	MSE	PSNR	Training [s]	Inference [s]
DeepAE	LSB plain	Sequential	6.36e-3	21.98	**4,231**	**0.0016**
		Rows				
		Squares				
	LSB variant	Sequential				
		Rows				
		Squares				
U-Net	LSB plain	Sequential	2e-4	36.95	13,648	0.0074
		Rows				
		Squares				
	LSB variant	Sequential				
		Rows				
		Squares				
U-Net+	LSB plain	Sequential	**1.6e-4**	**38.05**	18,646	0.0067
		Rows				
		Squares				
	LSB variant	Sequential				
		Rows				
		Squares				

4.2 Numerical Results

A preliminary analysis conducted to recover the malicious payloads showcases that all the considered neural architectures can sanitize the images. Indeed, the content extracted from the sanitized images does not contain even partially recognizable URLs. This is also confirmed by the results shown in Table 1 exhibiting a low value of MSE. This behavior can be observed regardless of the LSB technique and pattern employed. However, U-Net-based architectures produce higher quality images compared to the AE model. Notably, U-Net+ outperforms the baselines in both MSE and PSNR.

For all models, we also evaluated two timing behaviors. The first is the time required for the training phase. The second is the inference time, i.e., the duration of the sanitization process of a single image. As reported, the increased complexity of our approach accounts for higher training times. The inference phase exhibits similar behavior, but our solution preserves the original image quality, i.e., PSNR is equal to 38.05 compared to 21.98 of the DeepAE model.

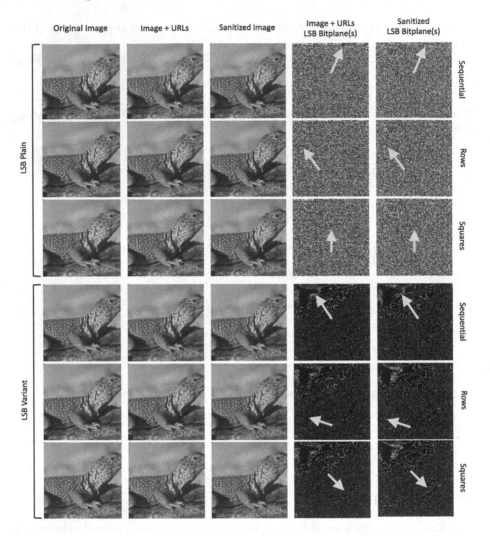

Fig. 4. Comparison among the original image, images with URLs embedded via different LSB methods/patterns, and the corresponding sanitized image. The comparison also includes the bitplane(s) for the modified and sanitized images.

To further quantify the effectiveness of our sanitization approach, Fig. 4 depicts a "visual" comparison among the original image, the images containing URLs embedded with the different LSB techniques and patterns, and the corresponding image sanitized through the U-Net+ architecture.

As shown, the sanitized image preserves the "quality" of the original counterpart while completely disrupting the malicious content. This can be viewed by inspecting the bitplanes. When the LSB plain mechanism is used, the bitplane refers to 1 bit, whereas for the LSB variant, it condenses three different layers. In

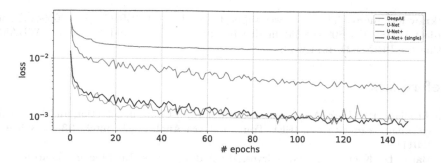

Fig. 5. Loss functions of the different architectures over the training epochs.

both cases, bits corresponding to the malicious content (indicated with arrows in the figure) are altered, thus leading to a complete disruption of the hidden information, which cannot be recovered by the threat actor, even partially (e.g., through a de-sanitization attack).

To highlight the advantages of our architecture compared to simpler models, we also conducted a convergence analysis of the training loss. As showcased in Fig. 5, skip connections lead to an accelerated convergence during the training process for both U-Net and U-Net+ architectures, i.e., fewer epochs are required to learn a reliable and effective model. This behavior allows to save time and reduces the overall energy footprint of the sanitization process. We point out that, the overall loss value of the U-Net+ architecture comprises four components. Hence, we also report the loss for the output layer responsible for reconstructing the image in its final form, i.e., see U-Net+ (single) in the figure.

5   Conclusions and Future Work

In this paper, we devised an approach for sanitizing images cloaking a list of URLs. Our mechanism exploited a U-Net-like model, which demonstrated its effectiveness in terms of quality of sanitized images and computational requirements. A main limitation of our solution is the need for a suitable amount of images containing hidden data to learn an effective functional mapping. Therefore, our ongoing research investigates the usage of semi-supervised methods to train reliable models on scarce data. Another possible refinement is to embed a regularization component in the loss function for estimating the probability that the secret has been destroyed. To make our sanitization technique suitable for realistic and large-scale scenarios characterized by tight privacy and computational constraints, a future development considers using a federated paradigm, e.g., to train the model in a decentralized fashion. Lastly, we are working to consider other cloaking mechanisms (e.g., Invoke-PSImage) and other hidden payloads, such as configuration files or small attack routines.

Acknowledgment. This work was supported by Project SERICS (PE00000014) and Project RAISE (ECS00000035) funded by the EU - NGE and by Project WHAM! within the PRIN2022 framework.

References

1. Arbelaez, P., Maire, M., Fowlkes, C., Malik, J.: Contour detection and hierarchical image segmentation. IEEE Trans. Pattern Anal. Mach. Intell. **33**(5), 898–916 (2011)
2. Bank, D., Koenigstein, N., Giryes, R.: Autoencoders. In: Machine Learning for Data Science Handbook, pp. 353–374 (2023)
3. Caviglione, L., et al.: Tight arms race: overview of current malware threats and trends in their detection. IEEE Access **9**, 5371–5396 (2021)
4. Caviglione, L., Comito, C., Guarascio, M., Manco, G.: Emerging challenges and perspectives in deep learning model security: a brief survey. Syst. Soft Comput. **5** (2023)
5. Caviglione, L., Mazurczyk, W.: Never mind the malware, here's the stegomalware. IEEE Secur. Priv. **20**(5), 101–106 (2022)
6. Chua, M., Balachandran, V.: Effectiveness of android obfuscation on evading anti-malware. In: ACM Conference on Data and Application Security and Privacy (2018)
7. D'Angelo, G., Ficco, M., Palmieri, F.: Malware detection in mobile environments based on autoencoders and API-images. J. Parallel Distrib. Comput. **137**, 26–33 (2020)
8. Gibert, D., Mateu, C., Planes, J.: The rise of machine learning for detection and classification of malware: research developments, trends and challenges. J. Netw. Comput. Appl. **153**, 102526 (2020)
9. Han, R., Yang, C., Ma, J., Ma, S., Wang, Y., Li, F.: IMShell-Dec: pay more attention to external links in PowerShell. In: ICT Systems Security and Privacy Protection, pp. 189–202 (2020)
10. Mazurczyk, W., Caviglione, L.: Information hiding as a challenge for malware detection. IEEE Secur. Priv. **13**(2), 89–93 (2015)
11. Mohaisen, A., Alrawi, O.: Unveiling zeus: automated classification of malware samples. In: International Conference on World Wide Web, pp. 829–832 (2013)
12. Paszke, A., et al.: PyTorch: an imperative style, high-performance deep learning library. In: NeurIPS (2019)
13. Robinette, P.K., Moyer, D., Johnson, T.T.: Monsters in the Dark: Sanitizing Hidden Threats with Diffusion Models. arXiv:2310.06951 (2023)
14. Robinette, P.K., Wang, H.D., Shehadeh, N., Moyer, D., Johnson, T.T.: SUDS: sanitizing universal and dependent steganography. In: ECAI, vol. 372 (2023)
15. Ronneberger, O., Fischer, P., Brox, T.: U-Net: convolutional networks for biomedical image segmentation, vol. 9351, pp. 234–241 (2015)
16. Rus, C., Sarmah, D.K., El-Hajj, M.: Defeating MageCart attacks in a NAISS way. In: 20th International Conference on Security and Cryptography, pp. 691–697 (2023)
17. Singh, A.K.: Data hiding: current trends, innovation and potential challenges. ACM Trans. Mult. Comp. Comms. Apps. **16**(3), 1–16 (2020)
18. Suarez-Tangil, G., Tapiador, J.E., Peris-Lopez, P.: Stegomalware: playing hide and seek with malicious components in smartphone apps. In: International Conference on Information Security and Cryptology, pp. 496–515 (2014)

19. Zhou, Z., Siddiquee, M.M.R., Tajbakhsh, N., Liang, J.: UNet++: redesigning skip connections to exploit multiscale features in image segmentation. IEEE Trans. Med. Imaging **39**, 1856–1867 (2019)
20. Zimba, A., Wang, Z., Chen, H.: Multi-stage crypto ransomware attacks: a new emerging cyber threat to critical infrastructure and industrial control systems. ICT Express **4**(1), 14–18 (2018)
21. Zuppelli, M., Manco, G., Caviglione, L., Guarascio, M.: Sanitization of images containing stegomalware via machine learning approaches. In: Proceedings of the Italian Conference on Cybersecurity, vol. 2940, pp. 374–386 (2021)

Digilog: Enhancing Website Embedding on Local Governments - A Comparative Analysis

Jonathan Gerber[✉], Bruno Kreiner, Jasmin Saxer, and Andreas Weiler

Institute of Computer Science, Zurich University of Applied Sciences, Obere Kirchgasse 2, 8400 Winterthur, Switzerland
jonathan.gerber@zhaw.ch
https://www.zhaw.ch/en/engineering/institutes-centres/init/

Abstract. The ability to understand and process websites, known as website embedding, is crucial across various domains. It lays the foundation for machine understanding of websites. Specifically, website embedding proves invaluable when monitoring local government websites within the context of digital transformation. In this paper, we present a comparison of different state-of-the-art website embedding methods and their capability of creating a reasonable website embedding for our specific task based on different clustering scores. The models consist of visual, mixed, and textual-based embedding methods. We compare the models with a base line model which embeds the header section of a website. We measure their performance in an off-the-shelf evaluation as well as after transfer learning. Additionally, We evaluate the models' capability of distinguishing municipality websites from other websites such as tourist websites. We found that when taking an off-the-shelf model, Homepage2Vec, a combination of visual and textual embedding, performs best. When applying transferred learning, MarkupLM, a markup language-based model, outperforms the others in both cluster scoring as well as precision and F1-score in the classification task. All mixed or markup language-based models achieve an F1-score and a precision over 97%. However, time is an important factor when it comes to calculations on large data quantities. Thus, when additionally considering the time needed, our base line model performs best.

Keywords: embedding evaluation · website embedding · content monitoring · cluster evaluation

1 Introduction and Motivation

From individuals seeking information to machine learning marvels like chatbots and trading algorithms, countless entities rely on the data ocean, known as the World Wide Web. In this landscape website monitoring plays a crucial role. By analyzing constantly changing internet data, these models handle diverse tasks ranging from event detection and price tracking to ensuring compliance with evolving policies. One noteworthy example is the European Union's 2016 Directive on website accessibility for public services (Directive (EU) 2016/2102). Monitoring tools can help to ensure these regulations are upheld, promoting an inclusive digital space for all.

When looking at the significance of website monitoring it becomes clear that tools are vital for understanding the broader landscape of digital transformation. We seek to analyze websites from local governments across Europe with the end goal of assessing their digitalization. While the assessment itself is not part of this paper, we set the foundation for an ongoing interdisciplinary research project between computer and political science called Digilog[1]. There is already work claiming to measure the level of digital transformation within local governments. Garcia-Sanchez *et al.* [9] presents an analysis of the development of e-governments of 102 Spanish municipalities and Pina *et al.* [22] conduct an empirical study about the effect of e-government on transparency, openness and hence accountability in 15 countries of the EU and a total of 318 government web sites. When analyzing websites over time, mutations such as domain changes or emergings of new websites frequently occur. To maintain an up to date list of municipality URLs, we propose a method to verify websites' authenticity, particularly distinguishing between governmental and tourism sites. Our research reveals that crawlers often mistake tourism sites for government ones. This classification task as well as all other downstream tasks (e.g. e-service detection, analysis of digital transformation, ...) require a numerical representation of the website. However, an accurate representation of websites using numerical embeddings derived from Natural Language Processing and Computer Vision models is challenging. We evaluate the performance of pre-trained model embeddings in two scenarios: using them directly and training a feedforward neural network (FNN) on top of them with domain-specific data, known as transfer learning (TL). This approach enhances generalizability for various downstream tasks.

2   Related Work

Websites use both visual (images, rendered HTML code) as well as textual (floating text) features to present content to users. To extract the full depth of information, an embedding model should be capable of processing both visual and textual data. Thus it is not surprising that Large Language Models (LLM) and Convolutional neural networks (CNN) are often used in recent work. There are also other classical machine learning approaches that rely more on feature engineering. However, they do not generalize as well as the state-of-the-art models, due to their lack of flexibility when it comes to structural changes of an HTML page. There exists a large amount of related work in the field of text-based-only embedding and classification of websites. Hashemi [11], Kowsari *et al.* [13] and Minaee *et al.* [19] provide surveys on past work and discuss different approaches on website embedding. We only mention a selection of the recent work which is

---

[1] https://www.digilog-project.org/.

related to our applied topic. Visual-only based classifications are in many cases applied to the detection of harmful content. Whether detection of propaganda of terrorism [12], alcohol, adult content, weapons [1,7] or just food, fashion and landscapes [17], the classes all have distinctive visual features. However, in many cases, these approaches cannot distinguish visually similar pages (e.g. municipality homepage vs. tourism page about the same municipality).

In the field of text-based website embedding/classification, there are approaches that rely on classical machine learning [2,18]. However, the majority are based on neural networks or transformers. There are several RNN and LSTM-based approaches [4,15,20,24] to embed websites. Lin et al. [15] and Zhou et al. [24] additionally combine their BiLSTM approach with a CNN. There are different Transformer based approaches [5,10,14,21]. We summarize the two most relevant approaches for our topic for each group (textual-/visual-based). Li et al. [14] propose MarkupLM, a pre-trained LLM for document understanding tasks based on the actual text as well as the Markup language. The model is based on the BERT architecture. They add the additional XPath embedding to the embedding layer which is based on various features to identify the target leaf. The pre-training objectives of the models are Masked Markup Language Modeling (prediction of text token of DOM tree leaf), Node Relation Prediction (e.g., child, sibling, ...), and Title-Page Matching. They compare their two models (base and large) with previous models such as FreeDOM-Full [15], Simp-DOM [24], and others on the SWDE dataset considering the F1-score. They also compare their models with BERT base, RoBERTa base, and ELECTRA large models from Chen et al. [5] on the WebSRC dataset. Their large model outperforms every other model in every aspect, while their base model outperforms the others in most cases. The proposed models are available only in English. Nandanwar and Choudhary [20] propose a classification model based on GloVe and a BiLSTM for categorizing. They test the model on the WebKB data set as well as the DMOZ dataset. They further compare their model against the ensemble model of Gupta and Bhatia [10] and a Support Vector Machine web page classification approach [2]. In most cases, the proposed model outperforms the other models.

There also exists research on mixed approaches. Bruni and Bianchi [3] introduce a procedure for website classification that leverages both textual and visual features. They compare different classification algorithms to identify e-commerce services on web pages. The classification approach they propose is highly sophisticated and may not align with our specific needs as they assume classes to have certain attributes such as those related to e-commerce services. Lungeon et al. [16] propose a language-agnostic website embedding for classification tasks. With their introduced homepage embedder "Homepage2vec" they create a multilingual embedding based on word embeddings of the textual content (the first 100 sentences), the metadata tags (title, description, keywords, ...), and also the visual appearance (screenshot) and other features such as domain name. The numerical features are concatenated and processed by a neural network. They are then used for classifying the website into 14 different classes (art, business, computers, games, ...). While the feature embeddings seem to effectively capture the essence of the homepage, the model is constrained by a narrow range of broad classes.

3   Methods

In this section, we clarify which pre-trained models we used for embedding, how we applied TL, and how we evaluated the models' embeddings.

As mentioned in Sect. 2 there are mainly three different approaches to embed websites: textual, visual, and combined methods. We apply two recently published methods and benchmark them on our dataset which is described in Sect. 4. We selected Homepage2vec [16] and MarkupLM [14] due to their performance and reproducibility. Both approaches leverage the deeper semantic understanding embedded within Markup Documents. MarkupLM incorporates the embedding of XPath and tags as features, while Homepage2vec integrates visual features alongside specific data from a Markup document, including keywords and descriptions found in the meta tag section. Notably, both authors provide a library or GitHub repository for applying their models. To accommodate the absence of a multilingual version for the MarkupLM model, text components were translated into English before being used for embedding.

Homepage2vec. We used the Homepage2vec [16] library[2] and its ready-made feature extractors. We slightly changed the way Homepage2vec retrieves websites. Namely, we allow for redirects using Requests. If requests cannot fetch a site, we use Selenium with a headless Chrome web driver. The library (see footnote 2) offers two options: Either use the visual embeddings using screenshots of the websites or simply leave them out. The library(See footnote 2) concatenates all the individual features and processes them using fully connected layers. We obtained 100-dimensional embeddings by accessing the last hidden layer.

MarkupLM. We used the MarkupLM [14] base model[3] and large model[4] to extract the text and XPath from the HTML. We limited the number of nodes to 512, which is the model's maximum processing capacity. We then translated each text node using the Libretranslate API[5]. We leveraged the MarkupLM model to embed each node and took the mean over all nodes of each HTML to obtain the embedding for the HTML. The final embedding has a dimension of 768.

Header Section Embedding. A website typically includes a header with the structure of a website including the main topics and subtopics of the website. Based on predefined rules we extracted this header. We then extracted the text and embedded it with a multilingual BERT-based sentence embedder. The embedding has 768 dimensions.

---

[2] https://pypi.org/project/homepage2vec/.
[3] https://huggingface.co/microsoft/markuplm-base.
[4] https://huggingface.co/microsoft/markuplm-large.
[5] https://libretranslate.com/.

ResNet Embedding. As a simple visual embedding method we used the pre-trained ResNet18 model for embedding screenshots of the websites. We retrieved the screenshot of each website and embedded it with this ResNet18 model. This resulted in an embedding vector with 512 dimensions.

Our TL approach involves using the models' embeddings and training a FNN on top of them. In the first hidden layer of the FNN, the embeddings are transformed into a 100-dimensional vector. The second hidden layer is the classification head. This architecture was also used for the classification in the original training of Homepage2vec, and we adopted the same activation functions and dropout rates.

After the website contents are processed by an encoder model unto latent space, we can analyze how well the ground-truth labels (municipality vs. non-municipality) are naturally clustered. To evaluate the two clusters we used the Silhouette Score [23], the Davies-Bouldin Index [6] as well as an additional score we call Separation Distance General Score (SDG-score). In the TL case, we simply take the 100-dimensional vector from the hidden layer.

The SDG-score evaluates each cluster individually and takes the mean of every cluster value. It assesses whether each cluster is separable from the rest of the dataset by comparing the third Quantile ($Q3$) of the within-cluster distance to its centroid with the first Quantile ($Q1$) of outside cluster distances to the centroid. A high outside distance (high separation) and low within distance (compact cluster) results in a high score which is preferable.

$$S_{SDG} = \frac{1}{C} \sum_{k=1}^{C} \frac{Q3(wcd(k))}{Q1(ocd(k))} \tag{1}$$

C represents the total amount of clusters. The function wcd returns a vector with length J_k, of distances of all observations whithin the k-th cluster to its centroid. The ocd function returns a vector of distances of observations outside the k-th cluster to the centroid of the $k - th$ cluster. The vector has the length $N - J_k$, with N being the number of all observations and J_k the number of observation within the k-th cluster.

3.1   Dataset and Infrastructure

Our dataset consists of 2901 municipality websites provided directly from the country administration and an additional 1349 municipality websites hand-labeled by domain exports. After dropping duplicates the remaining municipalities are used to retrieve 3813 non-municipality websites using DuckDuckGo by querying the municipality name + "tourism". We use touristic websites since they share various characteristics with municipality websites and act as difficult negatives. The dataset contains websites of municipalities from ten different countries: Albania, Azerbaijan, Bulgaria, Croatia, Cyprus, Hungary, Romania, Serbia, Slovakia, and the United Kingdom. Only the landing page of each website was used for the embedding. We generally used standard parameters from PyTorch for all the methods. We applied early stopping on the validation loss with a patience of 10 epochs for training. To ensure the robustness of the trained model, we implemented stratified K-Fold Cross-Validation with validation and test set. This involved dividing our dataset into 10 folds, which was split into a training, validation, and testing set with the proportions 60:20:20. Training and performance measurement was done on an 11th Gen Intel(R) Core(TM) i9-11950H CPU with 16 cores, 32 GB RAM, and an NVIDIA RTX A2000 Laptop GPU with 4GB dedicated RAM.

4 Case Studies and Results

The scoring of the embedding in Table 1 shows Homepage2vec with visual embedding to be the best embedding model when it comes to embedding domain-specific data without further adaptation. The combination of visual and textual embeddings seems to have an advantage. That is reasonable since certain distinctive features are only visually detectable by rendering images (e.g. municipalities tend to have a white background with an image of the municipality in the upper part of their websites). However, when only considering visual features, the model lacks the capability of distinctively building clusters. This is due to a lack of capability to understand the semantics of links and general text on the website. The light version of Homepage2vec without visual embedding performs worse than the other text-only-based approaches. The result of the ResNet embeddings, without further training, shows that it does not perform well in building clusters. The same is true for the pre-trained MarkupLM models and the header section embeddings. When using TL the MarkupLM models as well as the embedding of the header section perform best and have the biggest performance improvement. In both cases, the ResNet-based model is not able to compete with the other models. The high performance of Homepage2vec with visual embeddings does not translate to high performance with TL. This discrepancy may arise from the model's original training but crucially, the Homepage2vec embeddings are limited to 100 dimensions while the embeddings used

Table 1. Municipality vs non-municipality website clustering scores for each embedding method showing the score without (P) and with transfer learning (TL) using a FNN on top of the frozen pre-trained models. The mean processing time per observation is also added. The time (in seconds) includes scraping, feature engineering and embedding. Best score per column is written in bold.

Embedding Method	Silhouette		Davies-Bouldin		SDG		Time (s)
	P	TL	P	TL	P	TL	
Header section	0.169	0.631	2.443	0.511	1.133	3.197	4.287
ResNet	0.074	0.316	3.616	1.274	0.836	1.386	5.820
Homepage2vec	0.155	0.355	2.229	1.093	0.950	1.414	6.750
Homepage2vec with visual	**0.288**	0.446	**1.378**	0.880	**1.209**	1.757	12.815
MarkupLM base	0.168	0.742	2.259	0.337	1.002	**4.162**	7.531
MarkupLM large	0.172	**0.748**	2.203	**0.332**	1.069	4.107	8.050

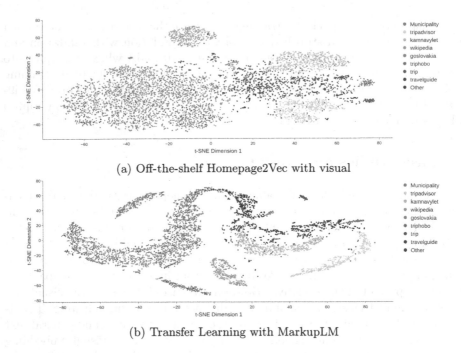

(a) Off-the-shelf Homepage2Vec with visual

(b) Transfer Learning with MarkupLM

Fig. 1. Embeddings of municipality (green) and non-municipality websites colored by most frequent domain. Both approaches are capable of creating reasonable clusters. (Color figure online)

from other models exceed 500. A higher-dimensional embedding vector leads these models to have a higher performance jump from pre-trained to TL. An alternative approach with Homepage2vec could involve extracting embeddings from a different layer of the model.

Figure 1 shows the separation of tourism websites from municipality websites. The embeddings with TL show less overlap of the clusters. The mean embedding time in seconds per URL is shown in Table 1. The time for the embedding includes fetching the HTML and screenshot if needed, followed by the embedding method. In the case of MarkupLM, the translation of the page content to English is also included in the time. The Header section embedding method performs best in terms of embedding time. The time correlates with the complexity of each model.

The TL results in discerning between municipality and non-municipality websites are shown in Table 2. We show the mean F1-Score and the Precision alongside their respective margin of error. We prioritize precision in this context to minimize false positives. We use the standard error of the mean (SEM) to calculate the margin of error of the 10K-Fold Cross-Validation. The MarkupLM large embedding as input to the TL step resulted in the best F1-Score and precision. There are different reasons why the MarkupLM models perform better than the Homepage2Vec model in this task. As mentioned before, the MarkupLM embedder offers embeddings 7 times larger, but it was also pre-trained differently on a broader and bigger dataset with 24 Million webpages making it potentially easier to use for TL and generalizability. Homepage2vec was trained on 886'000 webpages using multiple pre-trained models not trained on HTML content. The MarkupLM model was therefore exposed to much more HTML data during the training process than Homepage2vec. Furthermore the training dataset from MarkupLM which is a common crawl snapshot includes municipality websites, whereas the Hompage2Vec dataset does not.

Table 2. Transfer learning (TL) scores of the embedding methods using K-Fold Cross-Validation with 10 k in percentages, ± margin of error)

Embedding Method	F1-Score	Precision
Header section	98.10 ± 0.18	98.13 ± 0.27
Homepage2vec	97.52 ± 0.19	97.17 ± 0.16
Homepage2vec with visual	98.27 ± 0.21	98.33 ± 0.28
MarkupLM base	99.15 ± 0.12	99.21 ± 0.12
MarkupLM large	**99.18 ± 0.09**	**99.36 ± 0.09**
ResNet	91.62 ± 0.40	91.67 ± 0.50

5 Conclusions and Future Work

We compared different models on their capability of embedding HTML documents with high diversity and also tested their performance on classifying municipality websites in a binary classification task. We rated the embedding methods with several different clustering performance scores which reward the capability of separating websites within a binary classification system. We compared the performance of pre-trained models as well as models with TL. We have seen that, based on the clustering scores, Homepage2Vec with a combined approach of using textual and visual features outperforms visual or text-only based models in all the applied clustering scores. When applying TL and comparing the outputs of the last hidden layer as embeddings we have seen that markup language-based models had the biggest improvement and outperformed the mixed approach as well as the visual-based only. When comparing the models in a binary classification MarkupLM also outperforms the other approaches with a precision of 99.36% and an F1-score of 99.18%. However, when considering processing time, which is an important factor for classifying large datasets, the base line "Header section" model has the best balance between time efficiency and classification performance, achieving competitive clustering scores with TL.

Further research could be done on how these different sections correlate with the embedding or classification of a website focusing on explainability. Text-based embeddings seem to be the best choice when it comes to TL for the classification of websites in a binary classification. We could enlarge the embeddings to not only focus on one site but rather on subsites of the domain as well. Many features of government websites are not immediately visible at the top level but become apparent at deeper levels of crawling. A comparison of models that also consider linked sites could be conducted. Further research is needed to analyze the performance of models in a multi-class classification setting for website features such as e-forms, logins, payments, etc. To encourage the model to spread out the embeddings more effectively, one could apply triplet or contrastive learning approaches as seen in SimCSE [8]. This could be coupled with more sophisticated methods to handle outlier edge cases. One approach could be to crawl potentially hard-to-classify web pages as part of a dataset augmentation strategy. When it comes to the training of classification models labeled data is a valuable asset. Additional research could explore Semi-Supervised Learning and Active Learning in this specific context. The foundation of an efficient application is a reasonable embedding of a given website which we have demonstrated is achievable.

Acknowledgment. This work is supported by Grant No. GR 200839 of the Swiss National Science Foundation (SNF) and German Research Foundation (DFG) for the research project "Digital Transformation at the Local Tier of Government in Europe: Dynamics and Effects from a Cross-Countries and Over-Time Comparative Perspective (DIGILOG)".

References

1. Akusok, A., Miche, Y., Karhunen, J., Bjork, K.M., Nian, R., Lendasse, A.: Arbitrary category classification of websites based on image content. IEEE Comput. Intell. Mag. **10**(2), 30–41 (2015)
2. Bhalla, V.K., Kumar, N.: An efficient scheme for automatic web pages categorization using the support vector machine. New Rev. Hypermedia Multimedia **22**(3), 223–242 (2016)
3. Bruni, R., Bianchi, G.: Website categorization: a formal approach and robustness analysis in the case of e-commerce detection. Expert Syst. Appl. **142**, 113001 (2020)
4. Buber, E., Diri, B.: Web page classification using RNN. Procedia Comput. Sci. **154**, 62–72 (2019)
5. Chen, X., et al.: WebSRC: a dataset for web-based structural reading comprehension. arXiv preprint arXiv:2101.09465 (2021)
6. Davies, D.L., Bouldin, D.W.: A cluster separation measure. IEEE Trans. Pattern Anal. Mach. Intell. **2**, 224–227 (1979)
7. Espinosa-Leal, L., Akusok, A., Lendasse, A., Björk, K.-M.: Website classification from webpage renders. In: Cao, J., Vong, C.M., Miche, Y., Lendasse, A. (eds.) ELM 2019. PALO, vol. 14, pp. 41–50. Springer, Cham (2021). https://doi.org/10.1007/978-3-030-58989-9_5
8. Gao, T., Yao, X., Chen, D.: Simcse: simple contrastive learning of sentence embeddings. arXiv preprint arXiv:2104.08821 (2021)
9. García-Sánchez, I.M., Rodríguez-Domínguez, L., Frias-Aceituno, J.V.: Evolutions in e-governance: evidence from Spanish local governments. Environ. Policy Gov. **23**(5), 323–340 (2013)

10. Gupta, A., Bhatia, R.: Ensemble approach for web page classification. Multimedia Tools Appl. **80**, 25219–25240 (2021)
11. Hashemi, M.: Web page classification: a survey of perspectives, gaps, and future directions. Multimedia Tools and Appl. **79**(17–18), 11921–11945 (2020)
12. Hashemi, M., Hall, M.: Detecting and classifying online dark visual propaganda. Image Vis. Comput. **89**, 95–105 (2019)
13. Kowsari, K., Jafari Meimandi, K., Heidarysafa, M., Mendu, S., Barnes, L., Brown, D.: Text classification algorithms: a survey. Information **10**(4), 150 (2019)
14. Li, J., Xu, Y., Cui, L., Wei, F.: MarkupLM: Pre-training of Text and Markup Language for Visually-rich Document Understanding (2022). http://arxiv.org/abs/2110.08518. arXiv:2110.08518
15. Lin, B.Y., Sheng, Y., Vo, N., Tata, S.: Freedom: a transferable neural architecture for structured information extraction on web documents. In: Proceedings of the 26th ACM SIGKDD International Conference on Knowledge Discovery & Data Mining, pp. 1092–1102 (2020)
16. Lugeon, S., Piccardi, T., West, R.: Homepage2Vec: language-agnostic website embedding and classification. In: Proceedings of the International AAAI Conference on Web and Social Media, vol. 16, pp. 1285–1291 (2022)
17. López-Sánchez, D., Corchado, J.M., Arrieta, A.G.: A CBR system for image-based webpage classification: case representation with convolutional neural networks. In: The Thirtieth International Flairs Conference (2017)
18. Matošević, G., Dobša, J., Mladenić, D.: Using machine learning for web page classification in search engine optimization. Future Internet **13**(1), 9 (2021)
19. Minaee, S., Kalchbrenner, N., Cambria, E., Nikzad, N., Chenaghlu, M., Gao, J.: Deep learning-based text classification: a comprehensive review. ACM Comput. Surv. (CSUR) **54**(3), 1–40 (2021)
20. Nandanwar, A.K., Choudhary, J.: Semantic features with contextual knowledge-based web page categorization using the GloVe model and stacked BiLSTM. Symmetry **13**(10), 1772 (2021)
21. Nandanwar, A.K., Choudhary, J.: Contextual embeddings-based web page categorization using the fine-tune BERT model. Symmetry **15**(2), 395 (2023)
22. Pina, V., Torres, L., Royo, S.: Are ICTs improving transparency and accountability in the EU regional and local governments? An empirical study. Public Adm. **85**(2), 449–472 (2007)
23. Rousseeuw, P.J.: Silhouettes: a graphical aid to the interpretation and validation of cluster analysis. J. Comput. Appl. Math. **20**, 53–65 (1987)
24. Zhou, Y., Sheng, Y., Vo, N., Edmonds, N., Tata, S.: Simplified DOM trees for transferable attribute extraction from the web. arXiv preprint arXiv:2101.02415 (2021)

A Stream Data Mining Approach to Handle Concept Drifts in Process Discovery

Vincenzo Pasquadibisceglie$^{(\boxtimes)}$ ⓘ, Donato Lucente, and Donato Malerba ⓘ

University of Bari Aldo Moro, Bari, Italy
{vincenzo.pasquadibisceglie,donato.malerba}@uniba.it

Abstract. Process discovery algorithms discover process models from event logs recorded from the real-life processes. Traditional process discovery algorithms assume that logged processes remain in a steady state over time. However, this is often not the real-world case due to concept drifts. To continue using well-defined, off-line process discovery algorithms to process a stream of process execution traces, we propose an online approach that performs concept drift detection and adaption of process models discovered with traditional process discovery algorithms. Experimental results explore the effectiveness of the proposed approach coupled with several traditional process discovery algorithms.

Keywords: Concept drift · Process discovery · Event stream analysis

1 Introduction

An event log is a bag of process executions of a business process. A process execution is recorded as a trace, that is, an ordered list of activities invoked by a process execution from the beginning of its execution to the end. Process discovery is used to discover abstract process models from event logs. At present, several, powerful process discovery algorithms have been formulated in process mining literature [2] striking different trade-offs between the accuracy in capturing the behavior recorded in event logs and the simplicity of the derived process models. However, the majority of these algorithms perform offline analyses and assume processes are in a steady-state. On the other hand, few processes are truly in a steady-state. Due to changing circumstances (process concept drifts) process executions evolve and process models need to adapt accordingly.

In this study, we explore the performance of a stream data mining approach, called FAIRY (concept driFt detection and AdaptatIon pRocess discoverY), to integrate concept drift detection and adaptation in a traditional process discovery framework. FAIRY analyses a stream of traces and identifies possible concept drifts concerning how the recorded data conform to the process model. As soon as a concept drift is alerted, the discovery of a new process model is triggered. A contribution of the proposed approach is that it can wrap any process discovery

© The Author(s), under exclusive license to Springer Nature Switzerland AG 2024
A. Appice et al. (Eds.): ISMIS 2024, LNAI 14670, pp. 136–145, 2024.
https://doi.org/10.1007/978-3-031-62700-2_13

algorithm commonly used in the offline setting. For the concept drift detection, FAIRY adopts the ADWIN mechanism [3] to monitor how the upcoming traces conform to the online process model and vice-versa. ADWIN maintains an adaptive window of the newest traces performing a shrunk when a conformance change happens in context. In presence of conformance drifts, windowed traces (recorded by ADWIN) are used to discover a concept drift-tuned process model.

The paper is organised as follows. Section 2 illustrates the related work. Section 3 describes the basics of this study, while Sect. 4 describes the proposed methodology. Section 5 presents the experimental study. Finally, Sect. 6 sketched the future work.

2 Related Works

In the last decade, concept drift detection has attracted growing attention in the process mining literature. [11] presents an approach, named CDSF, that computes distances between streamed traces and a graph global model that represents the current state of the process. It processes distances through clustering and considers the emergence of a new cluster as the occurrence of a concept drift. However, this study mainly focuses on detecting and visualizing concept drift points as a means to perform conformance checking. [5] adopts ADWIN to monitor concept drifts in distances computed between pairs of activities and activity relations recorded in sub-windows. This study states that the adaptive window can, in principle, be used for the process discovery. However, it does not experiment this opportunity. Differently from this previous study, we verify the performance of ADWIN to monitor concept drifts occurring in the conformance of new traces to process models and we evaluate the performance of an active adaptation approach to adapt a process models to the newest windowed traces each time concept drifts are detected. ADWIN has been recently used in [9] to detect concept drifts in event streams and adapt a deep neural model trained for next-activity prediction to concept drifts. However, [9] investigates a predictive process mining problem, while this study explores a process discovery problem. The problem of detecting and handling concept drifts in process discovery has been recently investigated in [8]. This study describes an approach, named STARDUST, that uses a sampling technique to select the most representative, distinct trace variants to be considered for the process discovery. The proposed method alerts a concept drift as the histogram of the trace variants to be sampled changes over time, and triggers the discovery of a new process model as a drift is alerted. Notice that this previous approach do nor account for the effect of drifts on the process model. Differently, we monitor drifts in the conformance of process models to upcoming traces. Finally, [12] focuses on the online discovery of process models to handle large amounts of event data using finite memory. In this study, concept drifts may be recognized a-posteriori in the root-cause analysis of the changes observed in the process models continuously updated each time a new event is processed. Our study works in a different perspective as it explicitly detects concept drifts besides restarting the process model discovery in correspondence of detected concept drifts only.

3   Preliminary Concepts

Event and Trace Streams. Let $\mathcal{A}$ be the set of all activity names, $\mathcal{C}$ be the set of all case identifiers and $\mathcal{T}$ be the set of all timestamps. Let $\perp$ be the activity that denotes the completion of a case execution.[1] An event e is a triple $e = (c, a, t) \in \mathcal{C} \times \mathcal{A} \times \mathcal{T}$ that represents the occurrence of activity a in case c at timestamp t. Let $\mathcal{E} = \mathcal{C} \times \mathcal{A} \times \mathcal{T}$ be the event universe. Let us consider the functions: $\delta(e) = c$, $\pi(e) = a$ and $\gamma(e) = t$, respectively. An event stream Σ^* is an unbounded sequence of events $e_1, e_2, \ldots, e_i$ such that, for each $j = 1, 2, \ldots, i, \ldots,$ $\gamma(e_j) \leq \gamma(e_{j+1})$. A case c is an event sequence $c = \langle e_1, \ldots, e_n \rangle$, such that: (1) $c[i] \in \mathcal{E}$ for each $i = 1, 2, \ldots, n$, (2) $\delta(c[i]) = c$ for each $i = 1, 2, \ldots, n$ and (3) $\gamma(c[i]) \leq \gamma(c[i+1])$ for each $i = 1, 2, \ldots, n-1$. Let us introduce the function $len(c) = |c|$ that returns the number of events recorded in a case c. The case c is a full case if $\pi(c[len(c)]) = \perp$; a running case otherwise. In the remaining of this paper, we consider a case through its activity trace. Let c be a case, the trace σ of c is the activity sequence $\sigma = \langle a_1, \ldots, a_n \rangle$, such that $\pi(c[i]) = a_i$ for each $i = 1, 2, \ldots, n$ with $len(c) = n$. A trace stream $\Sigma = \sigma_1, \sigma_2, \ldots \sigma_i, \ldots$ is an unbounded sequence of traces of full cases so that, for each $i \in \mathbb{N}^+$, σ_i is the trace associated with the full case c_i with $\pi(c_i[len(c_i)]) = \perp$ and $\gamma(c_i[len(c_i)]) \leq \gamma(c_{i+1}[len(c_{i+1})])$. A trace σ_i is consumed by the trace stream Σ at time i. In this study, we adopt an adaptive, sliding window-based trace stream consumer. Let Σ be a trace stream, w be a window size, the window consumer $\mathbf{W}_{i-w+1 \longrightarrow i}$ keeps the traces $\sigma_j \in \Sigma$ with $i - w + 1 \leq j \leq i$.

Process Discovery. A process discovery task aims to discover a generative representation of a multiset of variant-traces (i.e. distinct trace of activities). This representation can generate multiple traces based on the optional paths it describes. We refer to the process model behavior as the set of variant-traces that can be generated from a process model. Let us consider a multiset of variant-traces $\mathcal{B}(\mathcal{A}^*)$, a generative representation M of $\mathcal{B}(\mathcal{A}^*)$ and a set of variant-traces $\mathcal{B}_M(\mathcal{A}^*)$ that can be generated by the execution of the model M. A process discovery algorithm is a function $\Delta \colon \mathcal{B}(\mathcal{A}^*) \mapsto \mathcal{B}_M(\mathcal{A}^*)$. Various robust process discovery algorithms are formulated in the process discovery literature [2]. They commonly discover Petri net-like process models from a multi-set of traces.

Conformance Checking. A conformance checker compares the observed behavior of real traces with the model behavior reflected by a process model [2]. The conformance of a model M, that can be formulated as a function $\epsilon_M \colon \mathcal{B}(\mathcal{A}^*) \mapsto \mathbb{R}^+$, is commonly computed as the FMeasure (i.e. harmonic mean) of fitness and precision of a trace to the process model M. The fitness measures how the process model parses observed traces, while the precision measures how the process model does not parse unobserved traces [2].

---

[1] The completion activity is commonly declared in various business processes such as ticket resolution in help desk processes. In any case, we are aware that business processes where the completion activity may not be explicitly declared require further investigation in future works.

Concept Drift Detection. Let us consider a process model M and a trace stream Σ. A concept drift is a change in the joint probability $P(\sigma, \epsilon)$ where σ is a trace fully recorded in Σ and ϵ is the conformance of σ to M. According to the definition formulated above concept drifts are mainly reflected in a deteriorating overall conformance of traces recorded in Σ to the process model M. Finding the exact points of change can be very challenging, as the change between two distributions can be gradual and has to be differentiated from transient noise that can affect the stream. In this study, we use an adaptive window strategy [2] that copes with infinite streams storing only data related to the most recent events and, periodically, analyzing them. The size of the window is adapted over time to the characteristics of the data. Let $\mathbf{W}$ be a sliding window populated with the latest traces recorded in Σ. Let $drift \colon \mathcal{B}(\mathcal{A}^*) \times \mathbb{R}^+ \mapsto \{0,1\}$ be a concept drift detection function. $drift(\mathbf{W}) = 0$ if no degradation is observed in the conformance $\epsilon_M(\sigma)$ of the sequence of traces σ actually recorded in $\mathbf{W}$, 1 otherwise. If $drift(\mathbf{W}) = 1$ then M is re-trained with the multi-set of variant-traces $\mathcal{B}(\mathbf{W})$ recorded in $\mathbf{W}$.

ADWIN. The concept drift detection mechanism ADWIN [1] is used to keep conformance statistics from a trace window consumer $\mathbf{W}$ of an adaptive, variable size, while detecting concept drifts on the sequence of conformance values associated with traces recorded in $\mathbf{W}$. The inputs to ADWIN are: (1) a confidence value $c \in [0,1]$ and (2) a (potentially infinite) sequence of conformance values. When each trace σ is added to the adaptive window $\mathbf{W}$, ADWIN analyzes the sequence of conformance values measured for the traces recorded in $\mathbf{W}$, in order to detect concept drifts. To this aim, ADWIN performs a search on all possible combinations of two sliding sub-windows in $\mathbf{W}$. If the distributions of conformance values in the two sub-windows are significantly different, then ADWIN alerts a concept drift and removes the oldest traces from the adaptive window.

4   The FAIRY Methodology

FAIRY is composed of three modules: initialization, conformance checking and process model discovery. Both the initialisation and process model discovery modules can wrap any process discovery algorithm Δ to perform the process discovery step. The initialization step sets the trace enumerator $i = 0$ and starts monitoring Σ, in order to discover an initial process model M. When a new trace σ_i is consumed by Σ at timestamp t_i (i.e. σ_i is recorded in $\mathbf{W}$), i is incremented by one. As soon as η traces have been recorded in $\mathbf{W}$, the process discovery module uses Δ to discover the initial process model M as a generative representation of the traces currently recorded in $\mathbf{W}$. Subsequently, the initialization step initializes ADWIN with the sequence of conformance values computed for M on the sequence traces actually recorded in $\mathbf{W}$. So, at the completion of the initialization step, $\mathbf{W}$ records η traces and M is the current process model that the online process model will use in the conformance checking module. The conformance checking module proceeds recording each upcoming trace in $\mathbf{W}$ and verifying the conformance of this trace to the current process model and adding this trace to

Table 1. Event log description: number of activities, events, traces and variant-traces recorded in the event logs, duration of the stream in days and average duration of traces in days

Stream	#Activity	#Trace	#Event	#Variant	Stream time	Trace time
Incidents (BPI13I)	13	7554	65533	2278	783	12.1
ControlSummary (BPI18C)	7	43808	161296	59	1051	57.3
ReferenceAlignment (BPI18R)	6	43802	128554	515	909	78.5
DomesticDeclarations (BPI20D)	17	10500	56437	99	889	11.5
PrepaiedTravelCost (BPI20P)	29	2099	18246	202	772	36.8
RequestForPayments (BPI20R)	19	6886	36796	89	941	12.0
Road	11	150370	561470	231	4917	341.6

the adaptive window of ADWIN according to its conformance value. If ADWIN alerts a conformance concept drift by estimating an adaptive widow with size w, then the window $\mathbf{W}$ is shrunken accordingly. In particular, the oldest traces are discarded from $\mathbf{W}$, while keeping the newest w traces only. Subsequently, the process discovery modules uses Δ to replace M with a new process model discovered as a generative representation of traces currently kept in $\mathbf{W}$. The new process model M will be used from now on in the conformance module to monitor the upcoming traces until a new concept drift will be detected by ADWIN. If new traces are consumed before the completion of a process discovery step, they are kept in queue until the process discovery step is completed. They are processed as the new process model is ready.

5   Experimental Results

We evaluated the performance of FAIRY[2] by conducting a range of experiments on various event logs handled as streams. The main objective of these experiments is to investigate the performance of the traditional process discovery algorithms in a dynamic setting by integrating concept drift detection and adaptation in the process discovery.

5.1   Event Logs, Experimental Setting and Metrics

We processed trace streams generated from seven benchmark event logs provided by the 4TU Centre for Research.[3] Table 1 reports the number of activities, events, traces and distinct variant-traces recorded in these event logs. For the process discovery, we ran Hybrid ILP Miner (ILP), Inductive Miner (IND) and Split Miner (SPL) as process discovery algorithms. To prevent the discovery of spaghetti-like models, we used the sampling algorithm [4] in advance to discovering process models with these algorithms. For each process discovery algorithm,

---

[2] The code is publicly available at https://github.com/vinspdb/FAIRY.
[3] https://data.4tu.nl/portal.

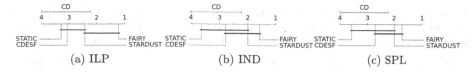

Fig. 1. Ranking of FAIRY, STARDUST, CDESF and STATIC according to the Nemenyi test performes on the conformance FMeasure of process models discovered with ILP (a), IND (b) and SPL (c). Groups of methods that are not significantly different (at $p \leq 0.05$) are connected.

we compared the performance of FAIRY to that of the static counterpart, named STATIC, that kept the process model discovered in the initialization step without updating it on concept drifts. In addition, we compared the performance of FAIRY to that of both CDESF that detected concept drifts with a cluster-based approach [11], and STARDUST that detected concept drifts monitoring changes in the histogram of variant-traces [8]. We also explored the sensitivity of FAIRY to the set-up of its parameters c (i.e. ADWIN confidence) and η (i.e. the number of events processed during initialization).

To measure the performance of discovered process models, we computed the average conformance of upcoming traces to the newer process model discovered in the stream, the average complexity of the process models discovered along the stream, and the total time spent processing the stream. The conformance was measured in terms of FMeasure of fitness and precision. This metric was computed on all the traces processed as the initialization step was completed. The average conformance FMeasure measured on all these traces is reported in this study. The complexity of process models was measured in terms of Extended Cardoso index [7]. This measures the complexity of a process model by its complex structures, i.e. Xor, Or and And components. The average extended Cardoso index was measured on the sequence of process models discovered on a stream. The total computation time spent processing a stream was measured in seconds by running all experiments on Intel(R) Core(TM) i7-9700 CPU, GeForce RTX 2080 GPU, 32 GB Ram Memory, Windows 10 Home.

5.2 Results and Discussion

Analysis of FAIRY, STARDUST, CDESF and STATIC. For this comparative study, we ran FAIRY with the default parameter set-up of $c = 0.002$ and $\eta = 10\%$ and remated methods with the parameter set-up reported in the reference studies. Table 2 reports the average conformance FMeasure of fitness and precision, the total time spent processing a stream, the number of concept drifts identified in a stream and the average extended Cardoso index of the sequence of process models discovered in a stream. Conformance results show that the adoption of a concept drift detection and adaptation strategy is crucial to ensure high conformance values of a process model on a stream. FAIRY achieves the highest conformance FMeasure on six out of seven streams with ILP, and four

Table 2. Average conformance FMeasure of precision and fitness, number of detected concept drifts, total computation time and average extended Cardoso index of FAIRY, STARDUST, CDESF and STATIC with ILP, IND and SPL. The best results are in bold.

Event log	Approach	Avg.FMeasure			Total time			Total drifts			Avg.ExtendedCardoso		
		ILP	IND	SPL	ILP	IND	SPL	ILP	IND	SPL	ILP	IND	SPL
BPI13I	FAIRY	0.51	0.58	0.51	107	111	91	11	8	11	6.75	11.44	8.42
	STARDUST	**0.52**	0.58	0.52	62	96	73	63	63	63	8.25	8.36	6.96
	CDESF	0.49	**0.60**	**0.53**	245	256	251	7	7	7	6.63	8.00	6.63
	STATIC	0.44	0.50	0.50	**46**	**50**	**49**	–	–	–	4	4	4
BPI18C	FAIRY	**0.99**	**0.99**	**0.99**	192	178	171	12	22	28	5.92	5.09	4.72
	STARDUST	0.97	0.97	0.97	198	160	169	8	8	8	5.22	5.78	5.89
	CDESF	0.97	0.97	0.97	4173	4136	4156	9	9	9	5.00	3.00	3.00
	STATIC	0.97	0.97	0.97	**188**	**147**	**157**	–	–	–	5.00	3.00	3.00
BPI18R	FAIRY	**0.98**	**0.97**	0.94	151	150	134	40	45	34	5.15	5.17	4.63
	STARDUST	0.94	0.96	**0.96**	132	130	120	11	11	11	4.00	4.83	5.17
	CDESF	0.94	0.94	0.94	3340	3342	3341	2	2	2	4.00	2.00	2.00
	STATIC	0.94	0.94	0.94	**128**	**109**	**110**	–	–	–	4.00	2.00	2.00
BPI20D	FAIRY	**0.85**	**0.85**	**0.85**	56	60	67	5	7	7	7.17	6.25	6.25
	STARDUST	0.83	0.83	0.83	47	47	48	4	4	4	6.13	5.75	5.88
	CDESF	0.64	0.68	0.68	1642	1643	1642	5	5	5	6.00	5.00	5.00
	STATIC	0.64	0.51	0.51	**43**	**43**	**44**	–	–	–	6.00	5.00	5.00
BPI20P	FAIRY	**0.69**	0.44	0.63	18	42	25	6	5	10	**12.29**	21.71	**16.82**
	STARDUST	**0.69**	**0.62**	**0.66**	18	25	**21**	68	68	68	13.44	**16.28**	17.08
	CDESF	0.50	0.21	0.33	459	462	460	3	3	3	13.50	19.75	24.75
	STATIC	0.50	0.26	0.29	**16**	**24**	25	–	–	–	13.00	17.00	25.00
BPI20R	FAIRY	**0.82**	0.81	**0.83**	35	39	36	3	8	4	6.75	6.33	6.20
	STARDUST	0.81	**0.82**	0.82	30	31	31	5	5	5	6.33	5.83	5.83
	CDESF	0.81	0.81	0.81	771	772	771	6	6	6	6.71	5.86	5.86
	STATIC	0.61	0.48	0.48	**27**	**28**	**28**	–	–	–	6.00	5.00	5.00
Road	FAIRY	**0.85**	**0.76**	**0.85**	892	941	802	487	270	300	**5.11**	**5.50**	**5.43**
	STARDUST	0.65	0.65	0.79	**659**	**727**	719	7	7	7	6.13	7.13	7.63
	CDESF	0.73	0.65	0.79	31400	31453	31395	1	1	1	7.50	9.50	7.00
	STATIC	0.79	0.50	0.79	704	1062	**672**	–	–	–	8.00	13.00	7.00

out of seven streams with both IND and SPL. To check whether the gain of conformance FMeasure is significant, we performed the Friedman's test to compare multiple methods over multiple data collections. The null-hypothesis states that all the approaches are equivalent. Under this hypothesis, the ranks of compared approaches should be equal. As we rejected the null hypothesis with p-$value \leq 0.05$, we performed the post-hoc Nemenyi test for pairwise comparisons. Results of Nemenyi tests reported in Fig. 1 show that FAIRY achieves the highest conformance with STARDUST as runner-up.

With regard to the number of detected drifts, this depends on the process discovery algorithm in FAIRY, while this is independent of the process discovery algorithm in both STARDUST and CDESF. In general, FAIRY discovers the higher number of concept drifts with a few exceptions in this study (i.e. STARDUST in BPI13 Incident and BPI20 PrepaiedTravelCost). However, we do not find any

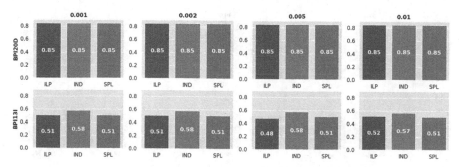

Fig. 2. Average conformance FMeasure of precision and fitness of **FAIRY** run in BPI20D and BPI13I by varying c among 0.001, 0.002, 0.005 and 0.05 with ILP, IND and SPL

specific relationship between the number of drifts and the conformance Fmeasure or the extended Cardoso Index of discovered process models.

With regard to the total time consumption, the use of a strategy to discover concept drifts and adapt process models to drifts is more time consuming than neglecting drifts and checking the conformance of a static process model. In any case, both **FAIRY** and **STARDUST** achieve comparable performances in terms of total time consumption. In addition, the total computation time of both **FAIRY** and **STARDUST** is significantly lower than the time consumption of **CDESF**. Notably, there are a few configuration of **STARDUST** which are less time consuming than **STATIC** (i.e. BPI20 PrepaiedTravelCost with SPL, as well as Road with ILP and IND). In these configurations, the average Extended Cardoso Index of the sequence of process models discovered with **STARDUST** is lower than the Extended Cardoso Index of the single process model discovered with **STATIC**. This condition may contribute to speed-up the conformance computation, despite more computation time is spent in the concept drift detection and adaptation.

Finally, with regard to the process model complexity, the average extended Cardoso index commonly increases with concept drift adaptations. However, there are a few cases were adapting a process model to concept drifts allows us to achieve a process model that is simpler than the initial process model discovered in the **STATIC** configuration (e.g. **FAIRY** with ILP and IND in BPI20 PrepaiedTravelCost, as well as **FAIRY**, **STARDUST** and CDESF with ILP, IND and SP in Road).

Analysis of Parameters c and η. We explored the sensitivity of the average conformance FMeasure of **FAIRY** to parameters c (sensitivity of ADWIN) and η (initialization size) in the streams BPI20 Domestic Declarations (BPI20D) and BPI13 Incidents (BPI13I). Figures 2 and 3 show the average conformance FMeasure of **FAIRY** by varying c among 0.001, 0.002 (default), 0.005 and 0.01, as well as η among 5%, 10% (default), and 15% of the total number of traces in the streams. In the experiment performed to evaluate the effect of η, we measured the average conformance FMeasure on the last 85% of traces in the stream. This

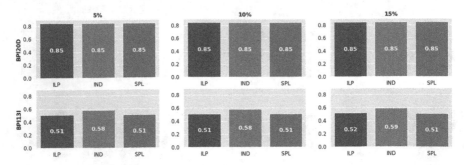

Fig. 3. Average FMeasure of precision and fitness of **FAIRY** run in BPI20D and BPI13I by varying η among 5%, 10% and 15% with ILP, IND and SPL

allows us to perform a safe comparison with average conformance FMeasure values measured on the same traces in the configurations with $\eta = 5\%$, 10% and 15%, respectively. The results show that the average conformance FMeasure does not depend on c. The only exception is observed in BPI13 Incidents with ILP, where average conformance FMeasure decreases from 0.51 ($c = 0.001$ and $c = 0.002$ to 0.48 ($c = 0.005$). Similarly average conformance FMeasure of **FAIRY** is approximately stable with η. We note a small gain in the average conformance FMeasure with $\eta = 15\%$ in BPI13 Incidents with ILP and IND wrapped as process discovery algorithms.

6 Conclusion

In this paper, we presented and evaluated the performance of a stream mining approach that can wrap any traditional process discovery algorithm in an online setting. It discovers a process model from an initial batch of activity traces, integrates a concept drift detector to monitor relevant changes in the conformance of the upcoming process execution traces to the current process model, and discovers a new process model as a concept drift is detected in the processed stream. In this study, we tested the ADWIN mechanism for concept drift detection. However, this choice is not the essential to the paper, and we plan to test other concept drift detectors formulated in the data stream literature in the future. Experiments on streaming event logs demonstrate that the proposed data stream approach improves process discovery by guaranteeing higher conformance over time.

One limitation of the proposed approach is that it is formulated for business processes that declare the completion activity of their full cases. Although this behaviour can be expected in several business processes, a future research direction includes exploring sequential pattern mining primitives [6] to handle sequence of events instead of sequences of traces. Another limitation is that the proposed approach forgets an old process model rather than updating it with the next newly observed traces. Updating a process model may help in filtering false drifts that may be detected when new traces are similar to old variant-traces forgotten with the old model. The proposed approach does not distinguish among sudden, abrupt and repetitive drifts. We plan to extend the experimentation exploring the performance of the proposed approach in artificial logs to analyse the ability to recognize different categories of concept drifts. Finally, the proposed solution could be adapted to update predictive process monitoring models to predict process evolution [10].

Acknowledgment. The work was partially supported by PNRR project FAIR - Future AI Research (PE00000013), Spoke 6 - Symbiotic AI (CUP H97G22000210007) under the NRRP MUR program funded by the NextGenerationEU and in partial fullfilment of the research objectives of PROMETEO project (CUP H93C23000900005).

References

1. Bifet, A., Gavaldà, R.: Learning from time-changing data with adaptive windowing. In: Proceedings of the Seventh SIAM International Conference on Data Mining, pp. 443–448. SIAM (2007)
2. Burattin, A.: Streaming process mining. In: van der Aalst, W.M.P., Carmona, J. (eds.) Process Mining Handbook. LNBIP, vol. 448, pp. 349–372. Springer, Cham (2022). https://doi.org/10.1007/978-3-031-08848-3_11
3. Ceravolo, P., Marques Tavares, G., Junior, S.B., Damiani, E.: Evaluation goals for online process mining: a concept drift perspective. IEEE Trans. Serv. Comput. 1 (2020)
4. Fani Sani, M., van Zelst, S., van der Aalst, W.: The impact of biased sampling of event logs on the performance of process discovery. Computing 1–20 (2021)
5. Hassani, M.: Concept drift detection of event streams using an adaptive window. In: Proceedings of the 33rd International ECMS Conference on Modelling and Simulation, ECMS 2019, pp. 230–239 (2019)
6. Hassani, M., van Zelst, S.J., van der Aalst, W.M.P.: On the application of sequential pattern mining primitives to process discovery: overview, outlook and opportunity identification. WIREs Data Min. Knowl. Discov. **9**(6) (2019)
7. Lassen, K.B., van der Aalst, W.: Complexity metrics for workflow nets. Inf. Softw. Technol. **51**(3), 610–626 (2009)
8. Pasquadibisceglie, V., Appice, A., Castellano, G., Fiorentino, N., Malerba, D.: STARDUST: a novel process mining approach to discover evolving models from trace streams. IEEE Trans. Serv. Comput. 1–14 (2022)
9. Pasquadibisceglie, V., Appice, A., Castellano, G., Malerba, D.: Darwin: an online deep learning approach to handle concept drifts in predictive process monitoring. Eng. Appl. Artif. Intell. **123**, 106461 (2023)
10. Pravilovic, S., Appice, A., Malerba, D.: Process mining to forecast the future of running cases. In: Appice, A., Ceci, M., Loglisci, C., Manco, G., Masciari, E., Ras, Z.W. (eds.) NFMCP 2013. LNCS (LNAI), vol. 8399, pp. 67–81. Springer, Cham (2014). https://doi.org/10.1007/978-3-319-08407-7_5
11. Tavares, G.M., Ceravolo, P., Turrisi Da Costa, V.G., Damiani, E., Barbon Junior, S.: Overlapping analytic stages in online process mining. In: 2019 IEEE International Conference on Services Computing, SCC 2019, pp. 167–175 (2019)
12. van Zelst, S.J., van Dongen, B.F., van der Aalst, W.M.P.: Event stream-based process discovery using abstract representations. Knowl. Inf. Syst. **54**(2), 407–435 (2018)

Explainability in AI

Enhancing Temporal Transformers for Financial Time Series via Local Surrogate Interpretability

Kenniy Olorunnimbe and Herna Viktor[✉]

School of Electrical Engineering and Computer Science, University of Ottawa,
Ottawa, Canada
{molor068,hviktor}@uottawa.ca

Abstract. The advent of Transformer architectures has ushered in a new era across various domains, including finance. These Transformer models, renowned for their scalability and efficacy, are considered opaque due to their black-box nature. The inherent complexity of these models, despite the partial interpretability afforded by the attention mechanism, poses a significant barrier to non-technical stakeholders in finance. This is challenging, not just for investors struggling to interpret the models' outputs, but also for technical experts who aspire to refine these models. In this research, we introduce a novel approach that leverages the interpretability of local surrogate models to enhance complex temporal Transformer models in financial forecasting. Utilizing insights from the surrogate models, we iteratively improve the model's predictive performance in both volatile and stable market conditions through informed feature selection. This contribution combines the simplicity of local surrogate models with the robust capabilities of temporal Transformers. Our experimental results demonstrate notable improvements in forecasting accuracy during volatile periods, with equally promising improvements in the non-volatile phases. These findings suggest that integrating explainability into the training process of financial models not only aids in demystifying their operation but also significantly bolsters their performance. Our solution leverages an emerging paradigm in machine learning where explainability becomes an integral component of model development, paving the way for more robust and intuitive financial forecasting models.

Keywords: Deep learning · Explainability in financial price prediction · Temporal Transformer · Explanation-guided learning

1   Introduction

Transformer architectures have indisputably revolutionized the landscape of artificial intelligence (AI), marking a paradigm shift including in the financial domain. At the forefront of this revolution are large language models (LLM) [17], whose ubiquity is anchored in the Transformers' ability to establish correlations across extensive sequences of data [16]. This characteristic is particularly advantageous in financial modeling, where leveraging historical data to predict future market trajectories is paramount. The superior performance of Transformer models in this context is well-documented, with some research integrating distinct financial characteristics into temporal Transformer models to further improve their performance for financial forecasting [8].

Despite these breakthroughs, Transformer models remain largely black-box models, i.e., their internal decision-making processes are not readily interpretable. The models' attention mechanism, while facilitating a degree of relationship mapping among encoded inputs, still presents a technical narrative tailored for a specialized audience, often out of the usual scope of stakeholders such as investors and market practitioners. This issue might be less pronounced during phases of consistent positive returns, but it becomes significantly more critical when these models underperform or when their outputs deviate from market expectations. The capability to intuitively interpret these models is not only crucial for correcting erroneous assumptions during experimentation but also plays a role in enhancing model performance. The need for comprehensible model interpretation and the ability to refine based on such insights falls under the broad domain of explainable AI (XAI) [3].

In a foundational work, the Temporal Fusion Transformer (TFT) model designed specifically for time series data, illustrated a degree of interpretability by highlighting long-term dependencies through its self-attention mechanisms [6]. While this represents a departure from conventional opaque deep learning (DL) models, it does not fully meet the desired XAI criteria. This is because it depends on complex architectural design and model-specific characteristics, demanding extensive machine learning (ML) expertise for full comprehension. XAI, in contrast, aims to make the outcomes of ML models straightforward and comprehensible to non DL experts.

In response to the need for enhanced explainability in DL models, a novel initiative, termed explanation-guided learning (EGL), has been proposed [2]. This approach seeks not only to augment the explainability of DL models but also to improve their generalizability, and overall performance. A pivotal element of EGL is the concept of interactive ML, which involves a feedback loop allowing ML experts to integrate insights from XAI directly into the training process. For instance, the study in [1] showcases a framework employing empirical mode decomposition (EMD) for data preprocessing, followed by artificial neural networks (ANN) for initial predictions and a random forest classifier in tandem with local interpretable model-agnostic explanations (LIME) for refining these predictions in subsequent stages. Recent works on EGL are summarized in the survey by Gao et al [2]. To our knowledge, there has been no research that combines EGL with Transformer models to enhance performance.

This research work is focused on making the application of Transformer architectures in financial forecasting clearer and more understandable while also improving the predictive performance. The paper is organized as follows. Section 2 describes local surrogate models for approximating Transformer models. Our experimental evaluation is presented in Sect. 3, and Sect. 4 concludes the paper.

2   Local Surrogate Model for Temporal Transformers

The Transformer model is fundamentally characterized by its attention mechanism, which computes the output as a weighted sum of values, where the weight assigned to each value is computed by a compatibility function of the query with the corresponding key. It is composed of Q, K, and V matrices, consisting of queries (q), keys (k), and values (v) vectors respectively. The q vector represents the information being queried, the k vector represents the information being compared against, and the v vector represents the query result to be generated [12]. In a time series scenario, the k vectors are generated from the historical data (e.g. the last 2 years), and the q vectors are generated from the input data (e.g. the last 30 days), representing the questions the model is asking about the historical data. The v vectors are the answers to those questions (i.e. the prediction horizon). In TFT, a long short-term memory (LSTM) sequence-to-sequence layer is used to establish short-term relationships within the encoder and decoder, while the self-attention network is used for long-term dependencies [6]. Similarity embedded temporal Transformer (SeTT) and run-similarity embedded temporal Transformer (r-SeTT) are further specialised for the financial domain by incorporating characteristics of financial time series into their architectures [9]. This involves adapting the attention mechanism to reflect the similarity of time series sections to the most recent window, employing Epps–Singleton (es) for SeTT and Hamming distance (d_H) for r-SeTT, as similarity functions. More details can be found in [9].

To satisfy the need for model interpretability and explainability (i.e., XAI) beyond what is possible with the attention mechanism, we use a *surrogate model*. Surrogate models are simpler, interpretable models that are used to approximate and interpret the predictions of complex models [7]. There are two main types of surrogate models: global surrogates, which provide an overall understanding by approximating the entire model behaviour; and local surrogates, which focus on explaining individual predictions or specific instances. In this work, we employ local surrogate models to interpret the decisions made by the complex temporal Transformer model in specific financial time series. Local surrogates provide insights at individual prediction points [11], which are essential for understanding the dynamic financial environments. Unlike global surrogates that offer a generalized perspective, local surrogates reveal specific, localized behaviors at the prediction points where data distributions or volatility may diverge from the general patterns. We use time series multivariate and univariate local explanations (TS-MULE) [13], an extension of LIME [11], a framework for interpreting black-box models by approximating them locally with interpretable surrogate models. Similar to LIME, TS-MULE is model-agnostic, and it focuses on providing local explanations for individual predictions, enhancing transparency and trust in complex models. The primary idea is to *perturb* (generate) new samples around the input to be explained, obtain predictions for these samples from the black-box model, and then learn a simple model on this new dataset. By using the temporal aware TS-MULE, we address the fundamental temporal dependencies in time series data, which is not a consideration in LIME [13].

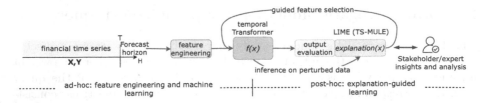

Fig. 1. An illustration of the process flow for financial forecasting with explanation-guided learning. The feature engineering and machine learning steps are followed by explainability feedback to update the model after analysis from stakeholders.

Figure 1 presents an end-to-end approach for augmenting the interpretability and performance of the Temporal Transformer model via EGL. Initially, feature engineering and exploratory data analysis (EDA) is conducted on financial time series data, establishing relevant features for the Temporal Transformer to generate forecasts over a specified horizon. However, the influence of these features on model's performance cannot be determined until the EGL loop's post-hoc analysis is conducted. As noted above, we employ TS-MULE, as it is designed to address the temporal structure associated with time series data, unlike LIME that perturbs features independently, making the generated explanations less reliable for time-dependent data in its default form. TS-MULE works in the following steps:

- Data segmentation: It first segments time series into interpretable components using segmentation techniques specific to time series. It employs novel strategies, including uniform, exponential, slope, and bin segmentation, to divide the time series into meaningful segments, either of equal length, increasing size, or based on local trends and patterns of related data regions, known as superpixels [13].
- Perturbation of segments: These segments are perturbed using zeros, inverse, or mean values as replacement strategies. This step aims to observe how changes influence the forecasting model's predictions, thereby identifying impactful data parts.
- Creation of surrogate model: A simpler, interpretable surrogate model is trained using the perturbed data to approximate the complex forecasting model's behaviour. The coefficients of the simple model provide local explanations for the model's predictions, enabling explanations into the complex model's decision-making [13]. As the simple interpretable model, we use Ridge regression, a form of regularized linear regression that addresses the potential for overfitting, a common concern in time series analysis due to autocorrelation and multicollinearity among variables [5]. By introducing a regularization term, $l2$ norm, Ridge regression imposes a penalty on the size of coefficients, thus discouraging overfitting on the noise in the training data at the expense of generalizability [5].
- Generate explanation: The coefficient generated from the segments is reverted to time series data points and is subsequently visualized to provide insight into

the effect of the individual features on the performance of the Transformer model. The process culminates in interpreting the surrogate model to identify and quantify the influence of specific time series segments on the model's predictions.

The generated explanations are integral to the post-hoc stage, where they are analysed and used in a guided feature selection feedback loop to iteratively refine the model's performance. Algorithm 1 depicts the pseudocode of this EGL guided process.

Algorithm 1. Enhanced temporal Transformer using EGL

Input:
1: X, y ▷ financial time series data
Output: $\hat{y}$ ▷ enhanced H future forecasts
2:
3: $X_f \leftarrow$ FEATURECONSTRUCTION(X) ▷ feature engineering and analysis
4: **repeat**
5:   $\mathtt{m} \leftarrow$ TRAINTEMPORALTRANSFORMER(X_f, y)
6:   $\mathcal{E} \leftarrow$ GENERATEEXPLANATION$(\mathtt{m}, X_f)$ ▷ generate explanation
7:   $X_f \leftarrow$ POSTHOCEGL$(\mathcal{E}, X_f)$
8:   $\mathtt{m} \leftarrow$ TRAINTEMPORALTRANSFORMER(X_f, y)
9: **until** satisfied
10:
11: **procedure** GENERATEEXPLAINATION$(X_f, \mathtt{m})$ ▷ via TS-MULE
12:   $\mathtt{m}_s \leftarrow$ SURROGATEMODEL   ▷ initialize surrogate model
13:   $\mathcal{C} \leftarrow \{\}$ ▷ initialize array for coefficients of surrogate model
14:   $\mathtt{S} \leftarrow$ SEGMENTTIMESERIES(X_f) ▷ divide X_f into segments
15:   **for each** $\mathtt{s}$ in $\mathtt{S}$ **do**
16:    $\mathtt{s}_p \leftarrow$ PERTURBSEGMENT$(\mathtt{s})$ ▷ Generate perturbed segment data
17:    $\mathcal{C} \leftarrow \mathcal{C} \cup$ EXPLAIN$(\mathtt{s}_p, \mathtt{m}_s, \mathtt{m})$
18:   $\mathcal{E} \leftarrow$ TOSERIES$(\mathcal{C})$ ▷ Convert segment coefficients to time series
19:   **return** $\mathcal{E}$
20:
21: **procedure** POSTHOCEGL$(\mathcal{E}, X_f)$
22:   Insights $\leftarrow$ ANALYZE$(\mathcal{E})$ ▷ expert stakeholder analysis of E
23:   $X_f \leftarrow$ REFINEFEATURES$(X_f,$ Insights$)$ ▷ Update features based on insights
24:   **return** X_f

---

3 Experimental Evaluation

Recall from Sect. 2, we discussed a class of Transformer models with temporal features, namely TFT, SeTT and r-SeTT. Our experiment consists of evaluating the predictive performance of these temporal Transformer models by comparing the quantile loss results from before the XAI-based interactive ML step with those from after its implementation.

Setup. For the temporal Transformer models, TFT, SeTT and r-SeTT, we used the following hyperparameters based on [9]: learning rate of 0.018, dropout rate of 0.3, and current window of 30 days. Also, we used es and d_H for the similarity function of the base learner SeTT and r-SeTT base learners respectively. For TS-MULE, we employed Ridge as the interpretable model, using the default settings of slope segmentation and zero values as the replacement strategy for perturbation within the segments.

Data. We used time series stock market data and financial data to evaluate our model. We leverage the stock market symbols for the companies listed in the Dow Jones Industrial Average (DJIA). DJIA is a stock market index, weighted by price, consisting of 30 prominent U.S. companies across 20 industries. We require both market and corporate results data across all timeframes for our experiments. Thus, we used the 20 stocks (Table 1a) for which we could obtain the complete data set, of the 30 companies available in the index, across 15 industries with a total weight of 65.02% of the overall index. Historical market and fundamental data are obtained from SimFin, an online financial data resource.

Table 1. Data and features.

Industry		Symbol	Weight(%)
TECH	Information technology	AAPL, CSCO INTC, MSFT	10.54
HEALTH	Managed health care	UNH	7.88
HOME	Home improvement	HD	6.65
FOOD	Soft drink, food industry	KO, MCD	5.52
PHARM	Pharmaceutical industry	JNJ, MRK	4.52
FIN	Financial services	V	4.34
AERO	Aerospace & defense	BA	4.01
CONST	Construction/mining	CAT	3.73
CONG	Conglomerate	MMM	3.38
ENT	Broadcasting & entertainment	DIS	3.18
APPAREL	Apparel	NKE	2.93
RET	Retailing	WMT	2.69
GOODS	Fast-moving consumer goods	PG	2.61
OIL	Petroleum industry	CVX	2.07
TELE	Telecommunication	VZ	0.97

Industry and stock symbols of 20 of the 30 DJIA companies, showing their collective industry weight (%) in the DJIA index.

Feature	Abbreviation	Description
Open, High, Low & Close prices	-	Daily open, high, low & close prices
Volume	-	Volume of shares traded in a day
Shares outstanding	-	All share available for trade
Earning per share (EPS)	-	Profit for outstanding shares
1-day log price change	p_logd	Log. change in close price over 1 day
1-day log volume change	v_logd	Log. change in volume over 1 day
Shares turnover	s_turnover	All share available for trade
Price variation	p_variatn	% change in open & close prices
Price-Earning ratio	per	Share price relative to EPS
Normalised High, Low Close	norm_h, norm_l, norm_c	High, Low & Close prices normalised with Open price
Year, Month, Weekday	year, month, weekday	Categorical representations of Year, Month & Weekday
Industry	industry	Market sector of company

Features used as input to the model. High, low and close prices are normalised against the open price for efficient learning [14]. A differenced price representation of the closing price, calculated as the logarithmic change in price over a single day is used as the prediction target [4,10].

As expected, features such as the normalised high, low, and closing prices (`norm_h`, `norm_l`, and `norm_c` respectively) are identified as strongly correlated during our EDA, because they are all normalised with the opening stock price (see Table 1). **Industry** and **ticker** are weakly correlated with other features in the financial time series dataset. The multicollinearity observed among these features is typically less problematic for deep learning models due to their ability to extract complex patterns and the effectiveness of regularization techniques such as dropout in mitigating overfitting concerns [15].

We constructed independent data sets consisting of features of the 20 companies, across two sets of independent periods, one volatile and the other non-volatile, based on the volatility index (VIX). The volatile periods, chosen for their high market volatility as indicated by the VIX, were from February 1, 2016, to February 1, 2018, and from February 1, 2015, to February 1, 2018. The non-volatile periods, characterised by lower VIX values, were from January 1, 2016, to January 1, 2018, and from January 1, 2015, to January 1, 2018. Each period comprises 2-year and 3-year intervals respectively. Our selection of these historical timeframes is informed by existing research, which indicates that such periods yield more optimal results [9]. All but the last 5 trading days of each period were used for training, with the model subsequently tested on the final 5 days, constituting the test set.

3.1 Experimental Results

In this section, we provide the details of the outcome of our experiments, conducted over four distinct periods. Following the initial training step of each period, we evaluate the importance and influence of the input features generated by TS-MULE. The corresponding feature importance and effect plots for the volatile market conditions are shown in Fig. 2. Following the insights gained from this analysis, we adjusted the input features for the models and conducted a subsequent round of modeling. The outcomes of this revised modeling are displayed in Fig. 3.

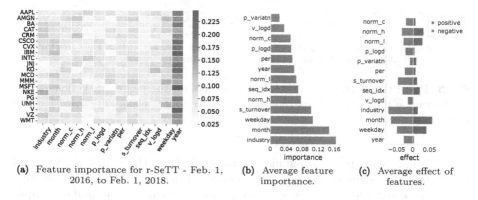

(a) Feature importance for r-SeTT - Feb. 1, 2016, to Feb. 1, 2018.   (b) Average feature importance.   (c) Average effect of features.

Fig. 2. Pre-EGL: Features importance and effects on the model predictions across all 20 symbols during volatile extrapolation periods before the explanation-guided features selection process.

During the volatile extrapolation period, Fig. 2(a) shows that the initial training of the r-SeTT model disproportionately focused on the year feature, overshadowing other critical features like normalised close (norm_c) or changes in trading volume (v_logd). However, the combined perspectives in Fig. 2(b,c) indicate a more balanced, yet still significant, emphasis on date-related features, as evidenced by the feature importance and effects plots, based on the output of TS-MULE. These features primarily exhibit a negative influence. While the direction of the effect (negative or positive) is not inherently indicative of quality, the prevalent focus on these features warrants consideration.

Informed by our findings, we eliminated the excessively weighted date-related features, and norm_c for volatile periods. The rationale for removing norm_c is that its primary information, the closing price, is already captured in other features like p_logd and p_variatn. For non-volatile periods, which exhibit a distinct relevance profile, we discarded the two least significant features, norm_h and v_logd. From the EDA step, we know that norm_h is highly correlation with other normalised price features, and v_logd's information regarding volume is sufficiently represented by s_turnover (volume relative to shares outstanding), which holds greater feature importance. Additionally, we opted to remove month and weekday features. However, we retained the year feature in this context, as preliminary tests indicate its significant negative impact on results when absent.

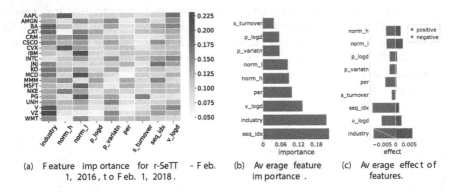

(a) Feature importance for r-SeTT - Feb.   (b) Average feature   (c) Average effect of
      1, 2016, to Feb. 1, 2018.   importance .   features.

Fig. 3. Post-EGL: Features importance and effects on the model predictions across all 20 symbols during volatile extrapolation periods after explanation guided features selection process.

Following the explanation-guided feature selection, the model is re-trained using the updated input features. Upon conducting a post-EGL explanability, we observed notable improvements in the model's focus on the updated input features, as indicated by Figs. 3. The feature representation in the r-SeTT models has significantly improved, particularly for the volatile period, which is evident from the altered feature importance in the average plot. Improvements in feature relevance for the non-volatile period have also been observed. However, further analysis is necessary to conclusively determine their generalizable impact on performance.

Table 2 presents the weighted loss results, contrasting the performance before and after the explanation-guided model updates. We employed the mean absolute scaled error (MASE) metric to assess the performance of each of the three models, comparing their pre and post-guided learning phases for every period examined. This approach enabled us to demonstrate significant improvements in performance using the same Transformer architecture and input features. This iterative improvement is not possible using EDA-based feature engineering. The influence of XAI in directing us towards varied input features for different periods further underscores this point. The table reveals a consistent improvement in the loss metrics for the volatile extrapolation period, ranging from 11% to 24%. While the improvements on three of the six non-volatile extrapolation periods are modest, ranging from 3% to 8%, they are still noteworthy. Furthermore, the

Table 2. Loss metric of six pairs of volatile and non-volatile extrapolation periods for two and three years timeframe, weighted for all stock symbols.

start	end	yrs	model	loss
2016-02-01	2018-02-01	2	rsett_hd	1.19 (-21.7%)
2016-02-01	2018-02-01	2	rsett_hd (post)	**0.932**
2015-02-01	2018-02-01	3	rsett_hd	0.922 (-31.2%)
2015-02-01	2018-02-01	3	rsett_hd (post)	**0.634**
2016-02-01	2018-02-01	2	sett_es	1.064 (-13.3%)
2016-02-01	2018-02-01	2	sett_es (post)	**0.922**
2015-02-01	2018-02-01	3	sett_es	0.949 (-21.9%)
2015-02-01	2018-02-01	3	sett_es (post)	**0.742**
2016-02-01	2018-02-01	2	tft	1.068 (-11.3%)
2016-02-01	2018-02-01	2	tft (post)	**0.948**
2015-02-01	2018-02-01	3	tft	0.924 (-23.9%)
2015-02-01	2018-02-01	3	tft (post)	**0.702**

Volatile extrapolation period

start	end	yrs	model	loss
2016-01-01	2018-01-01	2	rsett_hd	**0.523**
2016-01-01	2018-01-01	2	rsett_hd (post)	0.554 (-5.5%)
2015-01-01	2018-01-01	3	rsett_hd	0.468 (-8.3%)
2015-01-01	2018-01-01	3	rsett_hd (post)	**0.43**
2016-01-01	2018-01-01	2	sett_es	**0.508**
2016-01-01	2018-01-01	2	sett_es (post)	0.542 (-6.3%)
2015-01-01	2018-01-01	3	sett_es	**0.436**
2015-01-01	2018-01-01	3	sett_es (post)	0.446 (-2.3%)
2016-01-01	2018-01-01	2	tft	0.502 (-7.5%)
2016-01-01	2018-01-01	2	tft (post)	**0.465**
2015-01-01	2018-01-01	3	tft	0.472 (-3.7%)
2015-01-01	2018-01-01	3	tft (post)	**0.455**

Non-volatile extrapolation period

extent of deterioration in performance for the remaining three comparisons in this period is relatively lower, lying within the 2–6% range.

3.2 Discussion

Our experimental results underscore the significance of integrating XAI into the interactive ML pipeline for Transformer models. Despite thorough feature engineering before training, incorporating XAI into our methodology has enabled a clearer understanding of the impact of data features on prediction outcomes. Table 2 illustrates that as a result of the explainability feedback, we achieve substantial performance improvements, with gains up to 24% in volatile extrapolation periods, regardless of market complexity.

Moreover, the insights gained from this approach are valuable for non-technical market analysts. This is particularly pertinent when considering that such analysts often possess a deeper understanding of stock market fundamentals than a typical ML engineer or researcher. This methodology also lays the groundwork for a more comprehensive optimization strategy. It goes beyond adjusting the input hyperparameters of the ML model, but also learning from the XAI outputs to retrain and further enhance the model's performance in a guided reinforcing loop.

4 Conclusion

In this research, we introduce an approach that uses local surrogate models to enhance the predictive performance of temporal Transformer models across different financial market conditions through insights gleaned from a post-training XAI step. Our experimental evaluations spanned 2-year and 3-year timeframes, utilizing three distinct temporal Transformer models. The results notably indicate marked improvements in forecasting accuracy during volatile market periods, affirming the effectiveness of our approach and yielding promising outcomes in non-volatile periods as well. This research not only validates the integration of explainability into financial modeling but also establishes a promising direction for future financial forecasting methodologies. It signals a shift towards more robust, intuitive, and reliable forecasting models in the financial domain. Future work will include developing a hybrid EGL framework with automated XAI feedback and model optimization. Such a framework will allow for the potential enhancement of expert judgment by identifying input features with significant impacts on feature importance, thereby refining model interpretations and adherence to rules provided by the expert.

Disclosure of Interests. The authors have no competing interests to declare that are relevant to the content of this article.

References

1. Çelik, T.B., İcan, Ö., Bulut, E.: Extending machine learning prediction capabilities by explainable AI in financial time series prediction. Appl. Soft Comput. **132**, 109876 (2023). https://doi.org/10.1016/j.asoc.2022.109876
2. Gao, Y., Gu, S., Jiang, J., Hong, S.R., Yu, D., Zhao, L.: Going beyond XAI: a systematic survey for explanation-guided learning. ACM Comput. Surv. (2024). https://doi.org/10.1145/3644073
3. Gunning, D., Aha, D.W.: DARPA's explainable artificial intelligence program. AI Mag. **40**(2), 44–58 (2019). https://doi.org/10.1609/aimag.v40i2.2850
4. Hyndman, R., Athanasopoulos, G.: Forecasting: Principles and Practice, 3rd edn. OTexts, Melbourne (2021)
5. James, G., Witten, D., Hastie, T., Tibshirani, R., Taylor, J.: An Introduction to Statistical Learning: with Applications in Python. Springer, Cham (2023). https://doi.org/10.1007/978-3-031-38747-0
6. Lim, B., Arik, S.O., Loeff, N., Pfister, T.: Temporal fusion transformers for interpretable multi-horizon time series forecasting. Int. J. Forecast. **37**(4), 1748–1764 (2021). https://doi.org/10.1016/j.ijforecast.2021.03.012
7. Molnar, C.: Interpretable Machine Learning. https://christophm.github.io/interpretable-ml-book/
8. Olorunnimbe, K., Viktor, H.: Similarity embedded temporal transformers: enhancing stock predictions with historically similar trends. In: Ceci, M., Flesca, S., Masciari, E., Manco, G., Ras, Z.W. (eds.) ISMIS 2022. LNCS, vol. 13515, pp. 388–398. Springer, Cham (2022). https://doi.org/10.1007/978-3-031-16564-1_37
9. Olorunnimbe, K., Viktor, H.L.: Towards efficient similarity embedded temporal transformers via extended timeframe analysis (2024, accepted for publication)
10. De Prado, M.L.: Advances in Financial Machine Learning, 1st edn. Wiley, Hoboken (2018)
11. Ribeiro, M.T., Singh, S., Guestrin, C.: "Why should I trust you?": explaining the predictions of any classifier. In: Proceedings of the 22nd ACM SIGKDD International Conference on Knowledge Discovery and Data Mining. Association for Computing Machinery (2016). https://doi.org/10.1145/2939672.2939778

12. Russell, S., Norvig, P.: Artificial Intelligence: A Modern Approach, Global Edition, 4th edn. Pearson, Harlow (2021)
13. Schlegel, U., Vo, D.L., Keim, D.A., Seebacher, D.: TS-MULE: local interpretable model-agnostic explanations for time series forecast models. In: Kamp, M., et al. (eds.) ECML PKDD 2021. CCIS, vol. 1524, pp. 5–14. Springer, Cham (2021). https://doi.org/10.1007/978-3-030-93736-2_1
14. Soleymani, F., Paquet, E.: Financial portfolio optimization with online deep reinforcement learning and restricted stacked autoencoder-DeepBreath. Expert Syst. Appl. (2020). https://doi.org/10.1016/j.eswa.2020.113456
15. Srivastava, N., Hinton, G., Krizhevsky, A., Sutskever, I., Salakhutdinov, R.: Dropout: a simple way to prevent neural networks from overfitting. J. Mach. Learn. Res. 15(1), 1929–1958 (2014). http://jmlr.org/papers/v15/srivastava14a.html
16. Vaswani, A., et al.: Attention is all you need. In: Advances in Neural Information Processing Systems (2017). https://doi.org/10.48550/arXiv.1706.03762
17. Zhao, W.X., et al.: A survey of large language models (2023). https://doi.org/10.48550/arXiv.2303.18223

Explaining Commonalities of Clusters of RDF Resources in Natural Language

Simona Colucci[1]($\boxtimes$) (iD), Francesco M. Donini[2](iD), and Eugenio Di Sciascio[1](iD)

[1] Politecnico di Bari, Via Orabona 4, 70125 Bari, Italy
simona.colucci@poliba.it
[2] Università della Tuscia, Via S. Maria in Gradi, 4, 01100 Viterbo, Italy

Abstract. We introduce a system that provides explanations in Natural Language for individual clusters of RDF resources, where clusters are obtained using an external clustering tool. Our system is based on the theory of (Least) Common Subsumers (CS) in RDF. We propose an optimized algorithm for computing a CS, which allows us to compute the CS for up to 80 RDF resources (each with its own RDF-graph of linked data). We then generate a Natural Language sentence to describe each cluster. A unique aspect of our explanations is the use of relative sentences, including nested ones, to represent blank nodes in an RDF-path. We demonstrate the usefulness of our tool by describing the resulting clusters of a real, publicly available, dataset on Public Procurements.

Keywords: Explainable Artificial Intelligence (XAI) · Resource Description Framework (RDF) · Least Common Subsumer (LCS) · Natural Language Generation (NLG)

1 Introduction and Related Work

The success of subsymbolic approaches to the clusterization problem has increased the importance of describing the obtained clusters—*i.e.*, explaining why some items were put together—both to developers who tune parameters, and to end users [3]. For instance, a decision tree with k leaves has been recently proposed as an explanation of the clustering obtained by the well-known method of k-means [16], and by finding a minimal set of outliers to be excluded by the clusterization, one can approximate a "reasonably-sized" decision tree [4]. Yet, while decision trees and other data-structure-oriented descriptions could be well suited for Computer Science developers and practitioners, explanations using Natural Language—when available—are still among the most understandable ones for end users, given that an explanation is a *social interaction* [15].

A different perspective is given by the distinction between model-aware vs. model-agnostic explanation methods. Some clustering methods are self-explaining: conceptual clustering [9,14] was introduced in 1980 as the problem of returning, together with clustering results, a concept explaining the criterion for resources aggregation. A recent review [18] collected most influential proposals on the subject, but does not include any approach dealing with RDF resources. Again, while model-aware explanations describing the inner clustering mechanisms can be useful for developers' fine-tuning activities, model-agnostic explanations just describing the common characteristics of an obtained cluster seem to be preferable for communicating the results to lay users.

Motivated by the above considerations, in this paper we present a system that given any clusterization method for RDF resources in an RDF dataset, returns a description of the main commonalities in a cluster by a plain English sentence. Our system builds on a formal foundation of Least Common Subsumers of a set of RDF resources [6], but improves both on some computational aspects and the Natural Language generation (NLG) of descriptions.

After some preliminary notions (Sect. 2), we present in Sect. 3 the optimized algorithm computing the CS of several resources. An NLG tool for CS explanation is introduced in Sect. 4 and validated in Sect. 5 w.r.t. to a publicly available dataset for Public Procurement. Conclusion and future work close the paper.

2   Preliminaries

We assume the reader familiar with RDF and RDFS [17], fully covered by textbooks [11]. We recall here just some basic concepts and definitions about RDF Common Subsumers [6,7] to make the paper self-contained.

When comparing RDF resources, it is indispensable to select the portion of Linked Open Data [21] more suitable to describe them. In fact, taking into account all triples unboundedly linked to a resource (also in a limited set of datasets) would make unscalable most of applications. Thus we refer to an RDF resource as a pair $\langle r, T_r \rangle$—which we call *rooted* RDF-*graph* (in brief *r-graph*)— collecting the resource identifier r and a set of triples used for its description [6] with the condition that from r every other triple in T_r can be reached through an RDF-*path* (see below). Coherently, both Simple Entailment between r-graphs can be defined [6, Def.6], denoted as $\langle a, T_a \rangle \models \langle b, T_b \rangle$, and Common Subsumers (CS) in RDF [6, Def.7], which are themselves modeled as r-graphs of the form $\langle x, T_x \rangle$ (where x is a blank node unless $a = b$) such that both $\langle a, T_a \rangle \models \langle x, T_x \rangle$ and $\langle b, T_b \rangle \models \langle x, T_x \rangle$.

RDF-graphs have some peculiarities w.r.t. usual graphs (including Knowledge Graphs), which ask for the extension of some basic notions of Graph Theory. First, RDF admits also paths connected through the predicate (*i.e.*, the predicate of a triple stands as subject in another one). Thus, we define an RDF-*path* of length n from r to s as a sequence of triples $t_0, t_1, \ldots, t_n$ in which (1) the subject of t_0 is r, (2) for $i = 1, \ldots, n-1$, either the predicate or the object of t_i is the subject of t_{i+1}, and (3) *either* the predicate *or* the object of t_n is s [6]. Such a peculiarity changes also the notion of connectedness, which may occur also through the predicate; a resource r is RDF-*connected* to a resource s if there exists an RDF-path from r to s [6]. The RDF-*distance* between two resources is the length of the shortest RDF-path between them [6]. The distance between a resource r and a full triple t is the RDF-distance between r and the subject of t; note that triples whose subject is r have zero-RDF-distance from r itself.

The representation of RDF resources as r-graphs $\langle r, T_r \rangle$ asks for some criteria for the selection of triples to include in T_r describing a resource r. We encapsulate all selection criteria in a Boolean predicate ϕ [6]: only triples satisfying ϕ are selected. Our current implementation asks for the specification of three parameters determining the value of ϕ: i) the datasets to analyse; ii) the RDF-distance for exploration; iii) the list of so-called *stop-patterns* (analogous to stop-words to be discarded in search engines). In our proposal, ϕ aims at focusing on triples which are significant for the computation of a CS: tuning ϕ may lead to r-graphs which are more representative of the main commonalities in a CS. To further increase the significance of a CS, we iteratively eliminate from it triples that provide little information, called *uninformative triples*. Stop-patterns and uninformative triples include both general patterns/triples discarded in every application domain, and some domain-dependent patterns/triples.

3 An Optimized Algorithm for the CS of Several Resources

We recall [6,7] that the operation of computing a (Least) Common Subsumer is associative—that is, to compute the LCS of three or more r-graphs, one can start from computing the subsumer of any pair of them, and use the result to add the third one, etc. in any order. Hence below we first present Algorithm 1 for the computation of a CS of just two r-graphs, whose iteration to several r-graphs is just hinted at the end of this section.

Algorithm 1 is an optimized version of a previously published algorithm [6], which computes a CS $\langle x, T_x \rangle$ of two r-graphs $\langle a, T_a \rangle$, $\langle b, T_b \rangle$. Mimicking the depth-first search in n-ary trees, $\langle x, T_x \rangle$ is computed incrementally with respect to the triples (filtered by ϕ) directly outgoing from each resource, and recursively when the predicates and/or the objects of such triples are subjects of other triples. More precisely, for each pair of triples $t_1 = \ll a\ p\ c \gg, t_2 = \ll b\ q\ d \gg$, both satisfying $\phi(t_1), \phi(t_2)$, Algorithm 1 first recursively computes a CS $\langle y, T_y \rangle$ of the two predicates p, q, and a CS $\langle z, T_z \rangle$ of objects c, d, then it forms a provisional r-graph $\langle x, T_w \rangle$ with $T_w = \{\ll x\ y\ z \gg\} \cup T_y \cup T_z\}$. Then it adds the triples in T_w to T_x only if $\langle x, T_x \rangle$ does *not* entail $\langle x, T_w \rangle$ (Line 20).

Of course, Line 20 is only an optimization step, since in general it was proved [2] for Description Logics that the size of the *Least* Common Subsumer of several tree-shaped concepts can be exponential in the number of concepts, and that proof could also be repeated for r-graphs[1]. Also regarding time consumption, observe that Algorithm 1 could be launched with the two arguments equal to an r-graph $\langle x, G \rangle$ (with x any subject of a triple that reaches all other resources), and it is well known that finding a *lean*[2] equivalent of an RDF-graph G is NP-complete [19]. Since Algorithm 1 runs in time polynomial in the sizes of its arguments, we cannot expect it to yield a minimal r-graph (unless $P = NP$), but

---

[1] The same conclusion was recently reached [1] using cycles instead of trees.
[2] A lean graph G [17] is an RDF-graph which is $\subseteq$-minimal with respect to all other RDF-graphs logically equivalent to G.

only a reduced version of it. Nevertheless, on the real RDF datasets we tested, this optimization greatly reduces the size of the CS.

By iterating Algorithm 1—*i.e.*, starting from an initial pair of r-graphs and using its returned CS as the input of the next iteration with another r-graph—we can compute a CS of the whole cluster. By using the optimization of Line 20, we were able to compute in 462,78 s[3] a CS of clusters up to 80 resources—*i.e.*, the size of the largest cluster returned in our experiments – and cascade such a CS to the verbalization module.

4   Explaining Commonalities of Sets of RDF Resources

The triple set computed by iterating Algorithm 1 may be serialized in different formats and visualized as a graph, but none of these transformations is intelligible by unacquainted readers. Thus, we designed and developed a Natural Language Generation (NLG) tool able to explain in natural language the content of a CS.

Traditional NLG approaches for RDF are based on the application of rules and templates, *i.e.*, solutions highly domain-dependent and demanding manual intervention. Recently, the advancements in deep learning gave a boost to neural network-based NLG models (see [24] and [13], among others). In both approaches, the generated text considers only triples already present in the input RDF graphs and not including blank nodes.

We here propose a template-based approach, generating text from r-graphs logically computed from input RDF-graphs: CSs; such r-graphs, by construction, include blank nodes, that may also occur in positions other than the root. As an example, consider a group of contracting processes described in the dataset TheyBuyForYou [22] (further introduced in Sect. 5). The CS collecting their commonalities may include the following triples:

```
_:x <http://data.tbfy.eu/ontology/ocds#hasRelease>  _:y .
_:y <http://data.tbfy.eu/ontology/tbfy#releaseDate> "2019-01-14T00:00:00+00:00" .
```

which means that all contracting processes in the group (abstracted in the blank node _:x) have some release (_:y) dated 14 January 2019. Notably, _:y does not occur by itself in the input r-graphs but represents the fact that different contracting processes refer to different releases, yet all dated 14 January 2019.

To the best of our knowledge, the tool we present is the only NLG tool able to verbalize RDF triples involving blank nodes in any position. Other available tools[4] do not mention such a capability. Furthermore, Bouayad-Agha *et al.* [5] surveyed 11 NLG approaches working on RDF graphs in 2014; none of them was able to manage anonymous resources or generate text from triples not explicitly in the input RDF graph (*i.e.*, derived triples).

In the design of our NLG tool, we follow the approach of Gatt and Krahmer [10], synthesized in the following six tasks:

---

[3] Average execution time for running Algorithm 1 on 80 randomly selected resources—
 machine equipped with an Intel i7 processor at 3.60 GHz and 32 GB RAM.
[4] https://aclweb.org/aclwiki/Downloadable_NLG_systems.

Algorithm $Find\_ReducedCS(\langle a, T_a \rangle, \langle b, T_b \rangle, n)$

Input	:	$\langle a, T_a \rangle, \langle b, T_b \rangle$: a pair of r-graphs; n : the RDF-distance for RDF-graphs exploration;
Output	:	$\langle cs, T_{cs} \rangle$: r-graph s.t. both $\begin{cases} \langle a, T_a \rangle \models \langle cs, T_{cs} \rangle \\ \langle b, T_b \rangle \models \langle cs, T_{cs} \rangle \end{cases}$
Subroutine	:	Simple entailment between r-graphs $\langle x, T_x \rangle \models \langle y, T_y \rangle$ [6]
Global variables	:	ϕ : boolean predicate selecting triples; $uninf\_triples$: triple patterns to eliminate from the results S : set of records $[a, b, \langle w, T_w \rangle]$ s.t. $\begin{cases} (a, b) \text{ was already examined} \\ \langle w, T_w \rangle \text{ is their CS} \end{cases}$
Local variables	:	$\langle x, T_x \rangle$: the r-graph to be returned, incrementally built

```
 1  if there is already the record [a, b, ⟨w, Tw⟩] ∈ S then
 2  |    let ⟨x, Tx⟩ = ⟨w, Tw⟩
 3  else
 4  |    add [a, b, ⟨x, Tx⟩] to S;
 5  |    if a = b then
 6  |    |    let ⟨x, Tx⟩ = ⟨a, Ta⟩
 7  |    else
 8  |    |    let x be a new blank node not occurring in S;
 9  |    |    let Tx = ∅
10  |    end
11  |    if n > 0 then
12  |    |    foreach t1 = ≪a p c≫ such that φ(t1) = true do
13  |    |    |    foreach t2 = ≪b q d≫ such that φ(t2) = true do
14  |    |    |    |    let ⟨y, Ty⟩ = Find_ReducedCS(⟨p, Tp⟩, ⟨q, Tq⟩, n − 1);
15  |    |    |    |    let ⟨z, Tz⟩ = Find_ReducedCS(⟨c, Tc⟩, ⟨d, Td⟩, n − 1);
16  |    |    |    |    let ⟨x, Tw⟩ = ⟨x, {≪x y z≫} ∪ Ty ∪ Tz}⟩;
17  |    |    |    |    while ∃t ∈ Tw s.t. t matches a pattern in uninf_triples do
18  |    |    |    |    |    delete t from Tw
19  |    |    |    |    end
20  |    |    |    |    if ⟨x, Tx⟩ ⊭ ⟨x, Tw⟩ then
21  |    |    |    |    |    add Tw to Tx
22  |    |    |    |    end
23  |    |    |    end
24  |    |    end
25  |    end
26  end
27  return ⟨x, Tx⟩;
```

Algorithm 1: Optimized construction of a CS of $\langle a, T_a \rangle$ and $\langle b, T_b \rangle$

Content Determination. This step aims at identifying the portion of available informative content to include in the text to generate. In our tool, such a content coincides with a computed CS, which by construction holds a compact representation of all commonalities we consider significant. In fact, we: i) include in the input set only triple patterns which are relevant to the problem of finding similarities (Algorithm 1, Row 12); ii) exclude from the CS all triples too generic to be informative (Algorithm 1, Row 17).

Text Structuring. This task deals with deciding how to structure the content above in a readable text. In our case, the content is organized in an r-graph, in which different paths are RDF-connected to the root and include triples at variable RDF-distance from the root. The tool generates one sentence (with subject in the root) for each full path. Such sentences are presented in the order the paths appear in the CS construction, supposing they are equally informative.

Sentence Aggregation. The RDF-connection is a native criterion for sentence aggregation: each path RDF-connected to the root becomes one sentence.

Lexicalisation. This step aims at finding the right words to express information. Our approach exploits the inherent structure of paths for the lexicalization of triples, that depends on their RDF-distance form the root. At every RDF-distance, the subject of a triple is lexicalized with a (different) pronoun, the predicate with a verb in present tense and the object with a noun. Such nouns and predicates are collected in a dictionary. If the predicate and/or the object is a blank node, it is generated a phrase ("some generic resource"), further explained through a relative sentence when the resource has successors in the r-graph.

Referring Expression Generation. This task deals with selecting the words and phrases to identify domain objects. In our case, such entities are triple subjects at any level of the RDF path to consider, represented by blank nodes. We refer to them through generic pronouns, followed by (possibly recursive) relative sentences lexicalizing paths connected to them. All such information are organized around the root: the tool generates sentences whose main subject is the phrase corresponding to the root ("They all").

Linguistic Realisation. The final step consists in combining all words and phrases into well-formed sentences following a human-crafted grammar-based approach. We give below the main rules of the grammar (terminal symbols are quoted and vertical bar represents a choice between two forms of a rule).

$$CS \rightarrow \text{"They all" } Predicate \ (Noun \mid Noun \ RC)$$
$$RC \rightarrow \text{"which" } Predicate \ (Noun \mid Noun \ RC)$$

In the grammar above: RC is a nonterminal representing a relative clause; $Predicate$ (respectively $Noun$) is a nonterminal describing the text linguistically realizing a term in the position of triple predicate (respectively, subject/object). If the term is a blank node, the phrase "some resource" linguistically realizes it.

We notice that $Predicate$ and $Noun$ produce a finite number of terminals; thus, the substitution of all such terminals in the above rules yields a (very lengthy) right-recursive Type-3 Grammar. Such a grammar guides the implementation of the linguistic realisation, that follows a breadth-first strategy. In particular, if the CS includes multiple (say, n) RDF-paths, we first generate n phrases, one for the first triple of each path. Such phrases are complete if the path has length 0 (just one triple), or refer to a nonterminal RC when the path has length ≥ 1. The approach proceeds by browsing each path one triple further: their verbalization is added by substituting all non terminals RC once. The exploration ends when no path extends further.

5   Use Case: Explaining Clusters in TheyBuyForYou

We validate the NLG tool presented in Sect. 4 w.r.t. the public procurement knowledge domain, modeled in the "TheyBuyForYou" dataset [22]. TheyBuyForYou knowledge graph includes an ontology for procurement data, based on

```
The resources in analysis present the following properties in common:

1) They all have a release referencing some resource

        which  has publisher schema "Companies House"
                and  has publisher web page "https://openopps.com"
                and  has release date "14 January 2019"
                and  has type of release initiation"  "tender"
                and  has publisher name "Open Opps"
                and  has release publisher "TICON UK LIMITED"
                and  has release publication policy "legal"

2) They all have an award referencing some resource

        which  has an award date "14 January 2019"
                and   is emitted for a tender referencing some resource

            which  has tender status "complete"
                    and   has details of criteria for award referencing some resource

            and    require a specific item(s) referencing some resource

            which   has classification code referencing some resource

                    and  has classification schema "Common Procurement Vocabulary (CPV)"
```

Fig. 1. Explanation provided by our NLG tool of a cluster of 14 contracting processes, returned by running k-means (k=250) over the whole set (3,198 resources) of contracting processes emitted on 14 January 2019.

the Open Contracting Data Standard (OCDS) [23]. The OCDS data model is built around the concept of a contracting process, which models the procedures followed by a business entity when purchasing services or products.

Our use case shows the commonalities shared by different contracting processes, collected in the same cluster by the well-known clustering algorithm k-Means [12]. We implemented k-Means with $k = 250$ by *Scikit learn*[5] on the set of contracting processes released on 14 Jan. 2019[6], which includes 3,198 resources. The first returned cluster is made up of 14 resources, to which we iteratively applied Algorithm 1, finding a CS of them all. Figure 1 shows the text generated by our NLG tool to explain the commonalities in the above cluster.

We stress that we do not evaluate here clustering results: our approach is agnostic w.r.t. to the clustering process and aims at generating human-readable text corresponding to a set of RDF triples modeling the commonalities of a group of resources.

Intuitively, when finding commonalities in a set of resources, the addition of a new item may only reduce the CS, if the fresh resource does not share the full content collected up to its addition. In fact, in Fig. 2 we show the explanation of the commonalities of the first seven resources in the analyzed cluster. The reader may notice that the explanation in Fig. 2 is more informative (*i.e.*, specific) than the one in Fig. 1, because it describes a smaller set of resources.

We now discuss two distinguishing features of our NLG approach.

---

[5] https://scikit-learn.org/stable/modules/generated/sklearn.cluster.KMeans.html.

[6] The full Knowledge Graph is downloadable at https://tbfy.github.io/data/.

```
The resources in analysis present the following properties in common:

1) They all have a release referencing some resource

     which  has publisher name "Open Opps"
            and  has type of release initiation"  "tender"
            and  has publisher web page "https://openopps.com"
            and  has release publication policy "legal"
            and  has release publisher "TICON UK LIMITED"
            and  has publisher schema "Companies House"
            and  has release date "14 January 2019"

2) They all have an award referencing some resource

     which  has an award date "14 January 2019"
            and    has an award value referencing some resource

            and    is emitted for a tender referencing some resource

3) They all have an award referencing some resource

     which  is emitted for a tender referencing some resource

            which   require a specific item(s) referencing some resource

                     which  has classification schema "Common Procurement Vocabulary (CPV)"
                            and    has classification code referencing some resource

            and    require a specific item(s) referencing some resource

            and    has details of criteria for award referencing some resource

            and    require a specific item(s) referencing some resource

            which  has a schema for additional item classification"  "Common Procurement Vocabulary (CPV)"
                   and    has an additional item classification" referencing some resource

            and    has tender status "complete"

     and  has an award date "14 January 2019"
```

Fig. 2. Explanation provided by our NLG tool of the commonalities of 7 contracting processes selected in the cluster of 14 previously explained in Fig. 1.

First, the ability of managing blank nodes is crucial for abstracting several triples with common predicate/object. In both figures, the phrase "some resource" translates blank nodes with successors in the CS r-graph; triples rooted in such blank nodes are further explained in relative sentences whose common subject is the pronoun "which". As an example, consider Commonality 1) in Fig. 1. Each contracting process in the cluster has a release that references one resource, which, although different from all other contracting processes, shares with them 7 features (publisher name, release initiation, etc.). All the commonalities recursively expressed in relative sentences would be lost without considering blank nodes. Instead, our method uses blank nodes to chain triples reaching the same known object, until the maximum exploration depth.

Second, unlike NLG-based summarization approaches, our method does not just verbalize the set of triples explicitly included in input RDF-graphs; it generates human-readable text from a *newly computed* set of RDF triples that logically represent the commonalities shared by groups of resources. As a consequence, the informative potential of the returned explanation is rather significant and double-tied to the logic-based nature of computation.

6   Conclusion and Future Work

We implemented a *post hoc*, model-agnostic explanation system that provides natural language descriptions of single clusters of RDF resources, previously obtained by any external clusterization tool. Our system is based on the theory of (Least) Common Subsumers (CS) in RDF [6,7]. We presented an optimized version of an algorithm for computing a CS, that allowed us to compute the CS of the largest cluster in our experiments (80 RDF resources, each one with its own RDF-graph of linked data), to which we pipeline the generation of a Natural Language sentence describing it. An original feature of our explanations is the use of (possibly nested) relative sentences to represent blank nodes in an RDF-path. The application of our tool to a real dataset regarding Public Procurements shows both its usefulness in describing the obtained clusters and the possibility to use it in an interactive process to find more meaningful, fine-grained clusters.

The results of our tool show that although all the characteristics described by the CS are actually common to all resources, some of them could be considered more relevant than others for an end user. Hence we aim to filter by relevance the characteristics in the CS, considering both relevance to the general context, and to the user's previous knowledge [8]. Moreover, we plan to apply our tool to explain product commonalities [20].

Acknowledgements. We acknowledge support by project "LIFE: the itaLian system wIde Frailty nEtwork" founded by Ministry of Health (CUP D93C22000640001).

Data Availibility Statement. Data publicly available at https://tbfy.github.io/data/.

References

1. Amendola, G., Manna, M., Ricioppo, A.: Characterizing nexus of similarity within knowledge bases: a logic-based framework and its computational complexity aspects. https://arxiv.org/pdf/2303.10714.pdf
2. Baader, F., Küsters, R., Molitor, R.: Computing least common subsumers in description logics with existential restrictions. In: IJCAI, vol. 99, pp. 96–101 (1999)
3. Bae, J., Helldin, T., Riveiro, M., Nowaczyk, S., Bouguelia, M.R., Falkman, G.: Interactive clustering: a comprehensive review. ACM Comput. Surv. **53**(1) (2020)
4. Bandyapadhyay, S., Fomin, F.V., Golovach, P.A., Lochet, W., Purohit, N., Simonov, K.: How to find a good explanation for clustering? Artif. Intell. **322** (2023)
5. Bouayad-Agha, N., Casamayor, G., Wanner, L.: Natural language generation in the context of the semantic web. Semant. Web **5**(6), 493–513 (2014)
6. Colucci, S., Donini, F., Giannini, S., Di Sciascio, E.: Defining and computing least common subsumers in RDF. Web Semant. Sci. Serv. Agents World Wide Web **39**, 62–80 (2016)
7. Colucci, S., Donini, F.M., Di Sciascio, E.: Common subsumers in RDF. In: Baldoni, M., Baroglio, C., Boella, G., Micalizio, R. (eds.) AI*IA 2013. LNCS (LNAI), vol. 8249, pp. 348–359. Springer, Cham (2013). https://doi.org/10.1007/978-3-319-03524-6_30

8. Colucci, S., Donini, F.M., Di Sciascio, E.: On the relevance of explanation for RDF resources similarity. In: Babkin, E., Barjis, J., Malyzhenkov, P., Merunka, V., Molhanec, M. (eds.) MOBA 2023. LNBIP, vol. 488, pp. 96–107. Springer, Cham (2023). https://doi.org/10.1007/978-3-031-45010-5_8

9. Colucci, S., Giannini, S., Donini, F.M., Di Sciascio, E.: A deductive approach to the identification and description of clusters in linked open data. In: Proceedings of the 21st European Conference on Artificial Intelligence (ECAI 2014). IOS Press (2014)

10. Gatt, A., Krahmer, E.: Survey of the state of the art in natural language generation: core tasks, applications and evaluation. J. Artif. Int. Res. 61(1), 65–170 (2018)

11. Hitzler, P., Krötzsch, M., Rudolph, S.: Foundations of Semantic Web Technologies. Chapman & Hall/CRC (2009)

12. Jain, A.K., Dubes, R.C.: Algorithms for Clustering Data. Prentice-Hall Inc., Upper Saddle River (1988)

13. Li, J., et al.: Neural entity summarization with joint encoding and weak supervision. In: Proceedings of IJCAI-2020, pp. 1644–1650. ijcai.org (2020)

14. Michalski, R.S.: Knowledge acquisition through conceptual clustering: a theoretical framework and an algorithm for partitioning data into conjunctive concepts. Int. J. Policy Anal. Inf. Syst. 4, 219–244 (1980)

15. Miller, T.: Explanation in artificial intelligence: insights from the social sciences. Artif. Intell. 267, 1–38 (2019)

16. Moshkovitz, M., Dasgupta, S., Rashtchian, C., Frost, N.: Explainable k-means and k-medians clustering. In: Proceedings of the 37th International Conference on Machine Learning. Proceedings of Machine Learning Research, vol. 119, pp. 7055–7065. PMLR (2020)

17. Patel-Schneider, P., Arndt, D., Haudebourg, T.: RDF 1.2 semantics, W3C recommendation (2023). https://www.w3.org/TR/rdf12-semantics/

18. Pérez-Suárez, A., Martínez-Trinidad, J.F., Carrasco-Ochoa, J.A.: A review of conceptual clustering algorithms. Art. Intell. Rev. 52(2), 1267–1296 (2019)

19. Pichler, R., Polleres, A., Skritek, S., Woltran, S.: Complexity of redundancy detection on RDF graphs in the presence of rules, constraints, and queries. Semant. Web 4(4), 351–393 (2013)

20. Ruta, M., Colucci, S., Scioscia, F., Di Sciascio, E., Donini, F.M.: Finding commonalities in RFID semantic streams. Procedia Comput. Sci. 5, 857–864 (2011)

21. Shadbolt, N., Hall, W., Berners-Lee, T.: The semantic web revisited. IEEE Intell. Syst. 21(3), 96–101 (2006)

22. Soylu, A., et al.: TheyBuyForYou platform and knowledge graph: expanding horizons in public procurement with open linked data. Semant. Web 13(2) (2022)

23. Soylu, A., et al.: Towards an ontology for public procurement based on the open contracting data standard. In: Pappas, I.O., Mikalef, P., Dwivedi, Y.K., Jaccheri, L., Krogstie, J., Mäntymäki, M. (eds.) I3E 2019. LNCS, vol. 11701, pp. 230–237. Springer, Cham (2019). https://doi.org/10.1007/978-3-030-29374-1_19

24. Vougiouklis, P., et al.: Neural Wikipedian: generating textual summaries from knowledge base triples. J. Web Semant. 52–53, 1–15 (2018)

Shapley-Based Data Valuation Method for the Machine Learning Data Markets (MLDM)

Hajar Baghcheband[1,2]($\boxtimes$) (iD), Carlos Soares[1,2,3] (iD), and Luis Paulo Reis[1,2] (iD)

[1] FEUP - Faculty of Engineering University of Porto, Porto, Portugal
{h.baghcheband,csoares,lpreis}@fe.up.pt
[2] LIACC-Artificial Intelligence and Computer Science Laboratory (Member of LASI LA), Porto, Portugal
[3] Fraunhofer AICOS Portugal, Porto, Portugal

Abstract. Data valuation, the process of assigning value to data based on its utility and usefulness, is a critical and largely unexplored aspect of data markets. Within the Machine Learning Data Market (MLDM), a platform that enables data exchange among multiple agents, the challenge of quantifying the value of data becomes particularly prominent. Agents within MLDM are motivated to exchange data based on its potential impact on their individual performance. Shapley Value-based methods have gained traction in addressing this challenge, prompting our study to investigate their effectiveness within the MLDM context. Specifically, we propose the Gain Data Shapley Value (GDSV) method tailored for MLDM and compare it to the original data valuation method used in MLDM. Our analysis focuses on two common learning algorithms, Decision Tree (DT) and K-nearest neighbors (KNN), within a simulated society of five agents, tested on 45 classification datasets. results show that the GDSV leads to incremental improvements in predictive performance across both DT and KNN algorithms compared to performance-based valuation or the baseline. These findings underscore the potential of Shapley Value-based methods in identifying high-value data within MLDM while indicating areas for further improvement.

Keywords: Data Markets · Data Valuation · Machine Learning · Multi-Agent Systems

1 Introduction

The concept of data valuation (DV) has gained increasing attention in data markets [7]. A data market is a place where users can sell and buy data. An example of a data market is the Machine Learning Data Market (MLDM), introduced [2,3], which is based on multi-agent systems (MAS). A MLDM models a scenario of distributed agents that collect data and use it to develop models, which are then applied to new data. As data is distributed among all agents, they are encouraged to exchange data within budget constraints, emphasizing the significance of data valuation in optimizing their training sets and predictive performance. However, in data marketplaces, the monetization of valuable data is a complex challenge [4].

In this paper, we propose a Gain Data Shapley Value (GDSV) method for data valuation to be used in MLDM considering that agents collect data gradually and have a limited budget. We compare the Shapley-based DV with the one used originally with MLDM [3]. Agents will develop their models based on two common classification algorithms (KNN and DT) and test on 45 classification datasets.

2   Related Work

In this section, we provide a short review of Shapley Value (Sect. 2.1) and the Machine Learning Data Markets (MLDM) framework (Sect. 2.2).

2.1   Data Shapley Value

The Shapley Value, arguably one of the most popular ways of revenue sharing, is introduced based on game theory [9,10]. Its widespread acceptance is because it offers a distinctive profit allocation scheme that meets a range of desirable criteria, such as fairness, rationality, and decentralization [6,7]. In ML, the SV is a tool to put a value on data and is called "Data Shapley" [5]. Some approaches [5, 12] take a different view of the learning process: each training data point is perceived as a player in a cooperative game and these players (i.e. data points) work together through the learning process to build the model. In this case, the utility function is a measure of the predictive performance of the model being learned. Therefore, the SV of each data point measures its contribution to training an ML model. Consider N is the total number of data owners $D_i \in D = \{D_1, ..., D_n\}$ that has batch of data Bt_i, and $U(S)(S \subseteq \{Bt_1, .., Bt_n\})$ is the utility function representing the value calculated by coalition S. The SV for data point Bt_i as a player is defined as the average marginal contribution of data point Bt_i, to all possible coalitions formed by other data points [5,8,13]:

$$\phi_i = \sum_{S \subseteq \{Bt_1, .., Bt_n\} \setminus Bt_i} \frac{1}{N\binom{N-1}{|S|}} [U(S \cup \{Bt_i\}) - U(S)] \qquad (1)$$

The Shapley value (SV) can be used to value data and fairly compensate data owners for their contribution in data markets [1,8,13].

2.2   Machine Learning Data Market: MLDM

The MLDM is the data market platform based on multi-agent systems where each agent is motivated to exchange data to improve their predictive performance. Initially, each agent builds its learning model using a local training set.

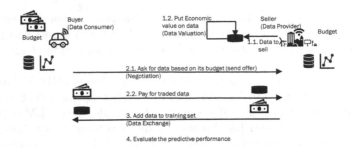

Fig. 1. Data Exchange in the MLDM

Subsequently, they value their recently collected data and advertise it within the platform. Agents select potential data providers based on perceived data value and engage in negotiations to acquire the desired data sets. Following the completion of the data trading process, the agent evaluates the impact and effectiveness of this data exchange on the overall performance of its learning model (Fig. 1).

More formally, The MLDM framework, (denoted as $MLDM = \{A, S, L\}$), consists of a society of n agents, $(A_i \in A = \{A_1, A_2, ..., A_n\})$ where agents utilize learning algorithms L, and negotiation strategy set S to improve their model performance [2,3]. Each agent A_i within the society undergoes specific operations at each time step t (see Fig. 2 and Algorithm 1):

Firstly, Agents learn the model, $M_{i,t} = L_i(TnB_{i,t})$ using available training data $TnB_{i,t}$ with a learning algorithm L_i (*The Modeling Component (MC)*). Next, They evaluate the model performance, $P_{i,t} = M_{i,t}(TsB_{i,t})$,using local test data $TsB_{i,t}$(*The Performance Estimation Component (PC)*). Agents then estimate the value of recently collected data to be traded, $TrB_{i,t} = TnB_{i,t} \subset Bt_{i,t}$, as $\phi_{i,t} = V(TrB_{i,t})$(*Data Acquisition and Valuation (DAV)*). Following this, agents engage in negotiation and data exchange to obtain data with the highest value ϕ_t based on its current budget $Bg_{i,t}$(*The Negotiation and Exchange Component (NEC)*). Each agent may act as a seller or buyer regarding the value of its data. Agent A_i is going to exchange data with another agent A_j, $NE_{i,t} = S_i(TrB_{i,j,t})$, pay data cost $C_{i,j,t}$ for traded batch of data $TrB_{i,j,t}$;

$$C_{i,j,t} = size(TrB_{i,j,t}) * BC, \tag{2}$$

where $size(Tr_{i,j,t})$ is the amount of traded data (number of records) and BC is the basic data cost regarding the value of data of agent A_j at time t, $\phi_{j,t}$. If the budget is not enough, $C_{i,j,t} > Bg_{i,t}$, agent A_i will request a portion of the data that is within its budget $(size(Tr_{i,j,t}) = Bg_{i,t}/BC)$. After negotiation takes place, agent A_i adds traded data to its training set, $Bt_{i,t} = (TnB_{i,t} \cup Tr_{i,j,t})$. All agents evaluate the improvement in the predictive performance after exchanging data and calculating the gain, $Gain_{i,t}$ (*The Evaluation Component (EC)*). Different functions can be used for $Gain$ (e.g. accuracy, computational time). Based on the mentioned components,the status of agents A_i at time t is defined as $A_{i,t} = \{Bt_{i,t}, M_{i,t}, P_{i,t}, Bg_{i,t}\}$.

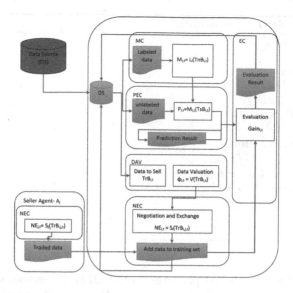

Fig. 2. Architecture of the Agents in the MLDM

3   Data Valuation in MLDM

The primary goal for agents is to identify high-quality data that improve their learning models. In this section, two data valuation methods are discussed: the original performance-based approach in MLDM and the proposed Shapley-based method.

3.1   Performance Method

The performance method is the simple, original data valuation method for MLDM [3]. This method is based on the informativeness of data [11]. It follows the idea that the quality of data is directly linked to the predictive performance. Having high-quality data leads to a high-performing prediction model.

$$\phi_{i,t} = max[P_{k,t}(Bt_{k,t})] \quad \forall k = 1, .., n \tag{3}$$

The performance method assesses the overall quality of the dataset by considering various predictive performance metrics. It takes into account the performance of the entire dataset but does not focus on the selected set that is being sold.

3.2   The Gain Data Shapley Value (GDSV)

An alternative data valuation method, proposed here tailored for MLDM, is based on Data Shapley Value (DSV) focusing on the training set as a player. This method evaluates the importance of each data point within the training set while considering the collective effort in the learning process. We utilize the DSV

Algorithm 1. MLDM Framework

Require: $T, Bt_i \forall i = \{1, .., n\}$
   $t \leftarrow 0$
  while $t \leq T$ **do**
     $M_{i,t} \leftarrow L_i(TnB_{i,t})$
     $P_{i,t} \leftarrow M_{i,t}(TsB_{i,t})$
     $\phi_{i,t} \leftarrow V(TnB_{i,t})$
     for A_i, A_j **do**
       if $\phi_{i,t} < \phi_{j,t}$ **then**
         $NE_{i,t} \leftarrow S_i(TrB_{i,j,t})$
         $Bg_{i,t} \leftarrow Bg_{i,t} - C_{ij,t}$
         $Bg_{j,t} \leftarrow Bg_{j,t} + C_{ij,t}$
       end if
     end for
  end while

(Eq. 1) adopted to the MLDM platform. In the MLDM scenario, the DSV is calculated for the recently collected training set. However, the DSV solely assesses the significance of the dataset concerning the impact on performance scores, which may not consider the relative improvement in performance accurately. To address this limitation, the Gain Data Shapley Value (GDSV) is proposed, which incorporates the relative improvement in performance, thereby offering an assessment of the value of data:

$$\phi_{i,t} = GDSV(TnB_{i,t}) = \frac{P_{i,t}(Bt_{i,t-1} \cup TnB_{i,t}) - P_{i,t-1}(Bt_{i,t-1})}{1 - P_{i,t-1}(Bt_{i,t-1})} \tag{4}$$

where $GDSV(TnB_{i,t})$ stands for the value of the training set at time t. We will often write $P_{i,t}(Bt_{i,t-1} \cup TnB_{i,t})$ and $P_{i,t-1}(Bt_{i,t-1})$ as $P_{i,t}, P_{i,t-1}$ to simplify notation. $P_{i,t}, P_{i,t-1}$ denote the performance score at time t and time $t-1$, respectively, over mentioned batch of data. The GDSV considers performance increase and normalizes it compared to previous performance. This method evaluates dataset value in relation to model performance enhancement.

4   Empirical Validation

This paper examines the effectiveness of data valuation methods on predictive performance in the MLDM platform. The real-world scenarios may be more complex, but the current investigation evaluates the principle of data valuation for MLDM.

The following are the experimental assumptions:

1. The MLDM framework was evaluated by choosing a random set of 45 classification data sets from the OpenML platform [14] with different sizes and properties to simulate diversity.
2. Each agent has a constant initial budget of $Bg_i = 1000$.
3. Batch size is the maximum proportion of data that is collected by each agent from the data source in each iteration.
4. The population of the society is the number of agents that exist in MLDM.
5. The single-agent scenario (SA) is a system in which data is not distributed and data exchange does not make sense. The performance of it is the maximum value for the performance.

Here, we consider batch size as $Bt = 0.01$ (i.e. every iteration, each agent selects one percent of the whole data set), and population size as five agents. Regarding this consideration, we can have a reliable number of iterations to analyze. We assess the predictive performance based on Global Evaluation [3]. In the Global Evaluation, $P_{i,t} = M_{i,t}(TsG)$, the agent estimates the predictive performance of the models on a common test set(TsG). It shows the behavior of the entire system. The results discussed in this paper are based on the global evaluation as GE_t, calculated by taking the average over all datasets $(D_j, j = \{1, ..., N\})$ for all agents $(A_i, i = \{1, ..., n\})$ at time t:

$$ GE_t = \frac{1}{n * N} \sum_{D_j=1}^{N} \sum_{i=1}^{n} P_{i,t}^{D_j} \tag{5} $$

where $P_{i,t}^{D_j}$ stands for the performance of agents A_i model at time t, $P_{i,t}$ applying data set D_j.

We study how GDSV affects performance improvement by comparing two learning algorithms in various scenarios (i.e. MLDM, NE, and SA). In Figs. 3 and 4, Performance-based data valuation and GDSV are evaluated in comparison to NE and SA scenarios using KNN and Decision Tree. Results show data exchange enhances model performance towards SA level, with GDSV slightly better at improving learning model efficiency. Therefore, we can conclude that GDSV has the potential to identify valuable data within the MLDM platform, consequently increasing predictive performance after data exchange.

In Fig. 5, we investigate the difference in performance between KNN and DT, when using GDSV. KNN with GDSV slightly outperforms DT with GDSV. Analysis explores GDSV effectiveness with various ML algorithms, offering insights on data valuation in MLDM platform.

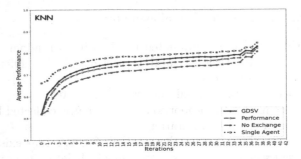

Fig. 3. Global Evaluation Average over all data sets for data valuation methods and baseline- Learning Algorithm: KNN

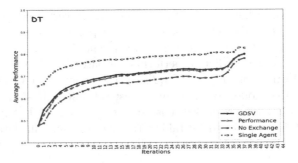

Fig. 4. Global Evaluation Average over all data sets for data valuation methods and baseline- Learning Algorithm: DT

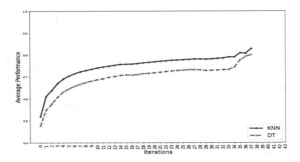

Fig. 5. Comparison of GDSV in Algorithm KNN and DT

5   Conclusion

In this paper, we introduced and examined GDSV, a method for data valuation in Machine Learning Data Markets. We aimed to see if Shapley value-based data valuation could improve predictive performance in MLDM platform. To evaluate GDSV, we analyzed 45 datasets and compared results to a scenario without data exchange. Results showed GDSV slightly outperforms performance-based data valuation. When comparing KNN and DT, GDSV with KNN performs slightly better. Overall, our study emphasizes the importance of data valuation in MLDM platform, showing improved predictive performance compared to baseline scenario. Future research in data valuation within the MLDM could explore more complex scenarios to enrich the understanding of data valuation.

Acknowledgments. This work was financially supported (or partially financially supported) by Base Funding – UIDB/00027/2020 of the Artificial Intelligence and Computer Science Laboratory - LIACC - funded by national funds through the FCT/MCTES (PIDDAC) and by a PhD grant from Fundação para a Ciência e Tecnologia (FCT), reference SFRH/BD /06064/2021.

References

1. Agarwal, A., Dahleh, M., Sarkar, T.: A marketplace for data: an algorithmic solution. In: ACM EC 2019 - Proceedings of the 2019 ACM Conference on Economics and Computation, pp. 701–726 (2019)
2. Baghcheband, H., Soares, C., Reis, L.: Machine learning data markets: trading data using a multi-agent system. In: 2022 IEEE/WIC/ACM International Joint Conference on Web Intelligence and Intelligent Agent Technology (WI-IAT), Los Alamitos, CA, USA, pp. 450–457. IEEE Computer Society (2022)
3. Baghcheband, H., Soares, C., Reis, L.P.: Machine learning data markets: evaluating the impact of data exchange on the agent learning performance. In: Moniz, N., Vale, Z., Cascalho, J., Silva, C., Sebastião, R. (eds.) EPIA 2023. LNCS, vol. 14115, pp. 337–348. Springer, Cham (2023). https://doi.org/10.1007/978-3-031-49008-8_27
4. Faroukhi, A.Z., El Alaoui, I., Gahi, Y., Amine, A.: Big data monetization throughout Big Data Value Chain: a comprehensive review. J. Big Data 7(1) (2020)
5. Ghorbani, A., Zou, J.: Data Shapley: equitable valuation of data for machine learning. In: Proceedings of the 36th International Conference on Machine Learning. Proceedings of Machine Learning Research, vol. 97, pp. 2242–2251. PMLR (2019)
6. Jia, R., et al.: Efficient task-specific data valuation for nearest neighbor algorithms. Proc. VLDB Endow. 12(11), 1610–1623 (2019)
7. Jia, R., et al.: Towards efficient data valuation based on the Shapley value (2023)
8. Liu, J., Lou, J., Liu, J., Xiong, L., Pei, J., Sun, J.: Dealer: an end-to-end model marketplace with differential privacy. Proc. VLDB Endow. 14(6), 957–969 (2021)
9. Shapley, L.S.: A value for n-person games. In: Kuhn, H.W., Tucker, A.W. (eds.) Contributions to the Theory of Games II, pp. 307–317. Princeton University Press, Princeton (1953)
10. Shapley, L.S.: The Shapley Value: Essays in Honor of Lloyd S. Shapley edited by Alvin E. Roth. Cambridge University Press (1988)
11. Sim, R.H.L., Zhang, Y., Chan, M.C., Low, B.K.H.: Collaborative machine learning with incentive-aware model rewards. In: 37th International Conference on Machine Learning, ICML 2020, vol. PartF16814(Ml), pp. 8886–8895 (2020)
12. Tang, S., et al.: Data valuation for medical imaging using Shapley value and application to a large-scale chest X-ray dataset. Sci. Rep. 11, 1–9 (2021)
13. Tian, Z., et al.: Private data valuation and fair payment in data marketplaces (2023)
14. Vanschoren, J., van Rijn, J., Bischl, B., Torgo, L.: OpenML: networked science in machine learning. SIGKDD Explor. 15(2), 49–60 (2013). https://doi.org/10.1145/2641190.2641198

Industry Session

ScoredKNN: An Efficient KNN Recommender Based on Dimensionality Reduction for Big Data

Seda Polat Erdeniz[1,2]([email]), Ilhan Adiyaman[1], Tevfik Ince[1], Ata Gür[1],
and Alexander Felfernig[2]

[1] Frizbit Technology S.L., Barcelona, Spain
{seda,ilhan,ata}@frizbit.com
[2] Graz University of Technology, Graz, Austria
{spolater,felfernig}@ist.tugraz.at

Abstract. E-commerce companies have an inevitable need in employing recommender systems in order to enhance the user experience, increase customer satisfaction, and drive sales. One of the most popular, intuitive and explainable recommender algorithm is the K-nearest neighbors (KNN) algorithm which is a well-known non-parametric collaborative filtering (CF) method. However, when dealing with big data, applying KNN poses computational challenges in terms of both time and space consumption. Several solutions proposed, but none of them could become a standard solution up to now. To address this issue, we propose a dimension reduction based approach with scoring functions which is applicable on all neighboring methods. With the help of this approach, similarity calculation is reduced into one dimension instead of two dimensions. The proposed approach reduces the KNN complexity from $O(n^2)$ to $O(n)$ and it has been evaluated on both publicly available datasets and also real-world e-commerce datasets of an e-commerce services provider company Frizbit S.L.. We have compared our method with state-of-the-art recommender systems algorithms and evaluated based on the criteria: time consumption, space consumption and accuracy. According to the experimental results, we have observed that our proposed approach ScoredKNN achieves a pretty good accuracy (in terms of MAE) and lower time/space costs.

1 Introduction

Recommender Systems algorithms are successfully applied on e-commerce domain. There are several examples of recommender systems in e-commerce that have been widely used by various platforms such as Amazon's recommendation system which is renowned for its effectiveness. It utilizes a combination of collaborative filtering, content-based filtering, and item-to-item collaborative filtering. It suggests items based on user browsing and purchase history, as well as similarities between items. Another important example can be Netflix which employs a sophisticated recommender system to personalize movie and TV show recommendations for its users. Netflix's system analyzes user viewing history, ratings, and interactions with the platform to generate tailored recommendations.

These systems frequently depend on collaborative filtering methods, which involves analyzing historical user interactions (browsing, purchase, etc.) to establish connections between users and products. As the number of users and items increases, recommender systems need to efficiently process and store large amounts of data. Managing and indexing user profiles, item catalogs, and interaction data can become challenging, requiring robust data storage and retrieval mechanisms. In many e-commerce and streaming platforms, recommender systems are expected to provide real-time recommendations as users interact with the system. As the user base grows, the recommender system needs to handle a high volume of requests and deliver recommendations in a timely manner. This requires designing efficient algorithms and infrastructure to support real-time recommendation generation and response. Traditional recommender algorithms, such as collaborative filtering, may face scalability issues when dealing with large datasets. The computation and memory requirements of these algorithms may become prohibitive, resulting in increased response times and resource consumption. Developing scalable algorithms and models that can handle large-scale data efficiently is a key challenge. Scalability often necessitates the use of distributed computing frameworks and infrastructure to handle large-scale recommender systems. Distributed storage, parallel processing, and efficient data sharing techniques are required to handle the increased workload and ensure high availability and fault tolerance.

A well-known collaborative filtering based recommender system algorithm is the K-nearest neighbors (KNN) which is a non-parametric model widely used. While KNN is effective, its exact implementation becomes impractical when dealing with big data due to the high computational requirements. To overcome this scalability limitation, various techniques and algorithms have been proposed to approximate or optimize the computation of pairwise similarities in KNN. These techniques aim to reduce the computational burden and improve the scalability of the KNN model, allowing it to handle larger datasets more efficiently.

To overcome this scalability issue while keeping the accuracy in an acceptable level, we developed our own method "ScoredKNN" which addresses the limitations by scoring users and items, so applies dimension reduction. This dimension reduction enables us to scale the implementations of both item-item and user-user approaches linearly, with the size of the data, which is $O(n)$.

In the rest of this paper, we first provide related work in Sect. 2. Afterwards, in Sect. 3 we describe our proposed method with providing pseudo codes and examples. Before concluding the paper, we evaluate the performance of our proposed method by comparing with state-of-the-art collaborative filtering methods of recommender systems (e.g., Matrix Factorization). Experimental tests are based on publicly available datasets (e.g., Movielens 100k, Movielens 1M) and also our real-world e-commerce transactions data.

2 Related Work

In this section, we provide the related work to solve the complexity issue of KNN. Aforementioned, KNN has $O(n^2)$ time and space complexity, so it is inapplicable on big amount of data even it is very intuitive and practical collaborative filtering algorithm in recommender systems. A computational complexity of $O(n^2)$ means that the running time of an algorithm increases quadratically with the size of the input data. In the context of big data, where datasets can be extremely large, a *quadratic* ($O(n^2)$) complexity becomes computationally expensive and inefficient. It is crucial to design algorithms with lower computational complexities, such as *linear* ($O(n)$), *logarithmic* ($O(log_n)$), or *sublinear* ($O(\sqrt{n})$)) time complexities. There are several research to scale KNN for big data usage.

First of all, *Approximate KNN* methods [1–3] aim to reduce the computational complexity of finding the exact nearest neighbors. They trade-off accuracy for efficiency by using techniques like locality-sensitive hashing (LSH) [4,5]. These methods enable faster retrieval of approximate nearest neighbors, making KNN more scalable.

High-dimensional data can significantly affect the performance and scalability of KNN algorithms. *Dimensionality Reduction* techniques, such as Principal Component Analysis (PCA) [6,7] or t-SNE [8], reduce the dimensionality of the data while preserving its structure. By reducing the number of dimensions, the computational complexity of KNN is reduced, improving scalability.

Clustering techniques [9] can also be applied as a "first step" for shrinking the candidate set in a nearest neighbor algorithm or for distributing nearest-neighbor computation across several recommender engines. While dividing the population into clusters may hurt the accuracy or recommendations to users near the fringes of their assigned cluster, pre-clustering may be a worthwhile trade-off between accuracy and throughput.

In large-scale datasets, working with the entire dataset can be computationally expensive. *Data Sampling* techniques can be employed to create smaller representative subsets of the data. By performing KNN on the sampled data [10,11], the computational load can be significantly reduced while maintaining reasonable accuracy.

Indexing and Data Structures can improve the scalability of KNN algorithms. The use of spatial indexing structures [12,13] like KD-trees, ball trees, or quadtrees allows for faster retrieval of nearest neighbors by organizing the data in a hierarchical manner. These structures help reduce the number of distance calculations required, improving scalability.

KNN computations can be parallelized across multiple processors or distributed across a cluster of machines. By utilizing *Parallelization and Distributed Computing* frameworks like Spark [14,15] or Hadoop [16,17], the computational burden of KNN can be distributed, allowing for faster processing and improved scalability.

Rather than recalculating distances and updating the entire model for every new data point, *Incremental Learning* techniques [14] can be employed. These techniques update the KNN model incrementally, incorporating new data while minimizing the computational overhead.

However, all these methods are neither available in standard libraries (e.g. python sklearn) nor applicable on recommender system problems. Therefore, scalable KNN is still a research topic in recommender systems domain. Researchers in the field of recommender systems continue to explore and develop novel scalable KNN approaches to address the challenges posed by large-scale datasets and increasing user and item populations. These advancements aim to improve the efficiency and effectiveness of KNN-based recommender systems and enable them to handle the demands of real-world applications.

3   Proposed Method: ScoredKNN

A typical problem in recommendation is that of rating prediction: given an incomplete dataset of user-item interactions which take the form of numerical ratings (e.g. on a scale from 1 to 5), the goal is to predict the missing ratings for all remaining user-item pairs [18]. To predict those missing ratings, our proposed method ScoredKNN combines both methods: KNN and GBE in order to improve the efficiency of KNN with the advantages of GBE's scoring technique. Moreover, it can be employed as item-based and user-based collaborative filtering algorithm, whereas GBE can be applied only as a user-based CF.

As shown in Algorithm 1, we defined a scoring function which is used to calculate similarity. Instead of computing the complete similarity matrix, we use these scores of users (or items) to find the k nearest neighbors of them. In other words, if score of an item is 0.81, the closest scored 3 nearest neighbors can have scores 0.78, 0.82, 0.87. Same strategy applies when calculating nearest neighbors of a user.

This scoring based similarity calculation decreases the computational dimensions from 2d-array (a complete similarity matrix) to 1d-array (scores array). Thus, ScoredKNN has linear ($O(n)$) complexity.

In the basic version, we use GBE for scoring. However, according to the evaluation metrics, we can adapt scoring function into many versions. Using GBE helps us to find similarly rating users or similarly rated items. In this scenario, nearest neighbors of a user would be users with similar rating behaviours (highly critical or giving always positive feedback).

4   Evaluation

In this section, we evaluate our proposed method by comparing with state of the art recommender system algorithms on the various datasets. Our target is to improve the computational performance without decreasing the accuracy dramatically. We have applied our tests on well-known public recommender systems benchmark Movielens with **100k** explicit ratings (943 users, 1682 items) which have a format as shown in Table 1 where ratings are in the scale of [1–5].

Algorithm 1. ScoredKNN

```
 1: Function Scoring():
 2:     μ = stdDev(all ratings)
 3:     For each u in n_users:
 4:         scoreᵤ = stdDev(ratings of u)
 5:         For each i in m_items:
 6:             scoreᵢ = stdDev(ratings of i)
 7:             rᵤᵢ = μ + scoreᵤ + scoreᵢ
 8: Function User_Based_Recommendations():
 9:     For each u in n_users:
10:         For each i in m_items:
11:             if ratingᵤᵢ == 0:
12:                 For each u2 in k_nearest_scored_users:
13:                     ratingᵤᵢ += rating(u2,i)
14:                 ratingᵤₗᵢ = ratingᵤᵢ/k
15: Function Item_Based_Recommendations():
16:     For each i in m_items:
17:         if ratingᵤᵢ == 0:
18:             For each u2 in k_nearest_scored_items:
19:                 ratingᵤᵢ += rating(u2,i)
20:             ratingᵤₗᵢ = ratingᵤᵢ/k
```

Table 1. Movielens Data: Ratings in the range of [1–5]

user_id	item_id	rating [1–5]
0	6	3
1	9	1
2	17	5

Moreover, to be able to test the computational performance more precisely, we have also used our real world e-commerce transactions data with **6.4M** (6,403,825) events. These events are holding implicit ratings, so we converted the implicit ratings into explicit ratings format to be able to use the algorithms which are implemented for explicit ratings data as in Table 2.

We have compared our proposed method with the state of the art recommender algorithms which are provided by a recommender systems library called "Surprise" [19]. We have also implemented our proposed method Scored-KNN in Surprise library using its new algorithm definition API. Surprise is a Python library for building and analyzing rating prediction algorithms. It was designed to closely follow the scikit-learn API, which should be familiar to users acquainted with the Python machine learning ecosystem. Surprise provides a collection of estimators (or prediction algorithms) for rating prediction.

Table 2. Frizbit Data: Events mapped into ratings in the range [1–3]

user_id	item_id	event	rating [1–3]
0	6	purchase	3
0	17	view	1
1	9	purchase	3
2	17	purchase	3
2	19	add to cart	2

We have evaluated the models in terms of both runtime and accuracy. Runtime and the memory consumption of the models are measured on a linux computer with 10 CPU and 60GB RAM. Accuracy is measured in terms of Mean Absolute Error (MAE) and Root Mean Squared Error (RMSE) as the formulas below:

$$MAE = \sum_{i=1}^{D} |x_i - y_i| \tag{1}$$

$$RMSE = \sqrt{\frac{1}{n}\Sigma_{i=1}^{n}\left(\frac{d_i - f_i}{\sigma_i}\right)^2} \tag{2}$$

MAE measures the average absolute difference between the predicted values and the actual values. MAE is advantageous because it provides a straightforward measure of the average magnitude of errors without considering their direction. RMSE is a commonly used evaluation metric to measure the accuracy of predictions or recommendations, particularly in regression tasks. RMSE quantifies the average magnitude of the differences between predicted values and actual values. Taking the square root of the mean squared error normalizes the metric back to the original scale of the data. A lower MAE and RMSE indicate better predictive accuracy, as it means the predictions are closer to the actual values on average.

We have evaluated ScoredKNN with the compared methods based on two datasets: Movielens's 1M observations public dataset with movie ratings (scaled in 1–5) and Frizbit's 6.4M observations private dataset with user transactions as view, add-to-cart, and purchase (converted into ratings respectively as 1, 2, and 3). Performance results are recorded based on 5-fold cross validation.

As observed in Table 3, the most efficient methods in terms of runtime are the $O(n)$ methods like GBE and ScoredKNN (except the Normal Predictor which is a kind of random recommender and used as a baseline for the experiment). ScoredKNN (both item based and user based) did not provide a better accuracy in terms of RMSE and MAE than GBE. This results was actually what we expected in the beginning of the experiment as well since the scoring function is not targeting the higher accuracy.

This experiment applied only to prove the computational efficiency of this method compared to other collaborative filtering methods. Its performance in

Table 3. Performance on Movielens data (100k ratings [1–5], 0.9k users, and 1.6k items)

	Movielens Data (100k)				
	Training Time (s)	Testing Time (s)	Memory (GB)	RMSE	MAE
Normal Predictor	0.09	0.15	0.013	1.51	1.21
GBE	0.21	0.10	0.012	0.94	0.74
KNN Basic	0.24	2.31	0.022	0.98	0.77
KNN WithMeans IB	0.60	2.77	0.033	0.93	0.74
KNN WithMeans UB	0.61	1.71	0.019	1.01	0.80
KNN Baseline	0.46	2.80	0.020	0.93	0.73
MF SVD	0.68	0.12	0.026	0.93	0.73
MF SVDpp	11.35	2.44	0.019	0.91	0.72
MF NMF	1.25	0.11	0.017	0.96	0.75
SlopeOne	0.60	1.91	0.054	0.94	0.74
CoClustering	1.30	0.11	0.012	0.96	0.75
ScoredKNN IB	0.20	1.02	0.011	1.21	0.82
ScoredKNN UB	0.21	0.17	0.012	1.45	1.01

terms of recommendation quality should be measured using different metrics and according to those target evaluation metrics, scoring function should be also adapted.

As seen, memory consumption of KNN WithMeans IB (item based) is higher than KNN WithMeans UB (user based) since number of items (1.6k) are higher than number of users (0.9k) in this dataset. For user based similarity, a similarity matrix with the shape 1.6kx1.6k is prepared wheras for item based similarity it is 0.9kx0.9k. However, actually in most of the real world scenarios, number of users are much more higher than number of items on an e-commerce system. Therefore, item based collaborative filtering methods become more "computationally efficient" for such systems. Therefore, we also tested the performance on our real world events data.

For the second test, we used 10 days transactions data of an e-commerce customer of Frizbit. In this small amount of time range, data includes 6.4M events with around 1.5M users and 20k products. For this second test, we have selected algorithms as in Table 4 because of the data size.

According to this error, we can understand that creating a user similarities matrix for around 1.5M users need 15.1 TB space. However, we used a computer with 60 GB memory. On the same computer, KNN WithMeans IB did not create a problem, for around 20k items it used 15 GB memory. Therefore, with this machine's capacity and the needed space allocation by KNN similarity matrix, we can only train KNN item based for maximum around 80k items, but user based similarity is impossible since the number of users in the data is much more than 80k. Therefore, our proposed method brings a solution for collaborative filtering

Table 4. Performance on Frizbit data (6.4M ratings [1–3], 1.4M users, and 20k items)

	Frizbit Data (6.4M)				
	Training Time (s)	Testing Time (s)	Memory (GB)	RMSE	MAE
Normal Predictor	61.83	34.64	2.9	0.50	0.34
GBE	63.27	45.35	3.2	0.38	0.25
KNN Basic	Error	Error	15098.02+	Error	Error
KNN WithMeans IB	1440.23	62.17	15.12	0.42	0.26
MF SVD	3246.28	124.35	12.24	0.39	0.23
ScoredKNN IB	78.34	63.87	4.1	0.43	0.28
ScoredKNN UB	82.13	56.34	5.2	0.45	0.31

on high amount of users (1M+) and ratings (6M+) data. Using ScoredKNN, both item and user based similarities can be calculated efficiently.

As observed also in this second test, it did not improve the accuracy in terms of RMSE and MAE. However, Ultimately, the success of a recommender system lies in its ability to satisfy users, engage them, and drive desired behaviors (e.g., increased purchases, longer session durations). Accuracy alone may not capture these user-centric aspects, and it is crucial to consider additional evaluation measures like user surveys, feedback, or business metrics to assess the overall performance and impact of the recommender system. Using ScoredKNN, the target metric can be changed based on the customer needs by adapting the scoring function.

5   Future Work

In this paper, we have evaluated the proposed method ScoredKNN using the basic scoring function which is based on GBE. With the results, we have observed that compared to KNN, ScoredKNN is increasing the computational performance significantly but not the accuracy, as expected. Therefore, as future work we would like to work on defining new scoring functions to increase the recommendation quality in terms of various evaluation metrics. For example, an alternative scoring function can be defined based on dimensions (features) of the items (products). Since in recommender systems, items are the products of the e-commerce company, dimensions of the items are simply their features such as: Price, Rating, Production Date, and etc.

If quality measure is *High Revenue*, the scoring function can be adapted to recommend high price products which can be interesting for the user. In this case, selected dimensions can be: Price and Rating where weights can be defined as $w_{Price} = 10$ and $w_{Rating} = 1$. With this approach, higher price items and the users who purchases those items will have higher scores.

For example, let's assume Tom and Bob are buying products on a camping products e-commerce company. If the recommender system of this store is using

Table 5. Example user acceptances for two recommendable items on an e-commerce company

	Camping Tent A (Price: 1000$, Rating: 4.8)	Camping Tent B (Price: 300$, Rating: 5)
Tom (High Score)	ACCEPT	ACCEPT
Bob (Low Score)	REJECT	ACCEPT

a standard collaborative filtering approach which is trained only according to Ratings, Camping Tent B is recommended to both Tom and Bob, so the revenue of the company will be 600$. However, if the recommender system of this store is adapted a "High Revenue" quality measure using ScoredKNN, Camping Tent A is recommended to Tom, and Camping Tent B is recommended to Bob, so the revenue of the company will be 1300$. Therefore, the *Opportunity Cost* of ignoring "High Revenue" as a quality metric is 700$ (Table 5).

6   Conclusions

KNN is an intuitive and widely-used recommender systems algorithm which has quadratic complexity. Running algorithms with quadratic complexity on big data can be expensive. It may require substantial computational infrastructure and resources, leading to increased costs for hardware, storage, and maintenance. In this research, we have proposed a novel approach "ScoredKNN" to scale up the KNN algorithm for the big data scenarios. We have evaluated ScoredKNN based on publicly available benchmark algorithms and datasets. As a result, we have observed that ScoredKNN makes KNN feasible on the big data since it reduces the complexity from $O(n^2)$ to $O(n)$. Moreover, the recommendation quality target can be changed by adapting the scoring function of ScoredKNN. This makes it very flexible for many recommendation scenarios.

Acknowledgements. This research is funded by ACCIO's Tecniospring INDUSTRY program for the project with the project number "ACE026/21/000108".

References

1. Yi, X., Paulet, R., Bertino, E., Varadharajan, V.: Practical approximate k nearest neighbor queries with location and query privacy. IEEE Trans. Knowl. Data Eng. **28**(6), 1546–1559 (2016)
2. Chen, J., Fang, H., Saad, Y.: Fast approximate KNN graph construction for high dimensional data via recursive Lanczos bisection. J. Mach. Learn. Res. **10**(9) (2009)
3. Anagnostou, P., Barbas, P., Vrahatis, A.G., Tasoulis, S.K.: Approximate kNN classification for biomedical data. In: 2020 IEEE International Conference on Big Data (Big Data), pp. 3602–3607. IEEE (2020)

4. Pan, J., Manocha, D.: Fast GPU-based locality sensitive hashing for k-nearest neighbor computation. In: Proceedings of the 19th ACM SIGSPATIAL International Conference on Advances in Geographic Information Systems, pp. 211–220 (2011)
5. Bagui, S., Mondal, A.K., Bagui, S.: Improving the performance of kNN in the mapreduce framework using locality sensitive hashing. Int. J. Distrib. Syst. Technol. (IJDST) **10**(4), 1–16 (2019)
6. Singh, A., Pandey, B.: An efficient diagnosis system for detection of liver disease using a novel integrated method based on principal component analysis and k-nearest neighbor (PCA-kNN). In: Intelligent Systems: Concepts, Methodologies, Tools, and Applications, pp. 1015–1030. IGI Global (2018)
7. Kamencay, P., Hudec, R., Benco, M., Zachariasova, M.: Feature extraction for object recognition using PCA-kNN with application to medical image analysis. In: 2013 36th International Conference on Telecommunications and Signal Processing (TSP), pp. 830–834. IEEE (2013)
8. Sakib, S., Siddique, Md.A.B., Rahman, Md.A.: Performance evaluation of t-SNE and MDS dimensionality reduction techniques with KNN, ENN and SVM classifiers. In: 2020 IEEE Region 10 Symposium (TENSYMP), pp. 5–8. IEEE (2020)
9. Sarwar, B.M., Karypis, G., Konstan, J., Riedl, J.: Recommender systems for large-scale e-commerce: scalable neighborhood formation using clustering. In: Proceedings of the Fifth International Conference on Computer and Information Technology, vol. 1, pp. 291–324 (2002)
10. Shokrzade, A., Ramezani, M., Tab, F.A., Mohammad, M.A.: A novel extreme learning machine based kNN classification method for dealing with big data. Expert Syst. Appl. **183**, 115293 (2021)
11. Deng, Z., Zhu, X., Cheng, D., Zong, M., Zhang, S.: Efficient kNN classification algorithm for big data. Neurocomputing **195**, 143–148 (2016)
12. Hering, T.: Parallel execution of kNN-queries on in-memory KD trees. Datenbanksysteme für Business, Technologie und Web (BTW) 2013-Workshopband (2013)
13. Barkalov, K., Shtanyuk, A., Sysoyev, A.: A fast kNN algorithm using multiple space-filling curves. Entropy **24**(6), 767 (2022)
14. Maillo, J., Ramírez, S., Triguero, I., Herrera, F.: kNN-is: an iterative spark-based design of the k-nearest neighbors classifier for big data. Knowl.-Based Syst. **117**, 3–15 (2017)
15. Geng, Y., Yan, X.: Research on improved k-nearest neighbor algorithm based on spark platform. In: 2017 2nd Joint International Information Technology, Mechanical and Electronic Engineering Conference (JIMEC 2017), pp. 553–557. Atlantis Press (2017)
16. Lu, S., Tong, W., Chen, Z.: Implementation of the kNN algorithm based on hadoop. In: 2015 International Conference on Smart and Sustainable City and Big Data (ICSSC), pp. 123–126. IET (2015)
17. Ma, C., Chi, Y.: kNN normalized optimization and platform tuning based on hadoop. IEEE Access **10**, 81406–81433 (2022)
18. Kaminskas, M., Bridge, D.: Measuring surprise in recommender systems. In: Proceedings of the Workshop on Recommender Systems Evaluation: Dimensions and Design (Workshop Programme of the 8th ACM Conference on Recommender Systems). Citeseer (2014)
19. Hug, N.: Surprise: a Python library for recommender systems. J. Open Source Softw. **5**(52), 2174 (2020)

Siamese Networks for Unsupervised Failure Detection in Smart Industry

Angelica Liguori[1,2(✉)], Ettore Ritacco[3], Giuseppe Benvenuto[4],
Salvatore Iiritano[4], Giuseppe Manco[2], and Massimiliano Ruffolo[4]

[1] University of Calabria, Rende, Italy
angelica.liguori@dimes.unical.it
[2] Institute for High Performance Computing and Networking, Rende, Italy
{angelica.liguori,giuseppe.manco}@icar.cnr.it
[3] University of Udine, Udine, Italy
ettore.ritacco@uniud.it
[4] Revelis S.r.l., Rende, Italy
{giuseppe.benvenuto,salvatore.iiritano,massimiliano.ruffolo}@revelis.eu

Abstract. In industrial production systems, detecting malfunctions or
unexpected behavior in devices early is crucial to avoid critical situations for both production plants and workers. In this context, we propose an unsupervised anomaly detection model that analyzes streaming
data from IoT sensors installed on critical devices to identify abnormal
behavior. Our model is based on a Siamese neural network, which embeds
time series windows into a latent space, generating distance-based clusters representing normal behavior. We evaluate our model in a real case
study focused on the predictive maintenance of elevators, where sensors
measure lift oscillations during daily use. Experiments demonstrate that
the model successfully isolates anomalous oscillations, correlating them
with potential malfunctions and preventing possible faults.

Keywords: Anomaly detection · Failure detection · Fault detection ·
Time-series analysis · Embeddings · Siamese networks

1 Introduction

The current industrial production systems need to face and adapt to market
dynamics continuously. There are essentially two significant triggers of innovation in modern industrial processes. On the one hand, we experience a demand for
constant improvement of competitiveness. On the other hand, compliance with
requirements of sustainability and safety drives continuous efforts to improve the
underlying models. Addressing both of these requirements necessitates the definition and implementation of automated control and intervention systems. This
poses a significant challenge for researchers yet simultaneously offers industries
an opportunity to surpass existing limitations.

Maintenance is one of the most important activities supporting all industrial production systems, serving as both opportunity and bottleneck. While it

A. Appice et al. (Eds.): ISMIS 2024, LNAI 14670, pp. 191–200, 2024.
https://doi.org/10.1007/978-3-031-62700-2_18

prevents machinery breakdowns, service interruptions, and potential monetary penalties, it can also slow down industrial processes since it requires starting activities not strictly correlated with the core business, committing the production machinery and numerous resources in terms of technicians, tools, time, and possible costs.

Fault detection and prevention [3] is one of the most critical parts of predictive maintenance, as it identifies anomalous system behavior and avoids sudden blocks and catastrophic machine failures. Typically, the focus is on developing solutions that produce alerts upon detecting anomalies. Machine learning-based fault detection methods involve analyzing historical data to devise models able to detect system failure, typically through supervised learning approaches. In this respect, the predictive maintenance process has largely benefited from the large-scale adoption of sensor devices. With the advent of the Internet of Things (IoT) technology, numerous sensors can be installed on production devices, which produce significant amounts of data based on their sampling frequency. Despite their inherent predictive value, managing these enormous data streams poses several challenges. Firstly, sensor streams are noisy temporal sequences with excessive dimensionality and are affected by specific issues, such as burst effects and seasonality. Automatic methods are needed to filter out noise and generate effective representations usable in predictive models. Additionally, the presence of streaming data often necessitates meeting real-time requirements. Maintenance solutions must be able to provide timely warnings to maintenance experts. Failure discovery tolerance times vary across application scenarios, but there are contexts in which the production machinery risks escalating damage within an exponential decay period following an initial failure event. In addition, predictive models must ensure high predictive accuracy to be effectively exploited in real production systems. These models are based on either supervised learning, where historical time sequences are labeled with failure notification variables, or unsupervised learning, which lacks such labels. Most available data are often unlabeled, and high-quality data with human labeling could be costly and time-consuming. Moreover, the disproportion of failure instances w.r.t. the normal ones, makes their detection challenging. Consequently, unsupervised methods are often considered more appropriate.

This paper studies a specific scenario where many of the above-mentioned issues occur. The primary focus of this study is to explore how integrating sensing technology with machine learning can accurately characterize the behavior of lifting systems. The goal is to devise effective predictive maintenance strategies that ensure the stability and efficiency of elevators while mitigating the risk of future breakdowns. For the case study, we propose an innovative methodology for quickly detecting failures in unsupervised contexts. Data streams generated by sensors equipped along the elevator allow training a neural network capable of making the identification of anomalous oscillations extremely easy and fast. The proposed methodology can be embedded in an overall maintenance system to be used by the expert for continuous monitoring of the underlying system.

The novelty of the proposed solution lies in its ability to address a classification problem without supervision by leveraging a classification sub-task.

The rest of the paper is organized as follows. First, an overview of the literature on failure detection is provided in Sect. 2. We next describe the details of the proposed methodology in Sect. 3. The description of the elevator case study and the application of the proposed methodology is proposed in Sect. 4 and finally in Sect. 5 we summarize our results and discuss the impact of the proposed approach.

2   Related Work

Traditional non-supervised approaches rely on one-class classification, e.g., One-class Support Vector Machines [2,19] or distance-metrics, such as e.g. Isolation Forest [13,14]. Recently, with the rise of Deep Learning (DL), there has been a growing interest in anomaly detection models exploiting Deep Neural Networks (DNNs) [17]. In this context, solutions based on the usage of autoencoders (AE) [4] have gained increasing attention among researchers and practitioners. For example, [1,6,8,15,16,18] use autoencoders to detect faults. Here, an autoencoder is exploited to identify faults by calculating the residual error in the reconstruction of the input provided by the decoder. Since most available data does not exhibit outliers and the autoencoder is trained on them, it tends to achieve lower errors for normal data and higher for outliers. Other works are based on variants of the autoencoder, such as [9,11,12] that propose an unsupervised fault detection based on a stacked and sparse autoencoder, adapted to cope with streaming data. Most of the unsupervised failure detection systems in the literature exploit the reconstruction error to discriminate between normal and anomalous data. Usually, when the reconstruction error is used as a measure of outlierness, a threshold is defined such that the data whose reconstruction error is above the threshold is marked as an outlier. Defining a threshold is very hard, especially when there is no knowledge background. Unlike these works, our proposal exploits the philosophy of the Siamese networks [10]. The intuition is to map the data into a latent space where data belonging to the same category are grouped while being separated from data belonging to different categories.

3   Methodology

Problem Statement. Let $\mathcal{D}$ be a set of devices characterized by functionality, structure, purpose, and/or working environment. Each device $d \in \mathcal{D}$ is equipped with a set sensors $\mathcal{M}_d$ emitting temporal sequences of events $S_d = \{e_d^{(1)}, e_d^{(2)}, e_d^{(3)}, \ldots\}$, where the superscript indicates the time step. Each event $e_d^{(t)}$ is a real vector of size $|\mathcal{M}_d|$ containing the values observed by the sensors. We further assume that S_d is labeled by a specific category that characterizes the process being monitored. Categories can also refer to the underlying device (i.e., each device can be seen as a distinct category); however, they can also describe the situations that the sensors are measuring, such as an elevator moving up with a load of two persons.

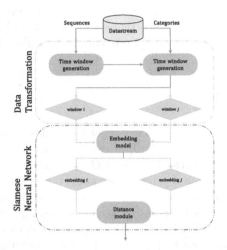

Fig. 1. Graphical representation of the proposed methodology.

Our research aims to detect anomalous situations within a sequence S_d associated with a device. The basic approach consists of building an ML-based model capable of characterizing each sequence's profile and marking anomalies whenever a given sequence does not fit that profile. However, we must face two main problems to apply such a methodology. Firstly, several different profiles can characterize each sequence. The categories encode different situations, and we can expect that events marked with a specific category are different from events labeled with a different one. Hence, we can assume that the expected number of categories is high. Further, we assume that $|\mathcal{M}_d|$ is large but the overall number of sequences is small compared to the number of categories. We can expect that each sequence is labeled with almost a different category. This clearly poses a problem in the learning stage since it is impossible to build specific profiles due to the lack of sufficient training data for each category.

We propose a model that reconstructs the different profiles by exploiting the length of the sequences and their associated category to map sequence fragments within an embedding space. As shown in Fig. 1, our methodology is divided into two parts. The first part consists of a data transformation approach that allows the definition of a classification problem. This classification problem can trigger the second part of the methodology, which consists of exploiting a modular Siamese network able to map the input (sequence fragments) into data points lying in a latent space. The latent space has a geometric interpretation: points that are close in the space correspond to devices that exhibit, in some time interval of their working process, similar behavior. The core concept is that clusters of these latent data points represent the different working modes of the target devices. Any element on the edge of a cluster or out of all of them can be highlighted to maintenance experts for further investigation. This allows approaching situations typical in many industrial processes, where critical anomalies (e.g., failures) are extremely rare.

Data Transformation. Since sequences may have different sizes, our methodology applies a sliding window extraction procedure to generate, for each sequence, a set of fixed-size observation windows. The sliding windows approach (both in the learning phase and in the prediction) allows to mitigate the noise of the data, enabling the calculation of aggregate statistics, and makes it possible to mark the presence or absence of faults in the time interval, offering a suitable margin of tolerance to the prediction. This procedure is characterized by the *size* and the *shift*. The former represents the number of events in each window, whereas the latter is the number of events between the beginning of a window and the beginning of the next. Contiguous windows partially overlap if the window shift is lower than the window size. We assume that the temporal interval between two events is almost constant in our setting.

Each window is a time frame that partially describes the target devices' behavior during the observation interval. From here on, $W = \{w_1, w_2, \ldots\}$ will be the set of all the time windows, and we will use the function $cat(w_i)$ for denoting the category relative to the device that generated the i-th time window.

Siamese Network for Embedding Generation. Given two arbitrary time windows w_i and w_j, the goal is to predict if they belong to the same category, i.e., whether $cat(w_i) = cat(w_j)$. To do this, we propose a Siamese Neural Network, introduced by [5] for addressing the problem of signature verification via image matching. Siamese Networks consist of two or more sub-networks processing two or more distinct inputs. The sub-networks are intended to be identical as they share not only their structure but also their weights. In this sense, supposing to have two inputs (and therefore two sub-networks) x_1 and x_2 (in our case, the two time windows w_i and w_j), if x_1 and x_2 are identical or very similar, their embeddings, named z_1 and z_2, will be similar as well since the two sub-networks share the weights and the two inputs have the same discriminative characteristics. The two embeddings, then, are fed into a distance module that calculates their distance, e.g., euclidean distance, to provide their similarity score. We exploit the philosophy of the Siamese network, shown in Fig. 2, and composed of two modules. The first is the *Embedding model* that maps a time window into the high-dimensional latent space. The Embedding subnet is a sequential model composed of a recurrent neural network (we exploited an LSTM layer [7]), for catching the time dependencies within each window and a feed-forward neural network (a dense embedding layer) that generates the data points in the latent space. The second module, the *Distance subnet*, computes the network output as the Euclidean distance between two embeddings.

The working flow of the whole architecture is described as follows. The network's input is a pair of time windows, w_i and w_j, randomly sampled from W. The set of all the built-up pairs called W_s, suitably represents each category, pro-

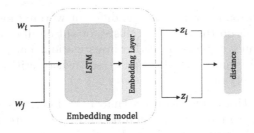

Fig. 2. Network architecture.

viding a sufficient number of positive (windows belonging to the same category) and negative (windows belonging to different categories) comparisons. Both w_i and w_j pass through the Embedding subnet that produces the two embeddings, respectively z_i and z_j. These two embeddings are the input of the Distance subnet that computes and returns their Euclidean distance. The loss function we chose is the following:

$$loss = \frac{1}{|\mathcal{W}_s|} \sum_{w_i,w_j \in \mathcal{W}_s} y_{i,j} \cdot f(w_i,w_j) - (1 - y_{i,j}) \cdot \log \left(1 - e^{-f(w_i w_j)}\right) \quad (1)$$

where $y_{i,j}$ is equal to 1 if $cat(w_i) = cat(w_j)$, 0 otherwise, and $f(w_i, w_j)$ is a function that computes the Euclidean distance between the embeddings z_i and z_j of the time windows w_i and w_j, respectively.

The proposed loss function encourages the network to generate pairs of embeddings close in the latent space if $y_{i,j} = 1$. Conversely, the network will produce distant embeddings if the two categories differ.

4   Experimental Assessment

Our methodology was empirically evaluated on a real industrial use case, whose objective is to monitor the health status and the working process of an elevator in an office building to detect anomalous behaviors, i.e., failures in the system. Our goal is to answer the following research question:

– **RQ1.** Is the model able to perfectly separate the different normality patterns?
– **RQ2.** Is the Siamese-based approach effective in detecting anomalous situations?

Industrial Use Case. Elevators are complex lifting systems that can account for different sophisticated architectures. A hydraulic elevator works through an oil injection system towards a cylinder. It has a pulley in the upper part of the rod on which cables are firmly anchored to the car's run. The pressure raises the head, and the descent occurs with the gradual emptying of the fluid. The electric elevator is made up of ropes tied to counterweights, and it operates by means of an electric winch that directs the actions from the cab to the ropes. In both cases, the elevator is guided by guides subject to oscillations, and the elevator car is connected to a rope or piston and is lifted thanks to a push. To safely use an elevator, the system must be subject to strict periodic maintenance rules, which will be entrusted to qualified personnel. The objective of maintenance is aimed at ensuring the stability and efficiency of the elevator, as well as preventing future breakdowns.

When a failure occurs, it usually coincides with non-optimal operation in the long run-up to failure, anomalies that are almost imperceptible. It is easy to understand how the first malfunction symptoms manifest in the oscillations of the guides. To ensure the operation of the elevator and avoid breakdowns, the elevator's progress must be constantly monitored. Good quality maintenance of elevators allows us to prevent any anomaly well in advance: from devices that collect sensor data and communicate it to whoever has to analyze it, it comes to more sophisticated tools that collect, process data and use them to indicate the need to act when necessary, but above all to learn from the data collected, diagnose possible scenarios and anticipate unwanted ones.

The objective is to propose installing a sensor system placed on the guides, capable of evaluating the status of the elevator and detecting/reporting problems and anomalies in real-time. Then, from the analysis of the data relative to the oscillations of the guides on which the elevator slides, we obtain an image of the general condition of the elevator. Leveraging Machine Learning and IoT technology facilitates the analysis of collected data, enabling the prediction and detection of failures, the planning of maintenance activities based on the actual condition of component usage, and the assessment of elevator health. Currently, maintenance plans are executed preventively to prolong component lifespan, yet they often overlook the current health status of the elevator. This oversight results in wasted time and resources, with no assurance against component damage.

Table 1. Sensor descriptions.

Sensor ID	Type	Location	Sampling frequency [Hz]
S1	Accelerometer	Ground	200
S2	Accelerometer	Shaft guide	200
S4	Accelerometer/Inclinometer	Shaft guide	200
S6	Accelerometer/Inclinometer	Shaft guide	200
S16	Accelerometer	Engine	4000
S21	Accelerometer	Cabin	200
S22	Magnetometer	Cabin	100

Dataset. The sensor system placed in the elevator (see Fig. 3 and Table 1 for details) records the movements in the x, y, and z axes of the oscillations of the

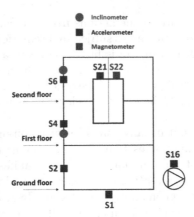

Fig. 3. Graphical representation of the elevator, highlighting the position of installation and the type for each sensor, according to the description in Table 1.

elevator guides and cabin, the inclination of the cabin and the magnetic field intensity in a relational data structure. Each record starts at a certain moment in nanoseconds and stops at a later point in time, collecting all the information that characterizes a certain set of signals in that time interval. Moreover, each record provides further information about the operation status of the elevator, highlighting its position, movement, and door status for each time step.

In this case study, we applied our methodology on a limited set of similar elevators in an office building. Thus, their sequences were split into fixed size time windows that were randomly paired up to be processed by the Siamese network. The category labels we used were the operational modalities: (i) Stationary; (ii) Moving up; (iii) Moving down; (iv) Opening doors; (v) Closing doors. Each of these modalities is further labeled by contour conditions representing the load (number of people) in the lift.

Results. Experiments show that, for each elevator, the network is able to map the different normal behaviors into clusters of the latent space. To allow a friendly visualization of the embeddings, we used the t-distributed stochastic neighbor embedding (t-SNE) library [20] which is a well-suited technique for the reduction of high-dimensional data into a low-dimensional space of 2 or 3 dimensions, helping to identify relevant patterns and preserving the local structure and data point relative distances: embeddings that are close in the high-dimensional space are modeled by nearby points in the low-dimensional space while embedding that are structurally different are modeled by distant points.

We visually show our results to answer to **RQ1**. Specifically, a 2D t-SNE plot of the learned behavior clusters is shown in Fig. 4a, where each point is related to the embedding of a time window, while colors indicate the categories the windows belong to. As can be seen, the network found 7 different behavioral clusters in which a specific color is dominant. The partial color overlapping is due to two factors. On the one hand, category labels were noisy since they were produced by humans with external chronometers, thus making the initial and final time windows imprecise for each category session. On the other hand, the sensors we used could not find appreciable differences in vibration when doors were opening or closing. To observe the model's capability to isolate anomalies and answer to **RQ2**, we performed new experiments in which passengers stopped and restarted several times the elevator movements or produced (weak) unexpected vibrations. We produced 30 anomalous time series and, as shown in Fig. 4b, the 2d t-SNE transformation of the embeddings, provided by the network, generated points, labeled as *Noise*, that are outside the clusters.

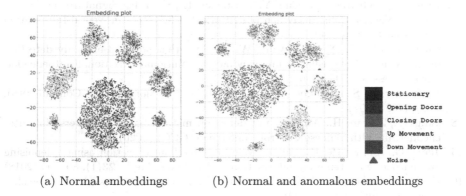

(a) Normal embeddings   (b) Normal and anomalous embeddings

Fig. 4. Embedding space

5 Conclusions

We proposed a new methodology for addressing the early detection problem of faults in critical devices. The methodology is designed to effectively work in settings where explicit information about previous failures is missing. Assuming that failures are rare events during the lifetime of a device, the proposed methodology supports a maintenance expert in easily identifying them as anomalous elements that are distant from all the clusters of normal behavior. Experiments on a real case study showed the capability of the proposal in effectively isolating anomalous time frames, suggesting that its application fields can span many different and more complex scenarios. The significant advantages of this methodology can be summarized as follows: (*i*) It leverages an unsupervised technique, enabling operation in environments with limited available information about past critical events; (*ii*) The inference phase of the model is highly efficient: the primary bottleneck lies in the sampling strategy used to pair up the time frames during the training phase, which can be executed offline; (*iii*) It demonstrates robustness even in noisy and imprecise environments.

References

1. Alfeo, A.L., Cimino, M.G.C.A., Manco, G., Ritacco, E., Vaglini, G.: Using an autoencoder in the design of an anomaly detector for smart manufacturing. Pattern Recogn. Lett. **136**, 272–278 (2020)
2. Mennatallah Amer, Markus Goldstein, and Slim Abdennadher. Enhancing one-class support vector machines for unsupervised anomaly detection. In *ACM SIGKDD*, 2013
3. Amruthnath, N., Gupta, T.: A research study on unsupervised machine learning algorithms for early fault detection in predictive maintenance. In: ICIEA (2018)
4. Bank, D., Koenigstein, N., Giryes, R.: Autoencoders (2020)
5. Bromley, J., Guyon, I., LeCun, Y., Säckinger, E., Shah, R.: Signature verification using a "Siamese" time delay neural network. In: Advances in Neural Information Processing Systems, vol. 6, (1993)
6. Chen, T., Liu, X., Xia, B., Wang, W., Lai, Y.: Unsupervised anomaly detection of industrial robots using sliding-window convolutional variational autoencoder. IEEE Access **8**, 47072–47081 (2020)
7. Hochreiter, S., Schmidhuber, J.: Long short-term memory. Neural Comput. **9**(8), 1735–1780 (1997)
8. Jian, W., Zhiyan, H.: A novel fault detection method based on adversarial auto-encoder. In: 39th Chinese Control Conference (CCC) (2020)
9. Jiang, G., Xie, P., He, H., Yan, J.: Wind turbine fault detection using a denoising autoencoder with temporal information. IEEE Trans. Mech. **23**(1), 89–100 (2018)
10. Koch, G., Zemel, R., Salakhutdinov, R.: Siamese neural networks for one-shot image recognition. arXiv (2015)
11. Liang, X., Duan, F., Bennett, I., Mba, D.: A sparse autoencoder-based unsupervised scheme for pump fault detection and isolation. Appl. Sci. **10**(19), 6789 (2020)
12. Lindemann, B., Fesenmayr, F., Jazdi, N., Weyrich, M.: Anomaly detection in discrete manufacturing using self-learning approaches. Procedia CIRP **79**, 313–318 (2019)
13. Liu, F.T., Ting, K.M., Zhou, Z.-H.: Isolation forest. In: IEEE International Conference on Data Mining (2008)
14. Liu, F.T., Ting, K.M., Zhou, Z.-H.: Isolation-based anomaly detection. ACM Trans. Knowl. Discov. Data **6**(1), 1–39 (2012)
15. Oliveira, D.F.N., et al.: Evaluating unsupervised anomaly detection models to detect faults in heavy haul railway operations. In: IEEE International Conference on Machine Learning and Applications (ICMLA) (2019)
16. Oliveira, D.F.N., et al.: A new interpretable unsupervised anomaly detection method based on residual explanation. IEEE Access **10**, 1401–1409 (2021)
17. Pang, G., Shen, C., Cao, L., Van Den Hengel, A.: Deep learning for anomaly detection: a review. ACM Comput. Surv. **54**(1), 1–38 (2021)
18. Principi, E., Rossetti, D., Squartini, S., Piazza, F.: Unsupervised electric motor fault detection by using deep autoencoders. IEEE/CAA J. Autom. Sin. **6**(2), 441–451 (2019)
19. Schölkopf, B., Williamson, R.C., Smola, A.J., Shawe-Taylor, J., Platt, J.C., et al.: Support vector method for novelty detection. In: NIPS, vol. 12 (1999)
20. van der Maaten, L., Hinton, G.: Visualizing data using t-SNE. J. Mach. Learn. Res. **9**, 2579–2605 (2008)

Adaptive Forecasting of Extreme Electricity Load

Omar Himych[1], Amaury Durand[1,3]([✉]), and Yannig Goude[1,2]

[1] Électricité de France R&D, Bd Gaspard Monge, 91120 Palaiseau, France
amaury.durand@edf.fr
[2] Laboratoire de Mathématique d'Orsay, Université Paris-Saclay,
307 Rue Michel Magat, 91400 Orsay, France
[3] Sorbonne Université, 4 Place Jussieu, 75005 Paris, France

Abstract. Electricity load forecasting is a necessary capability for power system operators and electricity market participants. Both demand and supply characteristics evolve over time. On the demand side, unexpected or extreme events as well as longer-term changes in consumption habits affect demand patterns. On the production side, the increasing penetration of intermittent power generation significantly changes the forecasting needs. We address this challenge in three ways. First, we consider probabilistic (quantile) rather than point forecasting; indeed, uncertainty quantification is required to operate electricity systems efficiently and reliably. The probabilistic forecasts are generated using both linear and non-linear quantile regressions applied to the residuals of the mean forecasting model. Second, our approach is **Adaptive**; we have developed models that incorporate the most recent observations to automatically respond to changes in the underlying process. Our adaptation methodology leverages the Kalman filter, which has previously been successfully employed for adaptive load forecasting, as well as Online Gradient Descent - a combination of an incremental strategy and pinball loss. Third, we extend the adaptive setting to **Extreme** scenarios by using the aforementioned methods to compute an adaptive threshold used as a reference in recently developed machine learning models targeting extreme values. Finally, we apply our different approaches on the french daily electricity consumption as use case.

Keywords: Electricity demand · renewable production · probabilistic forecast · adaptive · extreme values

1 Introduction

Forecasting electricity demand is fundamental in the process of maintaining supply-demand balance. This permanent equilibrium is necessary to maintain a reliable supply of electricity and to avoid damaging infrastructure. As electricity cannot be stored on a large scale, forecasts are crucial to informing production planning. This necessity explains why energy forecasting has gathered so

A. Appice et al. (Eds.): ISMIS 2024, LNAI 14670, pp. 201–215, 2024.
https://doi.org/10.1007/978-3-031-62700-2_19

much attention from the time series and forecasting community [12]. The recent increase in electricity prices in Europe, with prices rising to more than 400 euros per Mwh [28] further emphasizes the importance of demand forecast quality as small forecasting errors can cost millions euros to utilities.

Forecasting models are random estimates of quantities such as demand or renewable production. It is thus essential to provide insights on the distribution to take decisions such as positions on the energy market or grid optimization. Previous works on probabilistic forecasting were proposed in the electricity demand literature. Following the good performances of Generalized Additive Models (GAMs) for conditional mean forecasting (see e.g. [5,19]), [6,7] and [8] respectively proposed quantile GAMs and a variant of GAMLSS (Location Scale and Shape) to model the overall distribution. While these models perform well on stationary data, electricity data often evolve over time. For example, the COVID crisis have modified our electricity consumption habits [13] and the recent increase in electricity prices in Europe have entailed a drop in consumption [4]. These changes cannot be captured by traditional offline quantile regression methods. [25] proposed two methods to adapt probabilistic forecasts: the Kalman filter coupled with GAMs and Online Gradient Descent. However, due to the scarcity of data points exceeding extreme levels, these methods can fail at estimating conditional quantiles at extreme levels, which is an important task in risk assessment for rare events. Asymptotic results from extreme value theory can however be used to extrapolate beyond the range of the data. In [3], the Generalized Pareto Distribution (GPD) is proposed to forecast the deviation of the net demand (consumption minus renewable production) over a reference quantile using a GAMLSS approach (the parameters of the Pareto distribution depends on covariates via a GAM equation). In the same direction, several machine learning approaches have been introduced recently to model extreme quantiles as exceedance over a reference quantile, see [9] for random forest estimates, [17] for neural networks, [21] for gradient boosting machine. However, all these methods have been presented in an offline setting, assuming an implicit stationarity of the data generating process.

In this paper, we propose the first adaptation of extremal machine learning models to the non-stationary case. The novelty of the work lies in the combination of different existing tools (recent advances in adaptive quantile forecasting and extreme quantile regression) to achieve an adaptive extreme forecasting model. We naturally follow a three steps approach. The first two steps follow the ideas introduced in [25] and yield an adaptive quantile forecast at an intermediate level. In the last step, we use recent extremal machine learning models to fit a GPD on the exceedances over the intermediate quantile forecast. We then derive quantile forecasts at extreme levels. Section 2 gathers the theoretical background and models definitions for intermediate quantile forecasting. Extreme quantile forecasting is discussed in Sect. 3. Recent contributions from the extreme machine learning literature are gathered in Sect. 3.1 and Sect. 3.2 presents our main contribution, namely, the definition of adaptive extreme quantile models combining methods from the previous sections. Finally, in Sect. 4, we apply our method to french electricity load forecasting and compare it with relevant benchmarks used in operations by EDF (Electricité De France) forecasting team.

2   Intermediate Quantile Models

Assume we observe a stream of data $y_t \in \mathbb{R}$ and $x_t \in \mathbb{R}^d$ for $t \geq 1$ with $d \geq 1$. Let $\mathcal{F}_t = \sigma(x_1, y_1, \ldots, x_t, y_t)$ be the natural filtration (modeling the information contained in the observations up to time t). Then, we estimate the conditional distribution using the following two settings:

- **Offline** or *Batch* (Sects. 2.1 and 3.1). The model is learned on a training period, for instance on data up to time n_{train}. We estimate $\mathcal{L}(y_t \mid x_t, \mathcal{F}_{n_{\text{train}}})$, the conditional distribution of y_t given $x_t, \mathcal{F}_{n_{\text{train}}}$.
- **Online** or *Adaptive* (Sects. 2.2 and 3.2). The model is learned sequentially. We estimate $\mathcal{L}(y_t \mid x_t, \mathcal{F}_{t-1})$ at each time t.

2.1   Offline Quantile Regression Models

In this section we present three offline quantile regression approaches for probabilistic forecast: gam.qr, qgam and gam.qgam. The first one (gam.qr) is a two step approach proposed in [7] to solve the tricky issue of GAMs for quantile regression, at the cost of assuming that the smooth GAM effects are the same (up to a multiplicative factor) for the mean and all quantiles. The second approach (qgam) was developed in [6] and removes this restriction by minimizing a smoothed version of the pinball loss instead of a likelihood function. This approach achieves accurate quantile estimates and credible interval coverage. The third one is a combination of the others and is, to our best knowledge, novel.

We now formalize these approaches by first considering a GAM estimation of the conditional mean based on the following model.

$$y_t = \sum_{j=1}^{d} f_j(x_{t,j}) + \epsilon_t, \quad \epsilon_t \sim \mathcal{N}(0, \sigma^2), \tag{1}$$

where the smooth effects f_j are built using linear combinations of spline bases, see [27]. An optimization of the smooth effects on the training set yields an estimate $\hat{y}_t$ of $\mathbb{E}[y_t \mid x_t, \mathcal{F}_{n_{\text{train}}}]$. Similarly, the qgam method is based on a smooth additive representation of the conditional quantile instead of the conditional mean. Namely,

$$\mathbb{Q}_q(y_t \mid x_t) = \sum_{j=1}^{d} f_{j,q}(x_{t,j}), \tag{2}$$

where $\mathbb{Q}_q(y \mid \mathcal{F})$ denotes the conditional quantile at level $q \in (0, 1)$ of $\mathcal{L}(y \mid \mathcal{F})$. The smooth effects $f_{j,q}$ are obtained by minimizing a smooth version of the pinball loss called the Extended Log-F (see [6] for details). Again, an optimization of these effects on the training set yields an estimate of $\mathbb{Q}_q(y_t \mid x_t, \mathcal{F}_{n_{\text{train}}})$.

The two other approaches consist in estimating the conditional quantiles of the residuals $r_t := y_t - \hat{y}_t$ of (1). This takes into account the fact that r_t is not exactly equal to ϵ_t and may therefore also depend on the covariates x_t or could be non Gaussian. An estimate of the conditional quantiles of y_t are then obtained using the relation $\mathbb{Q}_q(y_t \mid x_t) = \hat{y}_t + \mathbb{Q}_q(r_t \mid x_t)$. In the `gam.qgam` method, we use a smooth additive model as in (2) for $\mathbb{Q}_q(r_t \mid x_t)$. In the `gam.qr` method, we assume the linear relation $\mathbb{Q}_q(r_t \mid x_t) = \beta_q^\top f(x_t)$ where $\beta_q \in \mathbb{R}^{d+1}$ and

$$f(x_t) := (\overline{f}_1(x_{t,1}), \dots \overline{f}_d(x_{t,d}), 1)^\top, \tag{3}$$

is the concatenation of the standardized smooth effects f_j of (1). The vector β_q is then estimated by minimizing the pinball loss on the training set [15], that is

$$\hat{\beta}_q \in \underset{\beta \in \mathbb{R}^{d_o}}{\operatorname{argmin}} \sum_{t=1}^{n_{\text{train}}} \rho_q(y_t - \hat{y}_t, \beta^\top f(x_t)) \text{ with } \rho_q(y, \hat{y}_q) := (\mathbb{1}_{\{y < \hat{y}_q\}} - q)(\hat{y}_q - y). \tag{4}$$

2.2 Online Quantile Regression Models

We present here the two adaptive models proposed in [25] to estimate the conditional quantile $\mathbb{Q}_q(y_t \mid x_t, \mathcal{F}_{t-1})$.

Kalman Based Model. The main idea is to plug a linear Kalman filter on the GAM effects in the same philosophy as for the `gam.qr` approach. We refer to this method as `gkal`. More precisely, we consider the following linear Gaussian state-space model.

$$\theta_t - \theta_{t-1} \sim \mathcal{N}(0, Q) \quad \text{and} \quad y_t - f(x_t)^\top \theta_t \sim \mathcal{N}(0, \sigma^2), \tag{5}$$

where $f(x_t)$ is defined in (3), Q is the state noise covariance matrix, σ^2 the space noise variance and we assume that the noises are independent. Estimation in a linear Gaussian state-space model with known variances has been optimally solved in [14] providing recurrent equation for θ_t. More formally, it is well known that we have $\theta_t \mid \mathcal{F}_{t-1} \sim \mathcal{N}(\hat{\theta}_t, P_t)$ where $\hat{\theta}_t$ and P_t are update online as

$$P_{t+1} = P_{t|t} + Q \quad \text{and} \quad \hat{\theta}_{t+1} = \hat{\theta}_t - \frac{P_{t|t}}{\sigma^2}\left(f(x_t)(f(x_t)^\top \hat{\theta}_t - y_t)\right),$$

with

$$P_{t|t} := P_t - \frac{P_t f(x_t) f(x_t)^\top P_t}{f(x_t)^\top P_t f(x_t) + \sigma^2}.$$

From the state posterior distribution and the observation distribution (5), we deduce an estimate of $\mathcal{L}(y_t \mid x_t, \mathcal{F}_{t-1})$.

As explained in [25], the variances σ^2 and Q play a crucial role in the performance of the Kalman filter similarly to the gradient steps in online gradient descent. In our case study, we use the iterative grid search approach implemented in the `viking` package [22] as in [16,23,25] providing a diagonal sparse matrix Q.

A more sophisticated Variance Tracking algorithm has however been developed in [24] where Q and σ are also time varying and could potentially switch automatically between stable periods and more volatile ones. For future work, we will leave this alternative open.

Online Gradient Descent. The Kalman filter is an optimal gradient descent solution for Gaussian data. To relax this assumption, [25] proposes to use Online Gradient Descent (OGD). This consists in updating the parameters of an offline linear quantile model (typically `gam.qr`) according to the gradient of the pinball loss over time. Suppose we have previously estimated an offline quantile regression model and we can reduce that model to a set of vectorial parameters $\hat{\beta}_q$ for each quantile; for example, the case `gam.qr` where our model is $\hat{y}_t + \hat{\beta}_q^\top f(x_t)$. To update $\hat{\beta}_q$ as a function of time, we use OGD to recursively estimate the parameter vector $\hat{\beta}_{t,q}$. At each time step, we update it with a step in the direction opposite to the gradient of the loss, that is

$$\hat{\beta}_{t+1,q} = \hat{\beta}_{t,q} - \alpha \frac{\partial \rho_q(y_t - \hat{y}_t, \beta^\top f(x_t))}{\partial \beta}\bigg|_{\beta=\hat{\beta}_{t,q}}.$$

To tune the gradient step size α online, [25] proposes to follow an expert aggregation strategy as in [30]. First, we run the OGD on a grid $(\alpha_k)_{1 \le k \le K}$. At each time t, it produces a forecast $\hat{y}_{t,q}^{(k)}$ for each step size α_k. Second, we combine these forecasts using the Bernstein Online Aggregation (BOA) method introduces in [26]. The principle of aggregation is to forecast $\hat{y}_{t,q} = \sum_{k=1}^K p_{t,q}^{(k)} \hat{y}_{t,q}^{(k)}$, where the weights $p_{t,q}^{(k)}$ are obtained sequentially. The properties obtained from the online learning literature guarantee that the total loss of the aggregation has a small regret compared to the total loss of the best expert. We refer to this method as `gam.ogd`.

Other Methods. The other methods aim to relax the Gaussian assumption by fitting offline quantile regression on the Kalman residuals similarly to the methods presented in Sect. 2.1. This approach results in two variants: `gkal.qr` and `gkal.qgam` which respectively fit a linear quantile regression and a QGAM on the Kalman residuals. Finally, we also include a last method, referred to as `gkal.ogd`, consisting in an online estimation (by OGD with BOA) of quantiles based on the Kalman residuals. The `gkal.qr` and `gkal.ogd` methods are introduced in [25] and, to our best knowledge, the combination of Kalman filtering with QGAM, i.e. `gkal.qgam`, is novel.

3   Adaptive Extreme Quantile Regression

To accurately estimate the extreme quantiles, we rely on extreme value theory since usual quantile regression methods may not be suitable for extreme quantiles where there are few or no training data points. The Pickands-Balkema-De Haan theorem [2,18] states that, under regularity assumptions on the tail behavior of a random variable, the law of its exceedances above a large threshold can be approximated with a Generalized Pareto Distribution (GPD). Extreme quantile models therefore rely on the assumption that, for a well chosen threshold $u_t \in \mathbb{R}$, the distribution of $y_t - u_t$ given $y_t > u_t$ and x_t is a GPD with parameters $\theta(x_t) = (\sigma(x_t), \xi(x_t))$, that is

$$\mathbb{P}\left(y_t - u_t > z \mid y_t > u_t, x_t\right) = \left(1 + \frac{\xi(x_t)}{\sigma(x_t)} z\right)_+^{-1/\xi(x_t)}. \qquad (6)$$

Under this assumption, we get that for any $y \in \mathbb{R}$,

$$\mathbb{P}\left(y_t > y \mid x_t\right) = \mathbb{P}\left(y_t > u_t \mid x_t\right) \left(1 + \frac{\xi(x_t)}{\sigma(x_t)}(y - u_t)\right)_+^{-1/\xi(x_t)},$$

which, taking $u_t = \mathbb{Q}_{q_0}(y_t \mid x_t)$ for an intermediate quantile level $q_0 \in (0,1)$, gives that for any $q > q_0$,

$$\mathbb{Q}_q(y_t \mid x_t) = \mathbb{Q}_{q_0}(y_t \mid x_t) + \frac{\sigma(x_t)}{\xi(x_t)}\left(\left(\frac{1-q}{1-q_0}\right)^{1/\xi(x_t)} - 1\right). \qquad (7)$$

Recently introduced Extremal Machine Leaning Models (EMLM) propose to estimate the conditional quantile at extreme levels by (7) where the intermediate quantile $\mathbb{Q}_{q_0}(y_t \mid x_t)$ is estimated using a standard quantile regression approach and the GPD parameters $\theta(x_t)$ are estimated by a machine learning model aiming at maximizing the GPD likelihood.

3.1   Extremal Machine Learning Models

In this section, we present briefly the different EMLMs used in our study to estimate the GPD parameters $\theta(x_t)$ of (6) based on a well chosen threshold u_t (typically based on one of the quantile regression models of Sect. 2.1). In the following, we let $z_t = y_t - u_t$.

- The benchmark method, referred to as gpd, provides a constant parameter $\theta(x_t) = \theta$ which is estimated from the z_t's with the maximum likelihood method of [20].
- The Extremal Random Forest (erf) of [9] uses similarity weights obtained by a Generalized Random Forest (GRF, see [1]) to fit a conditional GPD. Trained on the data up to n_{train}, the GRF provides weights $w_i(x)$, $i = 1, \cdots, n_{\text{train}}$ which indicate how relevant the i-th training sample is to estimate conditional intermediate quantiles at x. The GPD parameters $\theta(x)$ are obtained by minimizing the following weighted (negative) log-likelihood $\sum_{i=1}^{n_{\text{train}}} w_i(x)\ell_\theta(z_i)\mathbb{1}_{\{z_i>0\}}$ where ℓ_θ is the negative log-likelihood (or deviance) of the GPD. In our experiments we use the penalized version discussed in [9, Section 3.2] where a penalization is added to the weighted negative log-likelihood to control the variance of the shape parameter ξ of the GPD.

- The Gradient Boosting for Extremes (gbex) of [21] consists in minimizing the negative log-likelihood of the GPD model using stochastic gradient descent by iterating a tree-based learner in a gradient boosting fashion.
- The Generalized Additive Extreme Value Models (evgam) of [29] assumes that the GPD parameters follow a GAM equation similarly to (1).
- The Extremal Quantile Regression Neural Network (eqrn) of [17] proposes to model $\theta(x)$ with a neural network. In particular, the authors propose to use an LSTM or GRU architecture to take into account the sequential nature of the data. In this case, x_t is replaced by $(x_{t-s})_{0 \leq s \leq h}$ in (6) and (7) and the GPD parameters also depend on the lagged covariates up to lag h.

3.2 Adaptive Extremal Machine Learning Models

The methods listed in Sect. 3.1 rely on the implicit assumption that the data are stationary and therefore provide offline predictions. Although the EQRN method with LSTM or GRU architectures uses lagged covariates as input, it is not an online method since it only targets the quantiles of y_t given $x_{t-h}, \cdots, x_t$ and not given x_t and $\mathcal{F}_{t-1}$. It is the additional information provided by $y_1, \cdots, y_{t-1}$ which makes the strength of online methods. This motivates the need for online extreme quantile regression models which, to our up-to-date knowledge, have not yet been developed. To address this issue, we identify two approaches. The first one yields a partially adaptive EMLM. It is based on the intuition that the estimation of extreme quantiles using (7) depends strongly on the estimation of the intermediate quantile. Hence, we propose a simple method consisting in replacing $\mathbb{Q}_{q_0}(y_t \mid x_t)$ by an online estimator of $\mathbb{Q}_{q_0}(y_t \mid x_t, \mathcal{F}_{t-1})$ as provided in Sect. 2.2 while still relying on an offline estimation of the GPD parameters. The second, more involved, approach yields a fully adaptive EMLM based on the assumption that

$$\mathbb{P}\left(y_t - u_t > z \mid y_t > u_t, x_t, \mathcal{F}_{t-1}\right) = \left(1 + \frac{\xi_t(x_t)}{\sigma_t(x_t)} z\right)_+^{-1/\xi_t(x_t)} ,$$

where $\theta_t(x) = (\sigma_t(x), \xi_t(x))$ and u_t depend on $\mathcal{F}_{t-1}$. For example, taking $u_t = \mathbb{Q}_{q_0}(y_t \mid x_t, \mathcal{F}_{t-1})$, we get the online counterpart of (7),

$$\mathbb{Q}_q(y_t \mid x_t, \mathcal{F}_{t-1}) = \mathbb{Q}_{q_0}(y_t \mid x_t, \mathcal{F}_{t-1}) + \frac{\sigma_t(x_t)}{\xi_t(x_t)} \left(\left(\frac{1-q}{1-q_0}\right)^{1/\xi_t(x_t)} - 1\right).$$

In this paper, we only consider the first approach and leave the second one for future contributions. We summarize the adaptive EMLM framework in Algorithm 1. This encapsulates all the above approaches as follows.

- Offline EMLM: for all $t \geq 1$ we take $\hat{Q}_{t,q_0} = \hat{Q}_{q_0}$ and $\hat{\theta}_t = \hat{\theta}$ which are fixed over time and fitted on a training set using, for example, the methods of Sects. 2.1 and 3.1. In this case the adaptation phase of Algorithm 1 has no effect.

- Partially adaptive EMLM: for all $t \geq 1$, $\hat{\theta}_t = \hat{\theta}$ fixed over time and fitted on a training set using, for example, the methods of Sect. 2.1. On the contrary, the update of $\hat{Q}_{t+1,q_0}$ in Algorithm 1 is done using an adaptive method such as the ones of Sect. 2.2.
- Fully adaptive EMLM (not studied here): Both the estimators $\hat{Q}_{t,q_0}$ and $\hat{\theta}_t$ are updated online.

It should be noted that a fully adaptive EMLM can be achieved by simply updating $\hat{\theta}_{t+1}$ in Algorithm 1 with an offline model fitted on the data $(x_s, y_s)_{s \leq t}$. This approach is, however, usually too time consuming to be of any practical interest.

Algorithm 1: Adaptive EMLM

Input : Data $(x_t, y_t)_{t \geq 1}$, two quantile levels $0 < q_0 < q < 1$ and initial estimators $\hat{Q}_{1,q_0}$ and $\hat{\theta}_1$ for the intermediate quantile and GPD parameters.

for $t \geq 1$ **do**

    /* Prediction phase   */

    Observe x_t

    Estimate the intermediate quantile $\mathbb{Q}_{q_0}(y_t \mid x_t, \mathcal{F}_{t-1})$ by $\hat{Q}_{t,q_0}(x_t)$

    Estimate the parameters $\hat{\theta}_t(x_t)$ of the GPD for $\mathcal{L}(y_t - \hat{Q}_{t,q_0}(x_t) \mid x_t, \mathcal{F}_{t-1})$

    Deduce an estimate of the extreme quantile $\mathbb{Q}_q(y_t \mid x_t, \mathcal{F}_{t-1})$ using (7)

    /* Adaptation phase   */

    Observe y_t

    Update the estimator function $\hat{Q}_{t+1,q_0}$ using observations $(x_s, y_s)_{s \leq t}$

    Update the estimator function $\hat{\theta}_{t+1}$ using exceedances $(y_s - \hat{Q}_{s,q_0}(x_s))_{s \leq t}$

end

---

4 Case Study

In this section, we apply the proposed methods on the french daily electricity consumption, an open data set produced by RTE the french TSO. We evaluate their performances on the year 2020 to highlight the benefits of adaptive prediction in challenging situations like the COVID-19 crisis, see Fig. 1 where we observe an unusual break in the lockdown of March 2020. In the following the target variable y_t denotes the daily electrical load at day t and the covariates x_t depend on the model used. We split our data in three sets as illustrated in Fig. 1. The observations from 2012 to 2017 are used to fit the mean models gam and gkal. For the second model, the set is separated into two subsets: one to fit the gam model and one to fit the covariance parameters of the Kalman on the residuals of the GAM. Then the observations from 2018 and 2019 are used to fit the intermediate and extreme quantile models. Finally the models are evaluated on the year 2020.

For the mean estimation, we consider the Root Mean Squared Error (RMSE) defined on a test set $\mathcal{T}$ as RMSE $:= \sqrt{\frac{1}{|\mathcal{T}|} \sum_{t \in \mathcal{T}} (y_t - \hat{y}_t)^2}$. We evaluate the quantile forecasts based on calibration (also known as reliability) and sharpness (see

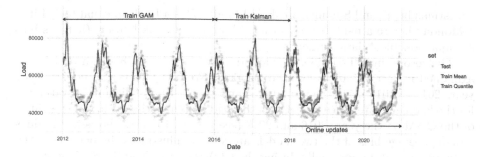

Fig. 1. Daily electricity load and 14 days moving average.

[11]). Calibration of the q-quantile forecasts $(\hat{y}_{t,q})_{t \in \mathcal{T}}$ is assessed by computing the coverage frequency

$$\frac{1}{|\mathcal{T}|} \sum_{t \in \mathcal{T}} \mathbb{1}_{\{y_t \leq \hat{y}_{t,q}\}}, \tag{8}$$

which, for a calibrated model, should be close to q. Sharpness measures the interval length provided by a probabilistic forecast. Among calibrated models, the one providing the sharpest intervals should be preferred. The pinball loss is known to target both calibration and sharpness and will be used to evaluate our models. To combine the evaluation at multiple quantile levels $q_1, \cdots, q_K$, we use the Average Calibration Error (ACE) and the normalized Ranked Probability Score (nRPS) proposed in [25] as a normalized discrete approximation of the CRPS [10]. These scores are defined as ACE $:= \frac{1}{K|\mathcal{T}|} \sum_{t \in \mathcal{T}} \sum_{k=1}^{K} |\mathbb{1}_{\{y_t \leq \hat{y}_{t,q_k}\}} - q|$ and nRPS $:= \frac{\sum_{t \in \mathcal{T}} \sum_{k=1}^{K} \Delta_k \rho_{q_k}(y_t, \hat{y}_{t,q_k})}{\sum_{t \in \mathcal{T}} |y_t - \bar{y}|}$, where we let $\Delta_k := q_{k+1} - q_{k-1}$ with $q_0 = 0$, $q_{K+1} = 1$ and $\bar{y} = \frac{1}{|\mathcal{T}|} \sum_{t \in \mathcal{T}} y_t$.

4.1 Estimation of the Mean

We compare two models for the estimation of the mean. The first one, referred to as gam, consists in fitting a GAM as explained in Sect. 2.1 on the whole training set (in green in Fig. 1). We use 6 smooth effects and 11 linear effects in the GAM Eq. (1). The formula is

$$\hat{y}_t = f_1(t) + f_2(toy_t) + f_3(Temp_t) + f_4(Temp95_t) + f_5(y_{t-1}) + f_6(y_{t-7})$$

$$+ \sum_{i=1}^{7} \alpha_i \mathbb{1}_{\{WD_t=i\}} + \beta_1 BH_t + \beta_2 CB_t + \beta_3 SB_t + \beta_4 P_t, \tag{9}$$

where toy denotes the normalized time of year from 0 (January 1st) to 1 (December 31st), $Temp$ and $Temp95$ are the average daily temperature and its exponential smoothing variant with parameter 0.95, WD indicates the day of the week from 1 to 7, BH, CB and SB are a boolean indicating bank holidays,

Christmas break and Summer break and finally P is a boolean equal to 1 if it is a Monday before a bank holiday or a Friday after a bank holiday (it takes into account possible 4 days weekends).

The second model considered is an adaptive GAM with Kalman updates as explained in Sect. 2.2 and referred to as `gkal`. We first fit a GAM model on the years 2012 to 2015 using the formula (9) and then estimate diagonal covariances for the Kalman state space model with the grid search method using the residuals of the GAM on the years 2016 and 2017 (see Fig. 1). Figure 2 shows the 14 days moving average bias of the two models and clearly illustrates the benefits of the Kalman adaptation, especially during the COVID crisis. The RMSE of `gam` goes from 856 in 2019 to 1926 in 2020 while `gkal` has much more stable scores with RMSE of 738 in 2019 and 812 in 2020.

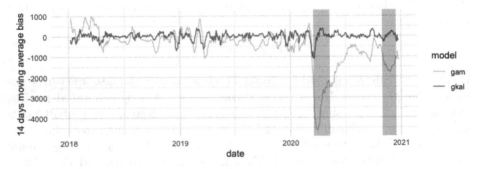

Fig. 2. 15 days moving average bias of the mean forecasts. Shaded areas correspond to the two lockdowns.

4.2   Estimation of Intermediate Quantiles

We estimate intermediate quantiles at levels $0.05, 0.1, \cdots , 0.95$ using the different methods introduced in Sect. 2.2 and summarized in Table 1. Figure 3 represents the coverage (8) at all quantile levels in the form of reliability diagrams. The online methods are clearly more calibrated than the offline ones since their reliability diagrams are much closer to the diagonal. It seems that better results are obtained with the methods based on `gkal` as confirmed by the scores of Table 2. This means that, for our case study, the adaptation of the mean forecast has the most impact on the performance of the quantile forecasts. Overall, the combination of an adaptive mean and non-linear quantile regression (`gkal.qgam`) gives the best results.

4.3   Estimation of Extreme Quantiles

Extreme quantiles forecasts are evaluated for both low and high quantiles. For high quantiles, we take the quantile forecasts at level $q_0 = 0.8$ provided by the methods of Sect. 4.2 as the intermediate threshold. Then we fit the EMLMs of

Table 1. Summary of intermediate quantile regression methods.

Name	Description
qgam	Fits smooth and linear effects using the same covariates as (9) to model the conditional quantile instead of the mean
gam.th	Theoretical quantiles from a normal distribution on the residuals of gam
gam.qr	Linear quantile regression on the residuals of gam using the effects of gam
gam.qgam	Similar to qgam but the target are the residuals of gam. The equation uses only smooth effects on toy, $Temp$ and $Temp95$ and linear effects on $\mathbb{1}_{\{WD_t=i\}}$ for $i = 1, \cdots 7$
gam.ogd	Same framework as in gam.qr but the coefficients of the linear quantile regression are updated online with gradient descent. We get 9 forecasts, one for each of the gradients steps $10^{-8}, 10^{-7}, \cdots, 1$ and we aggregate them online (using BOA) to compute the final forecast
gkal.th	Theoretical quantiles from the posterior normal distribution of gkal
gkal.qr	Linear quantile regression on the residuals of gkal using the effects of gam
gkal.qgam	Same as gam.qgam but on the residuals of gkal
gkal.ogd	Same as gam.ogd but on the residuals of gkal

Table 2. ACE, nRPS and pinball loss at levels 0.2, 0.5, 0.8 for the intermediate quantile forecasts in 2020. In bold: best per row.

	gkal.ogd	gkal.qgam	gkal.qr	gkal.th	gam.ogd	gam.qgam	gam.qr	gam.th	qgam
ACE	0.03	**0.02**	0.04	0.03	0.043	0.3	0.22	0.3	0.19
nRPS	0.048	**0.047**	0.048	0.049	0.07	0.12	0.10	0.13	0.09
20%	218	**213**	217	226	325	695	576	719	484
50%	279	**272**	279	274	372	644	559	718	515
80%	204	**194**	207	208	267	357	325	408	326

Sect. 3.1 on the exceedances over the intermediate threshold on the training set (years 2018 and 2019). We then compute the quantiles at extreme levels 0.99, 0.995 and 0.999 using (7). For low quantiles, we take the quantile forecasts at level $q_0 = 0.2$ as intermediate threshold and compute the quantiles at extreme levels 0.001, 0.005 and 0.01. To do so, it suffices to apply the same method as for high quantiles taking the opposite of the observations and intermediate quantile. The number of training samples, i.e. observations above (for high quantiles) or below (for low quantiles) the intermediate threshold represent between 119 and 144 samples, depending on the intermediate quantile method used. Due to the low number of training samples, we don't add categorical variables such as WD to the covariates. For all methods, we only consider the two covariates toy and

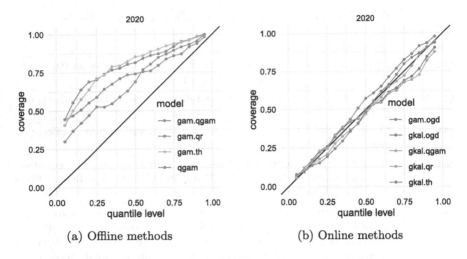

(a) Offline methods   (b) Online methods

Fig. 3. Reliability diagrams for the intermediate quantile forecasts in 2020

Temp. For each extreme quantile level, we also compare the EMLM estimates with the `base` estimate using the same model as the intermediate level.

An essential step in EMLMs is parameters selection. First, we make sure that the EMLMs yield constant shape ξ in the estimated GPD. For `evgam` and `eqrn`, this can be directly imposed in the R implementation of the models. For the `gbex` method, it suffices to set the depth of trees for the shape parameter to 0. For the `erf` method, we take a large shape penalty, here 1. To select the other parameters of the EMLMs, we use a 5-fold cross validation repeated 3 times and consider the GPD deviance (i.e. negative log-likelihood) as validation metric. For `erf`, The number of trees and the minimal node size of the `erf` are selected among $\{100, 500, 1000\}$ and $\{1, 3, 5, 10, 20, \cdots, 140\}$ respectively. For `gbex`, the depth of tree of the scale parameter σ is selected from 1 to 3 and the number of trees from 1 to 500. For `eqrn`, we consider 3 layers of GRU with size 256 and select the loss penalty among $\{0, 10^{-6}, 10^{-5}, 10^{-4}, 10^{-3}\}$. The maximum lag h (discussed in Sect. 3.1) for the input of `eqrn` is set to 1.

To evaluate the low and high quantile regression the ACE and nRPS metrics are not well suited as they put little weight to extreme levels. Instead, we use the individual coverages (8) and pinball losses gathered in Table 3. For low quantiles, we again observe that the online models outperform the offline models. We also almost always get better results by using an EMLM instead of the base model, especially at very low levels. The difference between the EMLMs is less striking. This may be explained by the small number of covariates used. An interesting observation is that the `base` variant of the `gkal.qgam` method also provides good results thus illustrating the effectiveness of QGAM even at extreme levels. For high quantiles, the results are more nuanced. A possible explaination is the fact that, during the COVID crisis the downward shift in electrical load makes the prediction of high quantiles particularly challenging. We also see the limitation of the calibration evaluation with (8) as, in our test data consisting of 366 samples, the best coverage are 0% and 100% for very low and very high levels respectively.

Table 3. Scores for extreme quantile regression. In bold: best score per line, underlined: best score.

Level		Coverage (%)						Pinball loss					
		base	eqrn	erf	evgam	gbex	gpd	base	eqrn	erf	evgam	gbex	gpd
0.1%	gkal.ogd	1.37	**0**	**0**	1.09	0.82	0.27	13.3	5.87	**3.53**	9.74	5.19	3.60
	gkal.qgam	**0**	0.27	**0**	**0**	0.27	**0**	3.72	4.33	4.18	**3.31**	4.82	4.19
	gkal.qr	3.01	**0.27**	**0.27**	1.09	0.55	0.55	16.6	4.36	**3.74**	8.04	5.01	3.97
	gkal.th	1.64	**0.27**	**0.27**	0.82	0.82	**0.27**	17	4.77	**4.03**	6.96	5.80	4.04
	gam.ogd	3.28	**0**	3.83	4.37	3.28	3.83	46.1	**9.92**	67.6	65.6	58.3	77.2
	gam.qgam	8.20	**0**	10.7	19.1	14.5	9.29	98.3	**6.39**	108	243	129	113
	gam.qr	33.1	**5.74**	13.1	12.8	12.3	12.3	416	**58.5**	164	164	141	179
	gam.th	14.7	**0**	7.92	35	9.02	9.84	229	**4.98**	86.03	402	121	124
	qgam	**1.09**	4.64	7.38	14.5	8.47	4.37	**5.98**	45.4	65.9	147	79.8	48.1
0.5%	gkal.ogd	1.37	**0.55**	1.37	1.37	1.37	1.37	21.3	**16.8**	17.6	24.2	22.1	22.3
	gkal.qgam	**0.82**	**0.82**	**0.82**	1.09	1.37	1.37	15.9	18	**15.1**	18.4	19.6	20.4
	gkal.qr	3.01	1.64	**1.37**	**1.37**	**1.37**	**1.37**	23.3	**20.7**	21.1	23.5	22.3	22.9
	gkal.th	3.01	**1.09**	**1.09**	1.37	**1.09**	1.37	31.6	**20.3**	21.3	23.4	22	23.1
	gam.ogd	3.83	0	5.19	5.19	4.64	5.19	61.9	**20.4**	96.9	94.1	91.6	104
	gam.qgam	19.1	**4.92**	19.1	27	22.7	16.4	241	**50.7**	239	316	267	237
	gam.qr	33.1	**13.1**	15.8	16.1	15.8	15.8	418	**156**	234	233	229	246
	gam.th	19.1	**3.28**	18.8	35.2	19.7	14.7	300	**24.8**	192	410	237	236
	qgam	9.02	**7.92**	15.6	18.8	14.5	8.47	99.1	**120**	163	196	171	123
1%	gkal.ogd	**1.37**	1.64	1.91	2.73	2.46	2.73	**30.1**	30.5	32	37.1	35	37.1
	gkal.qgam	**1.09**	2.19	1.91	2.73	2.46	1.91	**26.6**	32.5	30.5	32.4	31.8	35.2
	gkal.qr	**1.37**	2.46	2.46	2.46	2.46	2.73	**30.1**	36.3	35.9	36.3	35.4	37.6
	gkal.th	3.28	**1.04**	1.64	2.46	2.73	2.19	44	**34.9**	35	36.7	36.1	37.8
	gam.ogd	3.55	**2.73**	6.01	6.28	5.74	6.01	86.1	**47.5**	116	113	113	123
	gam.qgam	26.5	**12.3**	23.8	29.5	28.4	22.4	306	**144**	309	362	338	307
	gam.qr	32.8	**18.8**	19.1	19.7	19.4	**18.8**	414	**219**	273	274	277	285
	gam.th	23.2	**9.56**	22.4	35.8	25.1	19.1	344	**95.4**	268	419	306	296
	qgam	13.9	13.9	17.5	21	17.5	**13.7**	161	168	211	227	216	172
99%	gkal.ogd	98.6	99.2	99.2	99.2	**98.9**	99.2	24	22	24.5	23.3	22.7	**20.9**
	gkal.qgam	**99.2**	100	98.6	99.4	98.4	99.7	30.2	**24.3**	38.7	27.5	29.7	28
	gkal.qr	98.4	99.4	**99.2**	99.4	98.4	99.4	24	24.2	27.3	23.9	24.8	**21.2**
	gkal.th	96.7	100	99.4	100	**99.2**	100	**23.6**	35.6	35.2	26.9	26.4	27.4
	gam.ogd	99.7	99.7	**99.2**	98.4	98.6	99.7	26.3	27.4	28.9	26.5	27.8	**23.5**
	gam.qgam	100	100	**99.4**	100	**99.4**	100	34.4	35.7	33.6	**29.9**	30.9	33.5
	gam.qr	99.7	100	99.7	**99.2**	**99.2**	100	**26.3**	28	28.4	27.5	27.9	27.4
	gam.th	100	100	100	100	100	100	**31.2**	42.3	35.1	36.2	36.9	36.3
	qgam	100	**99.7**	**99.7**	**99.7**	**99.7**	**99.7**	36	32.7	36	30.6	32.9	**31.2**
99.5%	gkal.ogd	97	99.2	**99.4**	**99.4**	**99.4**	100	21	12.6	15.3	14.3	13.5	**12.3**
	gkal.qgam	**99.7**	100	99.2	100	99.2	100	17.4	**15.2**	25.1	17.2	18.2	17.9
	gkal.qr	95.1	100	**99.4**	**99.4**	99.2	100	28.2	15.6	17.3	14.9	14.2	**13.5**
	gkal.th	98.9	100	**99.4**	100	**99.4**	100	**13.2**	26.1	22.5	17.3	15.3	18.1
	gam.ogd	100	100	99.2	98.91	99.2	100	13.9	16.5	17.7	15.7	17.1	**13.3**
	gam.qgam	100	100	100	100	**99.7**	100	19.3	20.5	18.4	**15.6**	17	18.5
	gam.qr	100	100	**99.7**	99.2	**99.7**	100	**13.9**	15.9	15.9	15.3	15.7	15.2
	gam.th	100	100	100	100	100	100	**16.6**	24.5	18.7	19.3	20.5	19.4
	qgam	100	**99.7**	**99.7**	**99.7**	**99.7**	100	20.9	19	22.8	**17.9**	20	18.8
99.9%	gkal.ogd	97.8	**100**	**100**	**100**	**100**	**100**	11.1	**2.98**	4.12	3.43	3.24	4.02
	gkal.qgam	**100**	**100**	**100**	**100**	**100**	**100**	4.92	4.79	7.93	6.71	**4.38**	7.17
	gkal.qr	95.1	**100**	**100**	**100**	**100**	**100**	22.1	6.05	5.07	3.68	**3.28**	4.72
	gkal.th	99.2	**100**	**100**	**100**	**100**	**100**	**3.05**	12.58	7.59	5.90	4.38	6.58
	gam.ogd	**100**	**100**	99.2	99.2	99.2	**100**	**2.78**	4.98	5.24	6.15	5.92	3.42
	gam.qgam	**100**	**100**	**100**	**100**	**100**	**100**	5.76	5.48	4.46	**3.32**	4	4.49
	gam.qr	**100**	**100**	99.7	99.2	99.7	**100**	**2.78**	4.14	3.71	4.66	4.11	3.70
	gam.th	**100**	**100**	**100**	**100**	**100**	**100**	**3.74**	6.74	4.12	4.26	5.07	4.30
	qgam	**100**	**100**	**100**	99.7	**100**	**100**	6.52	**4.59**	7.32	4.83	6.10	6.49

5 Conclusion

In this work, we propose a new method for online extreme quantiles forecasting by combining recent advances in online adaptive probabilistic forecasting methods and EMLMs for GPD estimation. Our real life case study on the french electricity consumption illustrates the advantage of considering an adaptive quantile estimate as the intermediate threshold used in EMLMs during unstable periods like the COVID crisis. Our method is based on the implicit assumption that the GPD parameters do not need online adaptation. An interesting research topic, which we leave for future work, is to provide use cases where this assumption fails and derive fully adaptive EMLMs where the GPD parameters are also adapted online.

Acknowlegements. We thank our colleague Pr Olivier Wintenberger from Sorbonne Université and Wolfgang Pauli Institut, c/o Fakultät für Mathematik, Universität who helped us to initiate this project and provided insight and expertise that greatly assisted the research.

References

1. Athey, S., Tibshirani, J., Wager, S.: Generalized random forests. Ann. Stat. **47**(2), 1148–1178 (2019)
2. Balkema, A.A., de Haan, L.: Residual life time at great age. Ann. Probab. **2**(5), 792–804 (1974)
3. Browell, J., Fasiolo, M.: Probabilistic forecasting of regional net-load with conditional extremes and gridded NWP. IEEE Trans. Smart Grid **12**(6), 5011–5019 (2021)
4. Doumèche, N., Allioux, Y., Goude, Y., Rubrichi, S.: Human spatial dynamics for electricity demand forecasting: the case of france during the 2022 energy crisis. arXiv preprint arXiv:2309.16238 (2023)
5. Fan, S., Hyndman, R.J.: Forecast short-term electricity demand using semiparametric additive model. In: 2010 20th Australasian Universities Power Engineering Conference, pp. 1–6. IEEE (2010)
6. Fasiolo, M., Wood, S.N., Zaffran, M., Nedellec, R., Goude, Y.: Fast calibrated additive quantile regression. J. Am. Stat. Assoc. **116**(535), 1402–1412 (2021)
7. Gaillard, P., Goude, Y., Nedellec, R.: Additive models and robust aggregation for GEFCom2014 probabilistic electric load and electricity price forecasting. Int. J. Forecast. **32**(3), 1038–1050 (2016)
8. Gilbert, C., Browell, J., Stephen, B.: Probabilistic load forecasting for the low voltage network: forecast fusion and daily peaks. Sustain. Energy Grids Netw. **34**, 100998 (2023)
9. Gnecco, N., Terefe, E.M., Engelke, S.: Extremal random forests. J. Am. Stat. Assoc. (Jan), 1–24 (2024)
10. Gneiting, T., Raftery, A.E.: Strictly proper scoring rules, prediction, and estimation. J. Am. Stat. Assoc. **102**(477), 359–378 (2007)
11. Gneiting, T., et al.: Model diagnostics and forecast evaluation for quantiles. Ann. Rev. Stat. Appl. **10**(1), 597–621 (2023)

12. Hong, T., Pinson, P., Wang, Y., Weron, R., Yang, D., Zareipour, H.: Energy forecasting: a review and outlook. IEEE Open Access J. Power Energy **7**, 376–388 (2020)
13. Jiang, P., Van Fan, Y., Klemeš, J.J.: Impacts of COVID-19 on energy demand and consumption: challenges, lessons and emerging opportunities. Appl. Energy **285**, 116441 (2021)
14. Kalman, R.E., Bucy, R.S.: New results in linear filtering and prediction theory. J. Basic Eng. **83**(1), 95–108 (1961)
15. Koenker, R., Bassett, G., Jr.: Regression quantiles. Econ. J. Econ. Soc. 33–50 (1978)
16. Obst, D., de Vilmarest, J., Goude, Y.: Adaptive methods for short-term electricity load forecasting during COVID-19 lockdown in France. IEEE Trans. Power Syst. **36**(5), 4754–4763 (2021)
17. Pasche, O.C., Engelke, S.: Neural networks for extreme quantile regression with an application to forecasting of flood risk, April 2023
18. Pickands, J., III: Statistical inference using extreme order statistics. Ann. Stat. 119–131 (1975)
19. Pierrot, A., Goude, Y.: Short-term electricity load forecasting with generalized additive models. In: Proceedings of ISAP Power 2011 (2011)
20. Smith, R.L.: Estimating tails of probability distributions. Ann. Stat. 1174–1207 (1987)
21. Velthoen, J., Dombry, C., Cai, J.J., Engelke, S.: Gradient boosting for extreme quantile regression. Extremes **26**(4), 639–667 (2023)
22. de Vilmarest, J.: Viking: state-space models inference by Kalman or Viking (2022). https://CRAN.R-project.org/package=viking. R package version 1.0.0
23. de Vilmarest, J., Goude, Y.: State-space models for online post-COVID electricity load forecasting competition. IEEE Open Access J. Power Energy **9**, 192–201 (2022)
24. de Vilmarest, J., Wintenberger, O.: Viking: variational Bayesian variance tracking, November 2021
25. de Vilmarest, J., Browell, J., Fasiolo, M., Goude, Y., Wintenberger, O.: Adaptive probabilistic forecasting of electricity (net-)load. IEEE Trans. Power Syst. 1–10 (2023)
26. Wintenberger, O.: Optimal learning with Bernstein online aggregation. Mach. Learn. **106**(1), 119–141 (2017)
27. Wood, S.N.: Generalized Additive Models: An Introduction with R. CRC Press, Boca Raton (2017)
28. Youngman, B.D.: Electricity market report - update 2023. J. Stat. Softw. **103**(3), 1–26 (2022)
29. Youngman, B.D.: evgam: An R package for generalized additive extreme value models. J. Stat. Softw. **103**(3), 1–26 (2022)
30. Zaffran, M., Féron, O., Goude, Y., Josse, J., Dieuleveut, A.: Adaptive conformal predictions for time series. In: International Conference on Machine Learning, pp. 25834–25866. PMLR (2022)

Addressing Reviewers Comments: Explaining Voltage Control Decisions: A Scenario-Based Approach in Deep Reinforcement Learning

Blaž Dobravec$^{(\boxtimes)}$ [ID] and Jure Žabkar [ID]

Faculty of Computer and Information Science, University of Ljubljana,
Ljubljana, Slovenia
blaz.dobravec@elektro-gorenjska.si, jure.zabkar@fri.uni-lj.si
http://www.springer.com/gp/computer-science/lncs

Abstract. The electricity distribution system is evolving with increased electrification in transportation and heating, and the integration of distributed energy resources. This affects power quality in low-voltage networks. While Deep Reinforcement Learning (DRL) has been explored for voltage control, these methods lack the explainability, which reduces or even prevents their practical use. We introduce a Scenario-Based explanation (SBX) method to clarify the actions of the DRL agent in voltage control, enhancing operator understanding and ensuring system safety and reliability. Our method is based on the identification of the prototypical trajectories and the identified scenarios reflect comprehensible scenarios of the DRL agent behaviour in network management.

Keywords: Deep reinforcement learning · Explainability · Voltage control

1 Introduction

The European Union (EU) has set an ambitious goal of achieving a carbon-neutral economy by 2050. This goal necessitates a significant reduction in carbon emissions, requiring a substantial shift towards renewable energy resources (RES). The electricity sector is undergoing a radical transformation, with distributed photovoltaic solar systems at the helm of this new energy paradigm. Nonetheless, the anticipated electrification of transportation and heating is expected to augment electricity consumption [15]. This increase, alongside changes in consumption and production patterns, will significantly influence voltage profiles, especially within low-voltage distribution networks, thus, compounding network management complexities.

With advancements in technology, the controllability and observability of these networks have significantly improved, opening new possibilities for proactive distribution network management. Enhanced observability provides easy

© The Author(s), under exclusive license to Springer Nature Switzerland AG 2024
A. Appice et al. (Eds.): ISMIS 2024, LNAI 14670, pp. 216–230, 2024.
https://doi.org/10.1007/978-3-031-62700-2_20

access to both network conditions and household consumption data. Concurrently, the number of controllable components, such as OLTC transformers, battery storage systems, electric vehicles, and heat pumps, is on the rise, with a significant portion of these components owned by end users. Active consumers, who offer their components for use in flexibility services, bring versatility to the future electricity system. This versatility, coupled with the increasing number of active consumers and their flexibility potential, presents opportunities for distribution system operators to address voltage issues, including over- and under-voltage situations [34]. Although the active elements of consumers offer considerable flexibility potential, the role of the distribution system operator in voltage control on the low voltage network level is not designed to regulate individual consumers but rather optimise activations of multiple flexibility resources concurrently to achieve improvements in voltages over the entire network.

Contemporary strategies for voltage control predominantly rely on predefined or threshold-based control mechanisms, which often struggle to manage the intricacies of modern distribution networks. Deep Reinforcement Learning (DRL) presents a promising alternative and has demonstrated substantial potential in numerous complex domains:

- computer games (e.g. Chess [31], Go [32], Atari [23]),
- autonomous vehicles [18],
- and robot process automation in industrial settings [13].

Given the complexity of these domains, characterized by a large number of states and actions, DRL methods have recently been applied to optimize voltage control problems within simulated environments. Despite the promise of these methods in contrast to classical approaches, their adoption and implementation are hindered by a lack of explainability.

In this article, we introduce a Scenario-Based deep reinforcement learning eXplainability method (SBX) and apply it to a voltage control problem within a low-voltage distribution network. SBX integrates specific information about the observations and actions through the scenarios, which are derived from clustering sequences of observation-action pairs. These scenarios assist humans in understanding the correlation between a series of consecutive actions and the underlying observations of the environment. Our primary contribution is an explainability approach that clarifies actions based on the most representative time-based trajectories.

This article unfolds as follows: we review the current state-of-the-art voltage control and DRL explainability in Sect. 2. In Sect. 3, we detail the implementation of the deep reinforcement learning model. Section 4.1 elaborates on our scenario-based explainability approach. We provide simulation results and an analysis of explainability in Sect. 5. We examine the results in Sect. 6 and conclude by suggesting future research directions for our work in Sect. 7.

2   Related Work

2.1   Voltage Control Using Deep Reinforcement Learning

Voltage control is a standard practice in high-voltage distribution networks; it employs active elements like STATCOM to maintain voltage profiles at distinct network points. The majority of contemporary research on low-voltage distribution network control concentrates on optimizing consumption and generation at the individual consumer level [12,19,21,35]. Control of individual elements to improve and offer additional services to the DSO has also been explored in previous European projects [10] as well as integrated into existing business use-cases as proposed by Enedis [11]. Deep Reinforcement Learning (DRL) presents an appealing alternative approach, with recent studies lauding its superior performance in managing micro-grids and demand-side flexibility [24].

Specifically, a surge in research over the past year has explored the utilization of DRL for voltage control in low-voltage distribution networks [4,20,36,37,39]. Significant example includes the application of the Dueling Double Q-learning algorithm to mitigate over-voltages in distribution networks with solar power plants. A study conducted in Brazil demonstrated this approach to be both rapid and flexible [25].

Moreover, Mbuwir et al. [22] employed a model-free reinforcement learning model to devise optimal strategies for charging and discharging battery storage systems, aiming to enhance self-sufficiency. Their successful control strategy considered both consumption and generation within a micro-grid, taking into account electricity prices and distribution network constraints [2,5].

The incorporation of new elements at the household level-due to the electrification of transportation and heating-presents a significant challenge to existing control systems. This issue is further exacerbated by the increasing complexity of the low-voltage network. To tackle this problem, deep learning approaches are initially tested in simulation environments [9]. While some case studies concentrate on centralized micro-grid control, alternative decentralized methods can also be employed for active voltage control [6,36]. However, despite active research on applying DRL to voltage control issues, practical implementations of these algorithms remain scarce, potentially due to the lack of trust in these complex models. A recent survey concerning voltage control in low-voltage distribution networks emphasized the importance of user-oriented explainability in addition to superior performance for ensuring safety and reliability [27].

2.2   Explainability in Deep Reinforcement Learning

Explainability in deep reinforcement learning remains an active area of research. One widely employed explainability technique, primarily used in image classification, is the *saliency map*, which bases its explanations on pixel-wise feature attribution [33]. Building on this idea, Sequeira et al. [30] made the agent's *interactions with the environment* the focal point of their *Interestingness Framework*. Building on top of the feature attribution framework, the SHAP explainability method has also been applied in Reinforcement Learning tasks [40]. While the SHAP method offers individual steps or an average of multiple steps feature importance, other explainability methods focus on policy explanations. Hein et al. [14] propose a simplification of the RL policy using fuzzy particle swarm reinforcement learning (FPSRL) approach that can solve problems in domains where online learning is forbidden and which is human-readable and understandable. The final explanation is in the form of the logical statements (IF-ELSE). Coppens et al. [8] make the RL policy more interpretable using a Soft Decision Tree (SFT) which aims to mimic the original policy. The final explanation is given in the form of a graph. These approaches usually stem from the classical deep learning tasks, and the use of the sequential nature of the reinforcement learning process, we think that the explanations should be constructed with the component of time in mind. Time-based prototype learning similar to [7] has been explored by Ragodos et al. [28], but their approach is based on the prior definition of the prototypes. A similar approach was employed by Kenny et al. [17], where the authors propose a Prototype Wrapper network to generate human-friendly prototypes. Their approach is also hindered by the fact that the number of prototypes and how complete the explanation is depends on the number of prototypes defined at the beginning of the episode.

Despite the critical role of explainability in voltage control in low-voltage power systems, there is little research addressing this challenge. Zhang et al. [40] applied the SHAP explainability method to a deep reinforcement learning model for implementing proportional load shedding during under-voltage situations. They used Deep-SHAP to enhance the computational efficiency of their XAI model. The model's output presents its predictions through a visualization layer and a feature importance layer that addresses both global and local explanations.

Existing research on explainability in power systems, particularly regarding voltage control, predominantly focuses on feature-based explainability techniques. In contrast, our proposed approach aims to provide explanations for its predictions based on the prototypical behaviour of the agent. Compared to explanations for a single feature (individual voltage value) such as SHAP, our method considers the temporal component in the explanation process. To the best of our knowledge, this approach has not been applied to the reinforcement learning field in this specific manner.

3   Implementation of Deep Reinforcement Learning in Voltage Control

This section provides an overview of the deep reinforcement learning (DRL) model that we have implemented in this article. In this work, we have utilized the Proximal Policy Optimisation (PPO) algorithm, a DRL technique used in recent research [29]. While PPO has in previous work not provided the best performance [36], Yu et al. [39] improved the algorithm to be comparable to state-of-the-art approaches. We implemented our PPO algorithm in a single agent setting which produced good results.

3.1   Network Representation

The network representation follows the *pandapower* [16] format, where its components (e.g., lines, transformers, buses, loads, and generators) are implemented as tables. This simplifies the implementation of power flow algorithms required for voltage drop calculations. To emulate the flexibility activations, we categorise an active consumer as one who can voluntarily alter their consumption and production. Any modifications in their consumption or production patterns are quantified in the form of a percentile activation. Deep reinforcement learning is structured as a Markov decision process, with distinct definitions for states, actions, rewards, and a value function. The agent interacts with its environment (i.e., the distribution network) by executing actions that change the current state.

Definition of State. The state, s, is defined as an n-dimensional vector, where n is the number of buses in the network, and v_i represents the voltage at the ith bus within the distribution network $s = [v_1, v_2, \ldots, v_n]$.

Definition of Action. The action represents the agent's behaviour from a given state. Actions are represented as m-dimensional vectors, where m corresponds to the number of active consumers $a = [\alpha_1, \alpha_2, \ldots, \alpha_m]$.

Each $\alpha_i \in [-1, 1]$ denotes the activation level (in percentage) of the corresponding active consumer consumption decrease or generation decrease.

Definition of Reward. The primary objective of voltage control is to control the voltage across all network buses by making minimal adjustments to consumption and generation. To achieve this, we use a reward function proposed by previous research on multi-agent voltage control in distribution systems [36]:

$$r = -\frac{1}{|V|} \sum_{i \in V} l_v(v_i) - \alpha \cdot l_q\left(\mathbf{q}^{PV}\right),$$

where $l_v(\cdot)$ is a voltage barrier function and $l_q\left(\mathbf{q}^{PV}\right) = \frac{1}{|\mathcal{I}|} \left\|\mathbf{q}^{PV}\right\|_1$ is the reactive power generation loss (i.e. a type of power loss approximation easy for computation).

4   Generating Explanations with SBX

We consider an agent interacting with the environment over a sequence of steps; at each time t the agent receives a state or an observation $o_t \in \mathcal{O}$ from the environment and then performs an action $a_t \in \mathcal{A}$ in the environment. A sequence of consecutive steps is called a trajectory. A trajectory is a sequence of observation-action pairs, $\tau = ((o_0, a_0), \ldots, (o_{k-1}, a_{k-1})) \in (\mathcal{O} \times \mathcal{A})^k$. The trajectory represents the behaviour of the agent. We quantitatively compare the trajectories using Dynamic Time Warping (DTW) [1] distance. The notation $\tau^i \succ \tau^j$ signifies that trajectory τ^i is preferred over trajectory τ^j.

4.1   Scenario-Based eXplainability

The voltage control policy that the agent learns through Deep Reinforcement Learning is a complex model that is practically impossible to comprehend even by domain experts. To explain the actions of the agent, we propose a novel approach called *Scenario-Based eXplainability (SBX)* that aims to:

- aggregate trajectories with similar behaviours and abstract them into a common scenario,
- discover the number of meaningful scenarios in a given domain,
- visualize the scenarios to explain the actions of the agent.

We define a **scenario** as a prototypical trajectory, c_i, that captures characteristic behavioural patterns of the agent. Using the learned policy π of the agent, we generate a sequence of state-action pairs for the test data set. We use the non-seasonal cycle detection approach presented in [38] with time complexity of $O(n * \log(n))$ (n being the number of data points) on top of which we select the lengths of the cycles with the most amount of occurrences. We then split the sequence into i sub-sequences called trajectories τ_i all of the chosen length. Our goal is to then generate a meaningful set of scenarios using the generated trajectories $\tau_1, \tau_2, \ldots, \tau_m$. To this end, we use the K-means clustering algorithm on a set of trajectories. For each cluster, we obtain a centroid trajectory, which represents a scenario. The time complexity of the proposed approach is $O(n^2)$, where n is the number of data points. The number of clusters (scenarios), n, is determined using the Evidence Lower Bound method (ELBO) [3]. Dynamic Time Warping (DTW) [1] is used as a distance metric to measure the dissimilarity between trajectories at inference, accounting for possible temporal distortions:

$$\min_{c,\mu} \frac{1}{m} \sum_{j=1}^{m} DTW(\tau_j, c_{\mu(\tau_j)}) \tag{1}$$

Each trajectory τ_i is then assigned to the nearest cluster using DTW:

$$\mu(\tau_i) = argmin_{j-1,\ldots,n} DTW(\tau_i, c_j) \tag{2}$$

To generate the trajectories, let T denote the total number of steps computed for explaining the policy $\pi(o_i) = a_i$ from the training set (ignoring any episode limitations if they exist). The SBX currently presumes characteristic behaviour within a daily interval, which corresponds to trajectories of length $|\tau_i|$ based on the cycle detection algorithm. The agent performs control actions at each time step according to its policy; the observation-action pairs and trajectories of length $|\tau|$ are recorded. If the learning process terminates before reaching the trajectory length, we add padding as neutral states (1.03 p.u.) and zero actions to the end of the trajectory.

Algorithm 1. Scenario-Based eXplanation Method

Input: List of environment interactions containing the (o_i, a_i, r_i) pairs.
Input: Define the number β of different scenario types to view.
Procedure: Scenario-Based eXplainability
1. Calculate the lengths of the most common scenarios using the cycle detection algorithm. Limit the number of scenario types to be $\leq \beta$.
2. **for** each scenario length **do**
    a. Generate a list of trajectories $\tau_1, \tau_2, \ldots, \tau_i$.
    b. Calculate best n using the ELBO method.
    c. Initialize an empty set *Scenarios*.
    d. **for** each τ_i in $\{1, 2, \ldots, m\}$ **do**
        i. Find the first and the last index that is above λ.
        ii. Crop the trajectory between the first and last index.
    e. Apply K-Means using DTW distance to compute the pairwise similarity:
        $\min_{c,\mu} \frac{1}{m} \sum_{j=1}^{m} DTW(\tau_j, c_{\mu(\tau_j)})$
    f. Extract c_i of each cluster as an average of all elements in the cluster.
3. **end for**

---

5 Experiments

5.1 Simulation Environment and Voltage Control Policy

We examine a real-world low-voltage distribution network consisting of 26 consumers, of which 7 are active. Those active consumers are equipped with small solar plants (11kWp). The total yearly consumption in this network is negative, meaning that the solar plants are producing more electricity than is needed. An action in our model A visual representation of the network is displayed in Fig. 3.

The hyperparameters used in the PPO algorithm are the same as in the original Torch implementation [26]. The learning process extended over 1500 episodes, each containing 96 steps (representing a 15-minute interval across one day). We evaluated the model every 20 episode (1 epoch). In this network, we focus on handling mainly high voltages as those are a bigger problem in our example. We assessed the performance of the implemented algorithm using multiple deviation functions. The reward achieved during the training process is depicted in Fig. 1. In addition, we tested the model on a validation set, which consisted of two weeks' worth of real measurement data.

The Python library *pandapower* [16] was used to simulate the network states. The entire learning process, including the environment simulation, was conducted on a personal computer—a Macbook Pro equipped with an M1-Pro CPU. A single run of the process (1500 episodes), which includes 288 × 1500 load-flow calculations and the updating of network parameters, took approximately 45 min. During training, we tracked voltage deviations in the simulation based on the implemented PPO algorithm. The final model employs the *bump function* for voltage deviation in the reward function, as it consistently provides

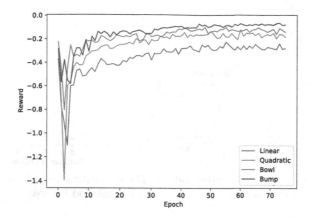

Fig. 1. The reward functions during the training process with different voltage deviation functions.

superior performance. We tested the model on a validation data set, comprising two weeks of consumer measurements, to mimic the actual environment the network might encounter.

5.2   Explaining a Simulation

To explain the actions of the agent, we created a visualization of the network. It displays actions and voltages (before and after) as streaming data. Following this, we present the three most prominent scenarios. We start by identifying the optimal length of the trajectories where we generate the cycle lengths and clustering the individual lengths to determine the ideal prototype length. The cycles and their lengths can be seen in Fig. 2a.

Then we split the training data into segments of length 95 based on the cycle detection algorithm. For these segments, we already have the corresponding actions generated by the agent according to the policy π. With the trajectories constructed, we determine the optimal number of clusters using the ELBO method [3]. In our example, the optimal number of clusters is determined to be three, as can be seen in Fig. 2b.

Activation curves in Fig. 4 serve as visual representations of the characteristic behaviours of the agent for each active consumer for each scenario. While the trajectories are clustered in observation-action space, we project the trajectories to separate action and observation sub-spaces for detailed analysis.

We observe three types of behaviour of activation curves (Fig. 3):

- **Scenario 1:** a jagged sequence of actions (blue),
- **Scenario 2:** a sequence of trivial actions (orange), and
- **Scenario 3:** a smooth sequence of actions (green).

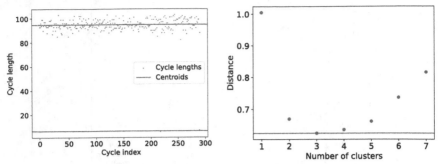

(a) Examples of the cycle lengths and their centroids of the clusters.

(b) The optimal number of clusters according to ELBO method is three.

Fig. 2. Deep-SHAP Explanations of the individual input feature (voltage value). Each active consumer represents a class and the colors of the class for each feature represent an average contribution/importance of the class on the voltage levels. When comparing the figures, we can see the correlation between the maximum impacts and the maximum voltages. (Color figure online)

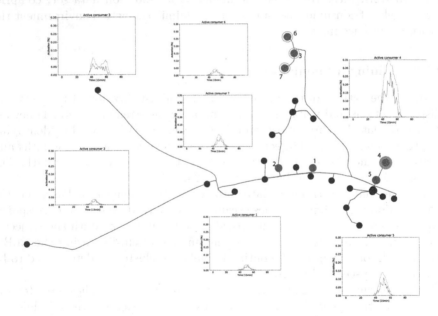

Fig. 3. The network is located in the north-western part of Slovenia. Active consumers (red), and their most representative activations are displayed with the corresponding graph. The width of the green circles denotes the most common and severity of over-voltage buses **before voltage control**. (Color figure online)

In the following, we analyse non-trivial scenarios, Scenario 1 and 3. By mapping these actions to corresponding observations, we analyse the relationships and patterns between observations and actions; we focus on the behaviour and voltages of active consumers.

Temporal profiles of individual active consumers reveal different amplitudes of activation and smoothness of the activation curves for each scenario (Fig. 3). The smoothness S of the activation curve denotes the intensity of the change in the activations in time. We quantify the smoothness by total variation; i.e. the sum of the absolute differences between adjacent activations. A smoother curve will have a smaller total variation

$$TV(X) = \sum_{i=1}^{n-1} |x_{i+1} - x_i| \qquad (3)$$

where $X = \{x_1, x_2, ..., x_n\}$.

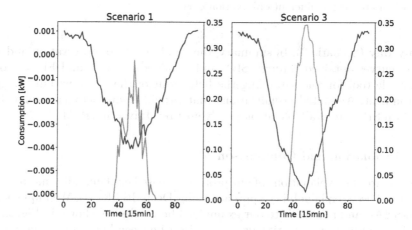

Fig. 4. Average consumption/generation activations of active consumer 4 and its activation curves for Scenarios 1 and 3.

The centroid activation curve for Scenario 1 has a total variation of 1.05 while the one for Scenario 3 is 0.70. The jaggedness of the consumption closely relates to the jaggedness of the representative activation curve. The comparison of the activation curves from Scenario 1 and 3 with the average historic consumption/generation over the same period is shown in Fig. 4. The second part of the explanation refers to the amplitude of the individual activation curves. Maximum consumption/generation in Scenario 1 is smaller than the one in Scenario 3, which implies a smaller amplitude of the activation curve in Scenario 1 compared to the one in Scenario 3.

The smoothness of the activation and observation curves (Fig. 4, Fig. 5) suggest that the detected scenarios depict three typical weather conditions. The typical **cloudy day** (Scenario 1), **in-between scenario** (Scenario 2), and the

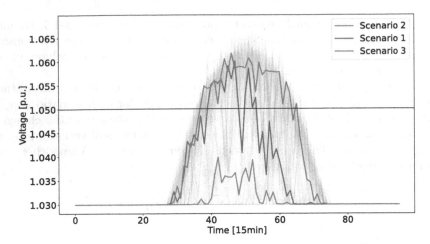

Fig. 5. Maximum observed voltages of all trajectories with the corresponding centroid. The colours represent different observation curves.

sunny day (Scenario 3). In summary, the combination of observation and activation curves obtained through SBX and depicted in Fig. 5 and Fig. 4 provide a valuable tool for understanding the behaviour of the agent within the given environment. Animated visualizations that provide more detailed explanations based on SBX are available at https://youtu.be/JMwaj9wvPhM.

5.3   Evaluation and Comparison

We compare the explanations of our prototype-based explainability method with a feature-based explainability approach using Deep-SHAP [41]. We implemented Deep-SHAP and applied it to our example. The results are shown in Fig. 6a. We found no appropriate quantitative comparison between both approaches; a general methodology for comparing feature-based and prototype-based explanations remains a challenging problem. We discuss the advantages and disadvantages of both approaches qualitatively.

Deep-SHAP provides insights into the importance of individual features at each step or their averages in the agent's decision-making process. In contrast, SBX provides a holistic time-dependent overview of the agent's decision-making. More concretely, at time t, we can, using SBX, provide the information on which scenario most closely resembles the current situation. The corresponding action curve indicates the explanation for the current action. Using SBX, we indirectly capture the influence of the agent's trajectory on its decision-making.

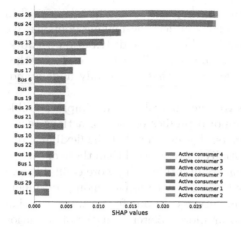

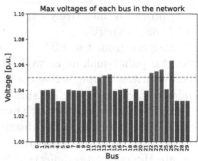

(a) The impact of the activation on the individual bus in the network.

(b) Maximum voltage data per each bus in the network.

Fig. 6. Deep-SHAP explanations of the individual input feature (voltage values). Each active consumer represents a class and the colors of the class for each feature represent an average contribution/importance of the class on the voltage levels. When comparing the figures, we can see the correlation between the maximum impacts and the maximum voltages. (Color figure online)

6 Discussion

The results of the proposed approach, applied in this domain, show promise; we obtained three distinct scenarios. Our results suggest that SBX has identified possible scenarios that closely correlate with different weather patterns. The actions executed in the environment, as determined by the observation-action pairs, are influenced by the proximity of the trajectory to the scenarios. This indicates that the distance metric employed in SBX is effective for differentiating and recognizing the scenarios based on their similarity to predefined patterns.

We consider the explainability of the obtained results as the most important contribution of this article. The scenarios identified by SBX align with common weather conditions, such as sunny, cloudy, and partly-sunny/partly-cloudy days, which are easily relatable to real-world situations and intrinsically explainable. This indicates that SBX not only provides accurate recognition of scenarios but also produces meaningful and explainable results that can be understood by domain experts or end-users.

7 Conclusion

In this article, we introduced SBX, a novel methodology for scenario recognition, predicated on the detection of characteristic scenarios. SBX has shown promising results in identifying and differentiating distinct scenarios. SBX stands out for its ability to provide meaningful, understandable explanations, which is critical in fostering trust in complex models, particularly in decision-making domains. We applied SBX to a voltage control problem in a low-voltage distribution network, where SBX identified three scenarios that aligned well with common weather patterns, providing intuitively explainable results that are readily understandable by domain experts.

The insights from the SBX in the voltage control domain imply possible direction for policy-making as well. Temporal profiles of control actions reveal notable disparities among active consumers; it is evident that the flexibility services offered by certain users are substantially more utilized than those of others. These insights could assist network managers in formulating more equitable and efficient policies, ensuring a fair distribution of service usage among consumers and optimizing the overall control of the network. In our future work, we plan to evaluate the correlation effects if two or more different lengths of scenarios are chosen. Furthermore, we aim to provide an interactive application for distribution system operators, which will enable them to play, pause, and review the trajectory prototype-based explanations.

Acknowledgements. This work was partially supported by the Slovenian Research Agency (ARRS), grant L2-4436: Deep Reinforcement Learning for optimisation of LV distribution network operation with Integrated Flexibility in real-Time (DRIFT).

References

1. Alizadeh, E.: An introduction to dynamic time warping (2022). https://builtin.com/data-science/dynamic-time-warping. Accessed 01 Feb 2024
2. Bahrami, S., Chen, Y.C., Wong, V.W.S.: Deep reinforcement learning for demand response in distribution networks. IEEE Trans. Smart Grid **12**, 1496–1506 (2021)
3. Bernstein, M.N.: The evidence lower bound (ELBO). https://mbernste.github.io/posts/elbo/
4. Cao, D., et al.: Data-driven multi-agent deep reinforcement learning for distribution system decentralized voltage control with high penetration of PVs. IEEE Trans. Smart Grid **12**(5), 4137–4150 (2021)
5. Cao, D., et al.: Model-free voltage control of active distribution system with PVs using surrogate model-based deep reinforcement learning. Appl. Energy **306**(Part A) (2021)
6. Cao, D., et al.: Deep reinforcement learning enabled physical-model-free two-timescale voltage control method for active distribution systems. IEEE Trans. Smart Grid **13**(1), 149–165 (2022)
7. Chen, C., Li, O., Tao, D., Barnett, A., Rudin, C., Su, J.K.: This looks like that: deep learning for interpretable image recognition. In: Wallach, H., Larochelle, H., Beygelzimer, A., d' Alché-Buc, F., Fox, E., Garnett, R. (eds.) Advances in Neural Information Processing Systems, vol. 32. Curran Associates, Inc. (2019)
8. Coppens, Y., Efthymiadis, K., Lenaerts, T., Nowé, A.: Distilling deep reinforcement learning policies in soft decision trees. In: International Joint Conference on Artificial Intelligence (2019)
9. Diao, R., Wang, Z., Shi, D., Chang, Q., Duan, J., Zhang, X.: Autonomous voltage control for grid operation using deep reinforcement learning. CoRR (2019)

10. Dumbs, C., et al.: Flexibility for DSOs on a local scale: business models and associated regulatory questions raised in the InterFlex project. CIRED - Workshop on Microgrids and Local Energy Communities (2018)
11. Enedis: Report on the integration of electric mobility in the public electricity distribution network (2019)
12. Fatima, S., Püvi, V., Lehtonen, M.: Review on the PV hosting capacity in distribution networks. Energies **13**(18), 4756 (2020)
13. Gomes, N., Martins, F., Lima, J., Wörtche, H.: Reinforcement learning for collaborative robots pick-and-place applications: a case study. Automation **3**, 223–241 (2022)
14. Hein, D., Hentschel, A., Runkler, T.A., Udluft, S.: Particle swarm optimization for generating interpretable fuzzy reinforcement learning policies. Eng. Appl. Artif. Intell. **65**, 87–98 (2016)
15. Iniciative, E.U.P.: Growing consumption in the European markets. https://knowledge4policy.ec.europa.eu/growing-consumerism
16. Institute, F.: Pandapower. https://pandapower.readthedocs.io/. Accessed 10 Nov 2024
17. Kenny, E.M., Tucker, M., Shah, J.: Towards interpretable deep reinforcement learning with human-friendly prototypes. In: The Eleventh International Conference on Learning Representations (2023)
18. Kiran, B.R., et al.: Deep reinforcement learning for autonomous driving: a survey. CoRR (2020)
19. Lan, Z., Long, Y., Rao, Y.: Review of voltage control in low voltage distribution networks with high penetration of photovoltaics. In: Proceedings of the 2nd International Conference on Information Technologies and Electrical Engineering, ICITEE-2019. Association for Computing Machinery, New York (2020)
20. Li, P., Shen, J., Yin, M., Zhang, Y., Wu, Z.: A deep reinforcement learning voltage control method for distribution network. In: 2022 IEEE 5th International Electrical and Energy Conference (CIEEC), pp. 2283–2288 (2022)
21. Ling Ai, W., Ramachandaramurthy, V., Walker, S., Ekanayake, J.: Optimal placement and sizing of battery energy storage system considering the duck curve phenomenon. IEEE Access **8**, 197236–197248 (2020)
22. Mbuwir, B.V., Spiessens, F., Deconinck, G.: Self-learning agent for battery energy management in a residential microgrid. In: 2018 IEEE PES Innovative Smart Grid Technologies Conference Europe (ISGT-Europe), pp. 1–6 (2018)
23. Mnih, V., et al.: Playing Atari with deep reinforcement learning. CoRR (2013)
24. Nakabi, T., Toivanen, P.: Deep reinforcement learning for energy management in a microgrid with flexible demand. Sustain. Energy Grids Netw. **25**, 100413 (2020)
25. del Nozal, A.R., Romero-Ramos, E., Trigo-Garcia, A.L.: Accurate assessment of decoupled OLTC transformers to optimize the operation of low-voltage networks. Energies **12**(11), 2173 (2019)
26. PyTorch: Reinforcement learning (PPO) with TorchRL tutorial. https://pytorch.org/rl/tutorials/coding_ppo.html
27. Qing, Y., Liu, S., Song, J., Song, M.: A survey on explainable reinforcement learning: concepts, algorithms, challenges. ArXiv (2022)
28. Ragodos, R., Wang, T., Lin, Q., Zhou, X.: ProtoX: explaining a reinforcement learning agent via prototyping. In: Koyejo, S., Mohamed, S., Agarwal, A., Belgrave, D., Cho, K., Oh, A. (eds.) Advances in Neural Information Processing Systems, vol. 35, pp. 27239–27252. Curran Associates, Inc. (2022)
29. Schulman, J., Wolski, F., Dhariwal, P., Radford, A., Klimov, O.: Proximal policy optimization algorithms. CoRR (2017)

30. Sequeira, P., Gervasio, M.T.: Interestingness elements for explainable reinforcement learning: understanding agents' capabilities and limitations. Artif. Intell. **288**, 103367 (2019)
31. Silver, D., et al.: Mastering chess and shogi by self-play with a general reinforcement learning algorithm. CoRR (2017)
32. Silver, D., et al.: Mastering the game of go without human knowledge. Nature **550**, 354–359 (2017)
33. Simonyan, K., Vedaldi, A., Zisserman, A.: Deep inside convolutional networks: visualising image classification models and saliency maps. CoRR (2013)
34. Strbac, G., et al.: An analysis of electricity system flexibility for Great Britain (2016)
35. Taczi, I., Sinkovics, B., Vokony, I., Hartmann, B.: The challenges of low voltage distribution system state estimation-an application oriented review. Energies **14**, 5363 (2021)
36. Wang, J., Xu, W., Gu, Y., Song, W., Green, T.C.: Multi-agent reinforcement learning for active voltage control on power distribution networks. CoRR (2021)
37. Wang, M., Feng, M., Zhou, W., Li, H.: Stabilizing voltage in power distribution networks via multi-agent reinforcement learning with transformer. In: Proceedings of the 28th ACM SIGKDD Conference on Knowledge Discovery and Data Mining (2022)
38. Witte, F., Kaldemeyer, C.: Cycle detection in time series: CyDeTS. In: Zenodo (2019)
39. Yu, L., Chen, Z., Jiang, X., Zhang, T., Yue, D.: Deep reinforcement learning for coordinated voltage regulation in active distribution networks. In: 2022 China Automation Congress (CAC), pp. 4005–4010 (2022)
40. Zhang, K., Xu, P., Zhang, J.: Explainable AI in deep reinforcement learning models: a shap method applied in power system emergency control. In: 2020 IEEE 4th Conference on Energy Internet and Energy System Integration (EI2), pp. 711–716 (2020)
41. Zhang, K., Zhang, J., Xu, P.D., Gao, T., Gao, D.W.: Explainable AI in deep reinforcement learning models for power system emergency control. IEEE Trans. Comput. Soc. Syst. **9**(2), 419–427 (2022)

Knowledge Graphs for Data Integration in Retail

Maxime Perrot[1,2(✉)], Mickaël Baron[2], Brice Chardin[2], and Stéphane Jean[3]

[1] Bimedia, La Roche-sur-Yon, France
[2] LIAS, ISAE-ENSMA, La Roche-sur-Yon, France
{maxime.perrot,mickael.baron,brice.chardin}@ensma.fr
[3] LIAS, Université de Poitiers, La Roche-sur-Yon, France
stephane.jean@ensma.fr

Abstract. Semantic web technologies are widely recognized for their utility in facilitating data integration tasks. While theoretical foundations have been extensively explored, few studies have displayed their practical implementation on real-world use cases and provided feedback on their scalability. This papers aims to address this gap by introducing a complete data integration framework tailored for Bimedia, a retail company. The framework is based on an integrated architecture of semantic layers and Knowledge Graphs (KGs), aiming to enhance data interoperability and provide a deeper understanding of retail dynamics by unveiling hidden relationships captured by the data. We present empirical evaluations of various architectural implementations, supported by quantitative analyses, to guide industry practitioners in effective decision-making. Furthermore, the case study of Bimedia showcases the practical application of knowledge graphs and semantic layers in the retail sector, bridging the gap between theory and practice. This study not only tackles Bimedia's specific challenges but also provides broader insights into the evolving landscape of retail technology.

Keywords: Knowledge graphs · virtual knowledge graphs · industrial context · knowledge graphs for data integration

1 Introduction

Over the past few decades, Semantic Web technologies have emerged among the most promising tools for data integration tasks. These encompass standardized languages for defining ontologies and knowledge graphs (such as RDF-Schema or OWL), database management systems (known as triplestores), query languages (like SPARQL), and mapping languages (for instance, R2RML) [12]. If, these technologies are well-defined due to substantial work on their theoretical aspects, the practical implementation of these technologies in real-world settings has received limited attention.

In this paper, we consider the case of Bimedia, a retail compagny. Technological innovations and new customer expectations have led Bimedia into developing

A. Appice et al. (Eds.): ISMIS 2024, LNAI 14670, pp. 231–245, 2024.
https://doi.org/10.1007/978-3-031-62700-2_21

a more intelligent and adaptive system, which has initiated a thorough examination of the potential offered by semantic technologies. The need is twofold: first, to establish a flexible and extensible semantic layer that can provide contextual meaning to raw data; and second, to construct a Knowledge Graph (KG) that can efficiently model intricate relationships within the retail landscape. In essence, Bimedia seeks a solution that not only handles the complexity of retail data but also translates this complexity into actionable insights for improved operational efficiency and customer experience.

While it is evident that semantic technologies can assist Bimedia in fulfilling these requirements, the practical implementation of these technologies in a real-world context is yet to be defined. Additionally, the capacity of these technologies to operate on a large scale requires further validation. Current studies assessing the scalability of semantic web technologies focus solely on these technologies in isolation – for example, benchmarks have been established to evaluate the scalability of triplestores. In an industrial context, a more comprehensive scalability study is necessary, one that considers the entire architecture used for implementing semantic technologies.

In this paper, we make two main contributions. The first is a comprehensive solution framework for implementing semantic technologies in a real-world scenario. We will outline the architecture, tools, and step-by-step implementation strategies customized for the retail domain. The second contribution involves providing empirical evidence through experimental results, assessing the performance of various architectural implementations. By presenting quantitative metrics and comparative analyses, we aim to offer valuable insights into the effectiveness of different approaches.

This article is organized as follows: We start by exploring the Bimedia business case (Sect. 2) and reviewing related work (Sect. 3). Next, we develop our methodology (Sect. 4), covering key steps. We then present our experimental evaluation (Sect. 5), followed by lessons learned (Sect. 6). Finally, we conclude with key insights (Sect. 7).

2   Motivation: The Bimedia Business Case

Bimedia is a leading French retail software publisher, part of the Orisha group. The company provides its services to diverse retail outlets such as tobacconists and bakeries. Its suite of solutions encompasses hardware, including cash registers and associated peripherals, as well as business software for tasks such as product ordering, stock management, or turnover and margin tracking. The company also offers integrated services, ranging from administrative tasks like car registration and fiscal stamps to financial services like money transfers and bank deposits. Bimedia serves over 6,000 businesses in France, handling over 60 million transactions per month, with more than 500,000 products referenced. Thus Bimedia accumulates a substantial amount of data pertaining to retail outlets, such as stock management and sales.

In retail, the capacity to make quick and informed decisions is critical to maintain competitiveness. This imperative emphasizes the central role of data analysis in corporate strategies. Internal data, such as sales history, customer preferences, and product performance, are crucial for understanding current trends and anticipating future needs. However, in an ever-changing environment, external data, such as those from open data source repositories – demographic, economic, and market trends – are equally indispensable. The integration of these diverse data sources is foundational for achieving a comprehensive and informed perspective on the market.

This paper focuses on a specific use case to illustrate our approach, employing semantic technologies for data integration, and to assess its scalability. The use case under consideration involves addressing a recurring analysis pattern within Bimedia: the extraction of a list of retail outlets based on specific criteria. This analysis is used, for instance, to select candidate outlets for targeted advertising, instead of broadcasting them across the entire network of stores. Some commonly employed criteria for searching retail outlets include: the volume of products sold belonging to a specific category (i.e. *press* or *confectionery*), the geographical location of the outlet (including factors such as proximity to points of interest) and the demographics (age, salary distributions, etc.) in the vicinity of the store.

In the experimental evaluation, our solution is tested with the following scenario: Bimedia's advertising department submits a request to the data analytics team, aiming to identify the most suitable retail outlets to promote a new edition of a magazine highlighting the role of women in the industry. This translates into identifying businesses that currently sell press products and are situated in areas with a high proportion of female executives.

In the past, accomplishing this task required a data scientist to manually consolidate data from diverse sources. The solution detailed in this paper streamlines this process by integrating all the data necessary through a semantic layer. This approach also allows for adjustment of analysis parameters based on business requirements through semantic queries. The proposed approach should also be applicable in an industrial context dealing with large volumes of data. Thus, we propose experiments conducted on real datasets of varying sizes to assess the performance and practical response times of our solution.

3 Related Work

In this section, we conduct a review of papers that explore the implementation of semantic web technologies within an industrial context. We classify these papers in two categories: those demonstrating the relevance of semantic web technologies in the industry and those offering technical insights into the utilization of semantic technologies.

The majority of the papers identified in the literature fall into the first category, addressing the question of why *Knowledge Graphs* (KGs) are beneficial in an industrial context. Thus, Li et al. [6] undertake a systematic survey of 119 papers, summarizing both technical and practical endeavors in the exploitation of KGs in industrial contexts. Yahya et al. [14] and Grangel-González [4] focus on the potential of these technologies for managing information related to maintenance, resource optimization, and production in industrial settings, and addressing interoperability issues arising from diverse entity representations and standards, respectively. Zou [15] and Abu-Salih [1] focus on the applications of KGs across various domains. They discuss the evolution of KGs, their impact in both academic and industrial fields, and how they enhance machine intelligence. Similar contributions have been made for *Virtual Knowledge Graphs* (VKGs), i.e., when the data remain in their original format and storage solution [2,7,11,13]. None of the identified papers explicitly evaluate possible architectures for integrating a semantic web layer. Additionally, none of these contributions involve quantitative performance comparisons of tools or technical architectures. Hence, papers falling within this first category rarely provide practical insights or feedback on the integration of semantic web layers and KGs in an industrial context.

We identified only two main contributions that fall within the second category of papers, providing insights into the utilization of semantic technologies in an industrial context [3,5]. Hubauer *et al.* [5] provide a very brief contribution, offering a concise and minimally detailed description of the approach used to establish a KG within the Siemens company. Similarly, Fishkin [3] presents a slide-based overview of the technical architecture employed for integrating knowledge graphs into Siemens' information system. These two contributions offer valuable insights into the tools and steps involved in implementing a KG within an industrial context. However, they do not offer a comparison of different architectures nor an evaluation of their performance.

Our contribution belongs to the category of papers offering technical insights into the utilization of semantic technologies in an industrial context. We aim at providing an overview of the solutions available for using semantic technologies in data integration in the retail sector, along with an experimental evaluation.

4   Semantic Data Integration Methodology for Industrial Applications

The use case presented in Sect. 2 requires the integration of the following data sources:

1. Bimedia's sales and business data that are stored in a PostgreSQL database. It encompasses a dozen of terabytes of information related to businesses, their sales, products, and other relevant details.
2. Data corresponding to the receipts of the businesses which are stored in a document-oriented NoSQL MongoDB database. This latter contains several hundred gigabytes of data.
3. Demographics open data provided by INSEE (a French public institute). These data are in the form of a CSV file, updated yearly.
4. Data from the Google Maps API on businesses. This data contains multiple pieces of information about Bimedia businesses, such as ratings of the outlets, opening hours, phone numbers, addresses, main activities, and more.

For the sake of simplicity, this paper only presents the integration of a single internal data source (Bimedia's sales data) and one external data source (INSEE's data).

Data integration can be achieved by adding a *semantic layer* on top of the integrated data sources. We define the notion of a semantic layer as the set of models, tools and languages that allow us to create KGs from a set of data sources. In this section, a four steps methodology is proposed to define this semantic layer. The originality of our proposition does not lie solely in the defined steps but rather in the practical implementation of each step, coupled with an evaluation of existing tools and models in an industrial setting.

4.1   Creation of the Bimedia Ontology

Ontologies serve the purpose of defining shared concepts and vocabularies for the integrated data sources. Many ontologies are available on the Web, and Bimedia's goal is to reuse these pre-existing ontologies whenever possible. Yet, in this case, they do not cover all the needs. Indeed, the concepts covering the internal data of Bimedia are specific to the retail domain and, to the best of our knowledge, not covered by existing ontologies. Conversely, the vocabulary covering the demographics open data is defined in an open ontology[1]. As a consequence, there is a need to define a new ontology on the specific concepts manipulated by Bimedia and to link it with open ontologies that encompass broader domains. We believe that a similar approach should be adopted for most industrial use cases as they involve specific and proprietary concepts.

To define the Bimedia ontology, we convened with all business experts within the company who possess knowledge relevant to the domains encompassed by the ontology. Together, we delineated the concepts and relationships to be represented within it. Reaching consensus on the definition of the ontology was a crucial aspect to consider.

The Bimedia ontology is intentionally crafted to adapt and evolve in response to shifting requirements, accommodating the incorporation of new concepts, relations, and other pertinent elements. For example, the definition of concepts such as *Tobacconist*, *Bakery*, or *Restaurant* may evolve over time. While, in a broad sense, a tobacconist is a store that sells tobacco-related products, a night club seldom selling those would not be considered a tobacconist by domain experts. Handpicked rules would typically define a tobacconist as a store that sells more than 1000 (or a given percentage of the stores' sales) tobacco-related items per month, although these thresholds may vary. This rule, subject to evolution over time – such as during periods of general market downturn – must be made explicit by domain experts. These kind of business rules can be represented in the ontology using SWRL (Semantic Web Rule Language), which combines OWL (Web Ontology Language) with RuleML (Rule Markup Language). The parameters of these rules can be easily modified to accommodate evolving requirements.

---

[1] https://github.com/InseeFr/Ontologies.

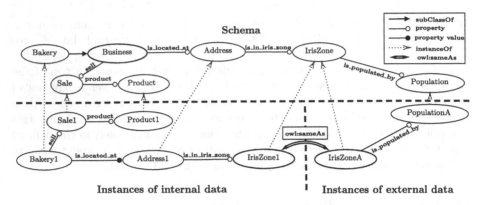

Fig. 1. Bimedia's ontology main classes

Figure 1 presents some of the main classes, properties and relationships of the ontology we designed for the Bimedia use case. This ontology contains 8 main classes: *Business, Sale, Product, Address, City, ZipZone, IrisZone,* and *InseeZone.* Each class has one or more properties. For instance, a business has a sign and a SIRET number. In this ontology, the relationships between classes are also expressed. For example: a business has an address, an address is linked to a city, a city is connected to an *InseeZone*, and so forth. Once the ontology has been setup, the next step involves integrating both internal and external data sources.

4.2   Semantic Integration of Internal Data

Semantic data integration involves the creation of a KG, which encompasses the defined ontology along with its instances derived from the integrated data sources. This KG may be virtual, referred to as a VKG when the instances remain in their storage solution. This helps avoid data duplication, which is crucial for Bimedia, especially given the large volume of sales data that are part of a broader data management process subject to various regulations.

Virtualization of a KG involves exposing the content of arbitrary relational databases as a KG. We demonstrate how this concept can be implemented in a real-world scenario using one of the internal data sources: the PostgreSQL relational sales database.

Different tools have proposed to build a VKG on top of various data sources. The ones we have identified as the most mature for inclusion in an industrial process are Ontop-VKG [8] and Virtuoso [9]. We choose to use Ontop-VKG due to its open-source nature, unlike Virtuoso, thereby enhancing the reproducibility of our approach.

Constructing a VKG with Ontop-VKG requires mapping rules to establish links between the ontology's concepts and the database's data. This is achieved using rules that can be expressed in two distinct languages, either

```
1  SELECT DISTINCT si.store_id, si.business_name, si.siret, addr.iris
2  FROM summary.store_ids si
3  LEFT JOIN summary.address addr ON addr.store_id = si.store_id
```

Fig. 2. Example of SQL query used in a mapping

OBDA (Ontology-based Data Access) or R2RML (RDB to RDF Mapping Language [10]). For this project, we opted for the R2RML language, since it is recommended by the W3C. More specifically, we employ SQL queries to (1) define how to retrieve instances of our ontology's classes and (2) create relations between instances by specifying which columns to use for joining.

We illustrate our approach by integrating a portion of the business data into our semantic layer, specifically focusing on legal data and addresses. This data is distributed across three tables within our relational database: the first contains legal information pertaining to businesses (such as trade name and SIRET number), the second stores the addresses of businesses, and the third contains city-related information.

Within our ontology (see Fig. 1 which illustrates a subset of the ontology), information concerning this data is represented under 6 classes: *Business*, *Address*, *City*, *ZipZone*, *IrisZone*, and *InseeZone*. To populate our VKG of businesses virtually, we need to express a mapping between the ontology and the relational table. This mapping is based on the SQL query presented in Fig. 2. This query returns four columns. Figures 3 and 4 present two examples of R2RML code, illustrating the mapping process of this data onto our ontology for the construction of the VKG. The business rules expressed in SWRL within the ontology (see Sect. 4.1) cannot be interpreted by the Ontop-VKG tool, as it doesn't support this type of inference. Therefore, these rules must be incorporated into the SQL queries within the R2RML mapping file.

In the excerpt of the mapping script depicted in Fig. 3, we link the results of this query to the concepts defined in our ontology. In the code section starting with `rr:subjectMap` (lines 5 7), we express the primary key of the instances to be populated and the class of our ontology to which these instances are attached. Defining the primary key in the mapping is crucial as it avoids what Ontop-VKG calls self-joins, a phenomenon that occurs during the rewriting by Ontop-VKG of SPARQL queries into SQL queries. The tables are then joined on themselves, drastically lengthening the response time. In this example, the class expressed in our ontology for the instances to be populated is *Store* and the primary key is *store_id*. Then the `rr:predicateObjectMap` (lines 8–16) specifies the values of instance properties by extracting data from columns of the query result.

We also need to define relationships between classes. This can be achieved with a certain form of `rr:predicateObjectMap`. The example depicted in Fig. 4 corresponds to the continuation of the previous example. In this mapping code, we express the relationship `is_in_iris_zone`, establishing the link between instances of the *Store* Class and instances of the *IrisZone* class (an IRIS zone

```
 1  @prefix sto: <http://ontologies.orisha.com/
         ↪ products_and_sales_ontology/> .
 2  <#Store>
 3    a rr:TriplesMap ;
 4    rr:logicalTable <#StoreTableView>;
 5    rr:subjectMap [
 6      rr:template "http://ontologies.orisha.com/
         ↪ products_and_sales_ontology/store/{store_id}" ;
 7      rr:class sto:Store; ];
 8    rr:predicateObjectMap [
 9      rr:predicate rdfs:label;
10      rr:objectMap [ rr:column "business_name" ]; ];
11    rr:predicateObjectMap [
12      rr:predicate sto:business_name;
13      rr:objectMap [ rr:column "business_name" ]; ];
14    rr:predicateObjectMap [
15      rr:predicate sto:siret;
16      rr:objectMap [ rr:column "siret" ]; ];
```

Fig. 3. Example of R2RML Code

```
 1  rr:predicateObjectMap [
 2      rr:predicate sto:is_in_iris_zone;
 3      rr:objectMap [ rr:template "http://ontologies.orisha.com/
         ↪ products_and_sales_ontology/IrisZone/{iris}" ]; ];
```

Fig. 4. Example R2RML Code for Relationship Mapping

denotes a small territorial unit in France, used in population statistics). In this mapping, we link the instances of these two classes by the IRIS column, explaining how to build the IRI of the corresponding *IrisZone* instance.

At this stage, the Ontop-VKG tool enables us to set up a SPARQL endpoint. Subsequently, this allows us to query the data from the ontology, even if it remains stored in relational tables.

4.3 Semantic Integration of External Data

In addition to sales data, the use case from Sect. 2 also requires external data. When integrating external data into the Bimedia semantic layer, various options are available in this integration process: storing the data in a native KG (managed by a triplestore), or storing the data in a relational database, followed by the virtualization of a KG on top of the relational data.

When using a VKG, the external and internal data can be kept separate, or stored by a single DBMS. These two options will be tested in our experiments. Since the process for implementing the loading of different external data sources into a relational database and virtualizing a KG above this relational database has already been covered, we will elaborate on the other solution: using a native KG.

The first step to integrate external data into a KG from files involves transforming tabular data into RDF triples. These triples should respect the previously defined ontology. The result of this step is a new file in the turtle format (Terse RDF Triple Language).

The second step is to serve this data using a triplestore, making them accessible through a SPARQL endpoint. Several technical solutions exist to manage the data. At Bimedia, we require a free triplestore that is regularly updated, efficient for our small volume of external data, compatible with the *owl:sameAs* property. Our testing has revealed that three triplestores meet these requirements : Apache Jena, Eclipse RDF4J and GraphDB by Ontotext. Amongst these three options, we have chosen GraphDB as it aligns better with Bimedia's IT architecture.

4.4   Federation of Knowledge Graphs for Mutli-Source Analysis

To enable the execution of queries across these different data sources (considering these sources are integrated into different databases), we need to setup a SPARQL endpoint capable of executing federated queries, and relationships between the different KGs.

SPARQL endpoints capable of executing federated queries can be easily identified by their compatibility with the SERVICE operator and, more generally, their compliance with the SPARQL 1.1 engine. There are also transparent federation tools such as FedX federation available in GraphDB or RDF4J. Given that we have previously chosen the GraphDB database management system for hosting data on the KG of external sources, compatible with the SERVICE operator, we will then use this SPARQL endpoint to launch our federated queries, thereby avoiding the complexity of the systems infrastructure.

To link multiple graphs, we used the **owl:sameAs** property on external data sources to match instances from our internal knowledge graph. This step is performed when populating the KG with data from external sources.

In our running example, for every *IrisZone* instance of the external data, we establish an equivalence relation with the corresponding instance in the VKG of internal data. Both use the same *IrisZone* class from the ontology, and we link them using the *owl:sameAs* property.

5   Experimentations

The goal of this experimental evaluation is to measure the impact of 1) virtualizing a KG compared to a native solution, and 2) separating internal and external data sources. Each of the six architectures – numbered A1 to A6 – is tested twice: once with a small-volume sales database with 29,409,768 rows (one month), and once with a larger sales databases with 111,315,470 rows (six months). This dataset is a summary of sales aggregated by day, commerce, and product. Other internal data includes information on businesses and products. For each architecture and amount of data, we execute three types of queries. Implementations of the queries are kept as similar as possible, but have their writing adapted to suit the architecture on which they are executed. For each configuration, we will comment on the execution time, the complexity of setting up the architecture, and the complexity of preparing the queries. The implementation of these experiments is available in a source repository[2].

To conduct these measurements, we defined three SPARQL queries, aiming at retrieving the following information:

- Query Q1 retrieves the list of SIRET numbers of businesses that are present in the French department code 17. To do this, the SPARQL query is a simple select with a filter on the first two digits of the zip code of the sales points' addresses. This query allows measuring the response time of different architectures on basic queries accessing a low amount of data (stores metadata).
- Query Q2 retrieves the list of SIRET numbers of businesses that have sold at least one product with coke (*coca* in french) in its name in a given month. The goal of this query is to compare the performances of the architectures on data volumes much larger than the previous one since, this time, the query relates to sales data. Having a solution that can efficiently filter sales data based on dates is important for the applications that Bimedia wishes to make of this KG. Optimization structures are present on the sales table in the underlying relational databases, so we expect adequate performances on VKGs, even with large volumes.
- Query Q3 retrieves businesses that sell more than 100 articles categorized as *press* and are located in an IRIS zone with more than 300 female executives, as recorded by the French state. This query corresponds to the case study presented in Sect. 2. A second version of this query (Q3'), intended for the first architecture only, does not consider external data and consists only in retrieving businesses that sell more than 100 articles categorized as *press*.

Table 1 provides a summary of the sizes of the different data graphs. For VKGs, we count all the triples of the virtual graph. We can observe slight variations in the number of triples between physical KGs and their virtual counterparts, even when the data sources are the same. This can be explained by the data transformations required during the construction of native KGs, which may lead to the consolidation of certain instances in the graph, thus reducing the number of triples.

The initial architecture (A1), illustrated in Fig. 5, includes a VKG layered over the sales and business relational database. This architecture functions as a foundational benchmark, enabling the assessment of the repercussions stemming from the incorporation of external data sources into the architectural framework in the subsequent architectures. Response times with this architecture are given in Table 2. We observe a significant influence of the amount of sales data on related queries (Q2 and Q3) execution times.

---

[2] https://forge.lias-lab.fr/retailkgintegration.

Table 1. Summary of Data Graph Volumes

Data Graph	Volume
Small Volume Internal KG	233,178,003
Large Volume Internal KG	638,033,173
External Population KG	25,183,519
Small Volume Internal VKG	233,026,330
Large Volume Internal VKG	638,024,436
External Population VKG	25,167,394
Small Volume Internal + External KG	258,338,227
Large Volume Internal + External KG	663,193,397

Fig. 5. A1 architecture schema

Table 2. A1 query execution times

Query	Small Volume	Large Volume
Q1	0.5 s	0.5 s
Q2	35.9 s	2 m 21 s
Q3' (w/o external)	52.3 s	4 m 31.2 s

The second architecture (A2), illustrated in Fig. 6, consists of a VKG over the relational database of sales and businesses, with the addition of external population data. This architecture serves as a baseline to measure the impact of separating external and internal data sources in subsequent architectures. This architecture is not of particular interest otherwise, as it is unconventional to include external data sources in a database built by aggregating internal data from a data lake, as is the case with the relational database of sales and businesses. Furthermore, this database is used to build customer reporting dashboards for clients, so it is prohibited to use it to store data that is not related to these purposes. The results obtained with this architecture are presented in Table 3. The response time for Q3 is lower compared with A1 when executed on a large volume of data. This is due to the addition of a second filter on the IRIS zone (population data), thereby reducing the number of rows returned. It is probable that the transformations performed on the data tables during the execution of the SQL query (here on a local PostgreSQL database on the machine) for graph virtualization are computationally expensive, hence the increased execution time.

The third architecture (A3), illustrated in Fig. 7, is composed of two VKGs, one over the relational database of sales and businesses, the second over another relational database containing population data. For this architecture, Q3 was executed from another SPARQL endpoint supporting the SERVICE operator (which is not the case with the SPARQL endpoint provided by Ontop-VKG). This query retrieves data from both VKGs by expressing the SERVICE operator

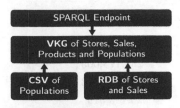

Fig. 6. A2 architecture schema

Table 3. A2 query execution times

Query	Small Volume	Large Volume
Q1	0.5 s	0.5 s
Q2	35.8 s	2 m 20 s
Q3	1 m 0.5 s	3 m 16.9 s

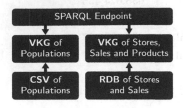

Fig. 7. A3 architecture schema

Table 4. A3 query execution times

Query	Small Volume	Large Volume
Q1	0.5 s	0.5 s
Q2	38.0 s	2 m 21 s
Q3	76 m 0.5 s	N/C

twice. The first use retrieves businesses that sold more than 100 *press* type articles over a month, the IRI of the business's *IrisZone*, and the corresponding IRI of this *IrisZone*. The second use of the SERVICE operator allows us to retrieve the IRIs of the *IrisZones* with more than 300 female executives recorded in their population, and the corresponding IRIs of this *IrisZone* in the sales and business graph (thanks to the *owl:sameAs* property). The SPARQL engine then calculates the intersection of these two sub-queries, yielding the expected results. The execution of this query allows us to measure the impact of multiplying VKGs in the technical architecture and to verify the proper functioning of joins in the case of VKGs. The results obtained for this architecture, presented in the Table 4, show a significant performance drop for Q3, likely due to the use of multiple SERVICE operators, in addition to the join that must be performed by the SPARQL engine to return the expected data. Moreover, we observe a true complexity increase in the query. For queries not requiring a join, no significant difference is observed compared to architecture A2, which was expected.

The fourth architecture (A4), depicted in Fig. 8, comprises a VKG atop the business's relational database, along with a native KG housing population data (previously transformed into triples and imported into a GraphDB KG). In this setup, Q3 was executed from the SPARQL endpoint of the population KG, leveraging its support for the SERVICE operator. The query initially retrieves *IrisZones* from the population graph, filtering for those with more than 300 female executives recorded. Subsequently, the query is performed on the VKG to fetch businesses that have sold over 100 "press" type articles within a month, with an additional filter on the IRI of the business's *IrisZone* corresponding to the IRI returned in the first step. Notably, the SPARQL engine avoids an intersection calculation between sub-queries, as there's only one, thus leading to a significant reduction in execution time, even compared to architecture A1

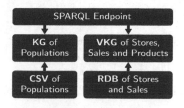

Fig. 8. A4 architecture schema

Table 5. A4 query execution times

Query	Small Volume	Large Volume
Q1	0.5 s	0.5 s
Q2	36.7 s	2 m 18.6 s
Q3	53.7 s	2 m 33.5 s

Fig. 9. A5 architecture schema

Table 6. A5 query execution times

Query	Small Volume	Large Volume
Q1	0.5 s	2.9 s
Q2	1 m 19 s	4 m 16 s
Q3	1 m 0.4 s	1 m 32.3 s

with a large volume. The results obtained for this architecture, presented in Table 5, demonstrate improved performance, particularly when compared to A3. We observe substantial time savings in executing Q3 on a large volume of data compared to architecture A2, underscoring the advantage of utilizing a native KG for external data coupled with the use of the *owl:sameAs* property.

The fifth architecture (A5), illustrated in Fig. 9, is composed of two native KGs, one containing population data and the other sales and business data. Although this architecture is unlikely to be used in production due to its cost in terms of storage and maintenance, it offers interesting performance and allows us once again to measure the impact of virtualizing all or part of the architecture compared to an entirely native architecture. It is particularly complex to transform a large volume of data into RDF triples and then load them into a database. This operation takes several hours. However, we observe in Table 6 a significant improvement in the execution time of Q3 on large volumes, explainable by the use of two native RDF storage solutions as opposed to A4 in which internal data was virtualized. Yet, query Q2 becomes slower compared with A4, showing some performance benefits of using a relational database for querying sales data exclusively.

The sixth architecture (A6), illustrated in Fig. 10, is composed of a single KG containing sales, business, and population data. This architecture allows us to measure the performance of an entirely native architecture. This solution includes the same disadvantages as A5 in terms of data duplication and setup complexity, only the architecture is less complex. Query Q3 does not require a join in this architecture; it is identical to that in architecture A2. Surprisingly, in the results presented in Table 7, the execution time of query Q3 on large volumes is increased compared with architectures A2, A4, and A5.

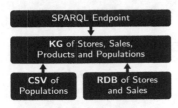

Fig. 10. A6 architecture schema

Table 7. A6 query execution times

Query	Small Volume	Large Volume
Q1	0.5 s	0.5 s
Q2	1 m 23.6 s	4 m 07 s
Q3	58 s	4 m 11 s

6   Lessons Learned

Considering the results obtained in the various experiments, we opt for the implementation of the technical architecture A4, with a VKG on top of internal data, and a native KG for external data. This architecture has several advantages: (1) it has reduced query complexity compared with A3; (2) it separates internal and external data sources, giving it an advantage compared with architectures A2 and A6; (3) since it avoids the prohibitive duplication of sales data, it offers a quick setup and ease of maintenance compared with architectures A5 and A6. In terms of cumulative response time, solution A4 outperforms all other solutions, achieving a 13% improvement compared with the second-best performing solution, A2. The only challenge in implementing this architecture is the construction of the mapping script, which remains a relatively accessible task for anyone familiar with the data sources and SQL queries development. When using the Ontop-VKG tool and an R2RML mapping file, the performance of VKGs can vary significantly. Some factors include: the construction of SQL queries expressed in the mapping file; the presence and use of indexes in the relational database containing the data to be virtualized; the construction of the mapping, including the use of primary and secondary keys to avoid self-joins when translating SPARQL queries into SQL queries by the Ontop-VKG rewriting engine.

7   Conclusion

In this contribution, we provide a practical insight into the implementation of a semantic web layer composed of KGs within Bimedia's information system. The objective was to provide Bimedia with access to their data in an enriched form, both through an extended semantic vocabulary and the seamless integration of external data sources. This enables advanced multi-criteria searches through simple SPARQL queries, a task that previously required the involvement and technical expertise of a data scientist. Our experiments have demonstrated that virtualizing a KG on top of a relational database containing large volumes of data is well-suited for this industrial context, where data duplication is not a viable solution in terms of storage and maintenance. We have also addressed the integration of external data sources by constructing a KG, a paradigm that opens up possibilities without complicating analysts work, thanks to the addition of explicit relationships between external and internal data sources.

This contribution has paved the way for several future directions to enhance the implementation of such solutions in industrial contexts. These include the integration of Natural Language Processing (NLP) models that facilitate the transformation of user textual queries into SPARQL queries suitable for this technical architecture. Additionally, there's the prospect of developing a rule transformation engine that can convert SWRL rules into R2RML mapping rules, allowing certain rules to be expressed at the ontology level rather than in the mapping file.

References

1. Abu-Salih, B.: Domain-specific knowledge graphs: a survey. J. Netw. Comput. Appl. **185**, 103076 (2021)
2. Chaves-Fraga, D., Corcho, O., Yedro, F., Moreno, R., Olías, J., De La Azuela, A.: Systematic construction of knowledge graphs for research-performing organizations. Information **13**(12), 562 (2022)
3. Fishkin, A.: Industrial knowledge graph at Siemens CERN Openlab Technical Workshop, Geneva (1999). http://web.archive.org/web/20080207010024/, http://www.808multimedia.com/winnt/kernel.htm. Accessed 01 Nov 2018
4. Grangel-González, I.: A knowledge graph based integration approach for industry 4.0. Ph.D. thesis, Universitäts-und Landesbibliothek Bonn (2019)
5. Hubauer, T., Lamparter, S., Haase, P., Herzig, D.M.: Use cases of the industrial knowledge graph at siemens. In: ISWC (P&D/Industry/BlueSky) (2018)
6. Li, X., Lyu, M., Wang, Z., Chen, C.-H., Zheng, P.: Exploiting knowledge graphs in industrial products and services: a survey of key aspects, challenges, and future perspectives. Comput. Ind. **129**, 103449 (2021)
7. Mendes de Farias, T., Wollbrett, J., Robinson-Rechavi, M., Bastian, F.: Lessons learned to boost a bioinformatics knowledge base reusability, the Bgee experience. GigaScience **12**, giad058 (2023)
8. Ontop-VKG. https://ontop-vkg.org. Accessed 01 Jan 2023
9. OpenLink Virtuoso. https://virtuoso.openlinksw.com/. Accessed 01 Jan 2023
10. R2RML: RDB to RDF Mapping Language W3C Recommendation. https://www.w3.org/TR/r2rml/. Accessed 27 Sept 2012
11. Vogt, L., Konrad, M., Prinz, M.: Knowledge graph building blocks: an easy-touse framework for developing FAIREr knowledge graphs. arXiv preprint arXiv:2304.09029 (2023)
12. Xiao, G., et al.: Ontology-based data access: a survey (2018)
13. Xiao, G., et al.: FHIR-Ontop-OMOP: building clinical knowledge graphs in FHIR RDF with the OMOP common data model. J. Biomed. Inform. **134**, 104201 (2022)
14. Yahya, M., Breslin, J.G., Ali, M.I.: Semantic web and knowledge graphs for industry 4.0. Appl. Sci. **11**(11), 5110 (2021)
15. Zou, X.: A survey on application of knowledge graph. J. Phys.: Conf. Ser. 012016 (2020)

Learning with Complex Data

Bayesian Approach for Parameter Estimation in Vehicle Lateral Dynamics

Fabien Lionti[1]([✉]) [ID], Nicolas Gutowski[2] [ID], Sébastien Aubin[3],
and Philippe Martinet[1] [ID]

[1] Université de Côte d'Azur - INRIA, 2004 Rte des Lucioles, 06902 Valbonne, France
{fabien.lionti,philippe.martinet}@inria.fr
[2] Université d'Angers - LERIA, 2 Bd de Lavoisier, 49000 Angers, France
nicolas.gutowski@univ-angers.fr
[3] Direction Générale de l'Armement - Techniques Terrestres, Rue de la Chédditière,
49460 Montreuil-Juigné, France
sebastien.aubin@intradef.gouv.fr

Abstract. Estimating parameters for nonlinear dynamic systems is a significant challenge across numerous research areas and practical applications. This paper introduces a novel two-step approach for estimating parameters that control the lateral dynamics of a vehicle, acknowledging the limitations and noise within the data. The methodology merges spline smoothing of system observations with a Bayesian framework for parameter estimation. The initial phase involves applying spline smoothing to the system state variable observations, effectively filtering out noise and achieving precise estimates of the state variables' derivatives. Consequently, this technique allows for the direct estimation of parameters from the differential equations characterizing the system's dynamics, bypassing the need for labor-intensive integration procedures. The subsequent phase focuses on parameter estimation from the differential equation residuals, utilizing a Bayesian method known as *likelihood-free ABC-SMC*. This Bayesian strategy offers multiple advantages: it mitigates the impact of data scarcity by incorporating prior knowledge regarding the vehicle's physical properties and enhances interpretability through the provision of a posterior distribution for the parameters likely responsible for the observed data. Employing this innovative method facilitates the robust estimation of parameters governing vehicle lateral dynamics, even in the presence of limited and noisy data.

Keywords: Bayesian Inference · Vehicle Dynamics · Parameter Estimation · Lateral Dynamics · Nonlinear Systems

1 Introduction

The parameter estimation in nonlinear dynamic systems is an extensively researched problem [1]. Traditional strategies for addressing this issue often involve minimizing an objective function to reduce the discrepancy between predicted model outcomes and actual observed data, typically employing gradient descent techniques [2]. However, these traditional methods can be limited by their tendency to converge to local minima and their sensitivity to the initial conditions of the system.

To circumvent these constraints, several alternative techniques have been proposed. Techniques such as genetic algorithms [3,4], simulated annealing [5], and particle swarm optimization [6] have been explored to more effectively navigate the parameter space towards the global minimum. These methods, however, face challenges as the computational demand increases exponentially with the number of parameters to estimate. Additionally, the reliance on integration methods like the *Runge-Kutta* method [7] for matching predictions with observations can lead to substantial computational efforts, particularly when integrating the system across various initial conditions.

Noisy observations further complicate the accurate determination of initial conditions for integration, potentially leading to biased parameter estimates in systems sensitive to such conditions. A strategy to mitigate this involves smoothing observations using splines [7], which enables the derivation of system observation derivatives and subsequent parameter estimation from the differential equation residuals. This concept has been extended to include neural networks, specifically Physics-Informed Neural Networks (PINN) [8], which interpolate observations from processes described by partial differential equations (PDEs) while also estimating the parameters of these equations using automatic differentiation to optimize parameters through gradient descent.

Despite the advancements, these methods generally do not offer a probabilistic interpretation of the estimated parameters, a gap that recent efforts within the Bayesian framework aim to bridge. By incorporating priors about parameter distributions, these approaches not only limit the search space but also enhance interpretability by yielding a posterior distribution over the parameters. Techniques leveraging Gaussian processes for interpolation and modeling observation noise uncertainty [9,10], where the derivative of a Gaussian process remains within the Gaussian framework [11], facilitate the definition of a likelihood function. This function, combined with sampling strategies like the Metropolis-Hastings algorithm [12] or Gibbs sampling [13], evaluates the posterior distribution over parameters, assuming a normal error between observed and predicted derivatives.

The complexity of parameterizing these sampling methods, such as selecting the number of samples for approximating the posterior distribution or the transition proposal, poses significant challenges. The efficiency and effectiveness of these approaches heavily depend on the precise specification of transition proposals, with inadequate choices leading to low acceptance rates or poor exploration of the parameter space. Furthermore, ensuring the convergence of samples towards the target distribution is crucial, especially as the target distribution's complexity increases.

In parallel, *likelihood-free* approaches like the *Approximate Bayes Computation (ABC)* method [14] have emerged, approximating the posterior distribution without explicitly defining a likelihood function. This simulation-based sampling strategy filters parameters likely to have produced the observed data using a similarity measure. The *ABC-SMC* method [15], a *Sequential Monte Carlo* technique, refines this process by reducing rejection rates through targeted sampling in parameter spaces more consistent with the observed data.

Despite a comprehensive review, no direct applications of these methods for parameter estimation in nonlinear dynamic vehicle systems were identified, which is the purpose of our work. Nevertheless, their utility in dynamic system parameter estimation, particularly in biological contexts [16–18], is well-documented. The proposed methodology, tested on simulated data using a bicycle model incorporating the Pacejka tire model [19], focuses on estimating Pacejka coefficients and physical model parameters, such as center of gravity location and vehicle moment of inertia. Our research contributes to broader efforts to model ground vehicles to enhance stability and safety, potentially reducing the costs associated with extensive physical testing across varied environments and configurations.

2 Approximate Bayes Computation

The *ABC* method [14] estimates the posterior distribution $P(\theta|\mathcal{D})$ without having to specify the likelihood function and without having to calculate the marginal probability of the data. The only prerequisite is the ability to simulate data $\mathcal{D}^*$ from a model corresponding to different parameterizations of θ.

This approach remains inefficient in the choice of θ samples over iterations, as it explores the parameter space without favoring regions most likely to result in sample acceptance. The *ABC-SMC* approach [15] (See Algorithm 1) improves sampling efficiency by effectively using prior information. In *ABC-SMC*, samples from the previous iteration are used to generate new samples. This leverages already acquired information on the posterior distribution and focuses computational efforts on the most promising regions of the parameter space. Thus, by improving the quality of sampling, it is possible to progressively improve the approximation of the posterior distribution using a sequence of progressively decreasing thresholds. The generation of new samples is based on an importance sampling mechanism of previous samples by weighting drawing probabilities. Each drawn sample is then perturbed by a perturbation function K, often uniform or Gaussian, which models the transition probability $P(\theta|\tilde{\theta}_{t-1})$. It can be parameterized to control the amplitude of perturbations and thus influence the diversity and exploration of the parameter space. After generating new samples, they are evaluated using the sampling adequacy measure ρ, and the samples generating data $\mathcal{D}^*$ closest to the observations $\mathcal{D}$ are selected according to the threshold ϵ_t for the next iteration of the sampling process. This iterative step is repeated until convergence of the posterior distribution is achieved or other convergence criteria are satisfied. Note that we provide a *Python* implementation of the *ABC-SMC* algorithm in the following GitHub repository:

Algorithm 1: ABC-SMC Algorithm

Data: Observed data $\mathcal{D}$, number of iterations T, sample size N, thresholds
$\quad \epsilon_1 > \epsilon_2 > ... > \epsilon_T$
Result: Approximation of the posterior distribution
Initialize the sample $\theta^1, \theta^2, ..., \theta^N$;
for $t \leftarrow 1$ **to** T **do**
    while $i \leq N$ **do**
       if $t = 1$ **then**
          | $\tilde{\theta} \leftarrow P(\theta)$;
       end
       else
          Draw a sample $\tilde{\theta}_{t-1}$ from the previous iteration's sample set θ_{t-1}
          weighted by w;
          $\tilde{\theta} \leftarrow P(\theta|\tilde{\theta}_{t-1})$;
       end
       if $P(\tilde{\theta}) \neq 0$ **then**
          Generate simulated data $\mathcal{D}^*$ from $\tilde{\theta}$;
          Calculate the distance $\rho(\mathcal{D}, \mathcal{D}^*)$;
          if $\rho < \epsilon_t$ **then**
             $\theta_t^i \leftarrow \tilde{\theta}$;
             if $t \neq 1$ **then**
                | $w_t^i = \dfrac{P(\theta_t^i)}{\sum_{j=1}^N w_{t-1}^i K(\theta_t^i, \theta_{t-1}^j)}$
             end
             else
                | $w^i = 1$
             end
             $i = i + 1$;
          end
       end
    end
    for $j \leftarrow 1$ **to** N **do**
       | $w_t^j = \dfrac{w_t^j}{\sum_{i=1}^N w_t^i}$
    end
end

---

https://github.com/fabien-lionti/abc-smc-vehicle-dynamic. This repository also includes extended results.

3 Modeling of Vehicle Lateral Dynamics

3.1 Two Degrees of Freedom Bicycle Model

The two degrees of freedom bicycle model is a simplified representation of a vehicle's lateral dynamics. It is widely used in the field of vehicle dynamics to analyze driving behavior, stability, and handling. Applying the fundamental principle of dynamics to the force model described in Fig. 1 yields the following system of DEs:

$$\begin{cases} \dot{v}_y = \frac{F_{yf}}{m}\cos\delta_f + \frac{F_{yr}}{m} - v_x r \\ \dot{r} = \frac{l_f}{I_z}F_{yf}\cos\delta_f - \frac{l_r}{I_z}F_{yr} \end{cases} \tag{1}$$

The vehicle's yaw rate is represented by r, m represents the mass of the vehicle, the distance of the front axles relative to the center of gravity (CG) is represented by l_f, the distance of the rear axles relative to the CG is represented by l_r, and the vehicle's yaw inertia is represented by I_z. The lateral forces exerted by the vehicle's front F_{yf} and rear F_{yr} tires are modeled using the nonlinear Pacejka model [19].

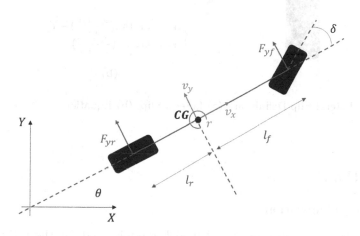

Fig. 1. Bicycle model.

3.2 Pacejka Model

The Pacejka model is a hybrid model since it is based on physical laws assuming a parabolic distribution of forces on the tire/ground contact area but also relying on phenomenological observations from test results. Thus, it represents a trade-off between computation time and accuracy. The Pacejka model is used to model the lateral front F_{yf} and rear F_{yr} forces exerted by the road on the tires.

$$F_{y_i} = D \cdot \sin\left(C \cdot \arctan\left(B \cdot \left(\alpha_i - E \cdot \left(\alpha_i \right.\right.\right.\right.$$
$$\left.\left.\left.\left. - \arctan\left(B \cdot \alpha_i - \arctan\left(B\alpha_i\right)\right)\right)\right)\right)\right) \tag{2}$$

With $i \in \{f, r\}$ denotes the front force and the rear force. B denotes the stiffness coefficient, C the shape coefficient, D is the maximum value reached by F_y, and E is the curvature factor.

The Pacejka model inputs the slip angles of the front α_f and rear α_r tires, corresponding to the difference between the real speed vector v of the tires and v_y (Fig. 2). This slip angle is influenced by factors such as the lateral speed of the vehicle, the rotation speed, the distance between the front/rear axle and the vehicle's center of gravity, as well as the steering angle δ_f of the front wheels.

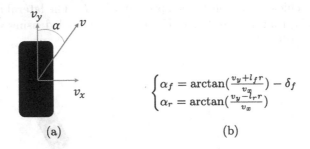

$$\begin{cases} \alpha_f = \arctan(\frac{v_y + l_f r}{v_x}) - \delta_f \\ \alpha_r = \arctan(\frac{v_y - l_r r}{v_x}) \end{cases}$$

(a)   (b)

Fig. 2. Lateral Slip Definition - (a) Lateral Slip, (b) Equations for Lateral Slip

4   Method

4.1   Data Generation

Based on the previously described model, data is used for the estimation of parameters $\theta = \{B, C, D, E, l_f, l_r, I_z\}$. The DE system is integrated using the Runge-Kutta method with a sampling period of 0.01 seconds for a duration of 60 seconds. Gaussian noise is added to the obtained trajectories with a mean of zero and variances $\sigma_{vy}^2 = 2.5e^{-3}$ for v_y and $\sigma_r^2 = 6.25e^{-6}$ for r. The integration parameters used are as follows: $B = 3$, $C = 2$, $D = 1$, $E = 1$, $l_f = 1.5$, $l_r = 3$, $I_z = 1.9$, $m = 1500$, $v_x = 30$.

4.2   Method Settings

The developed method (Fig. 3) is carried out in two stages. Initially, the obtained observations for the system's state variables v_y and r are smoothed using a cubic spline. To filter the noise from the observations, the spline is regularized by penalizing its curvature to avoid interpolating the noise. Two functions $S_{vy}(x)$ and $S_r(x)$ are obtained, allowing the approximation of the derivatives of the two state variables of the system $\dot{S}_{vy}(x)$ and $\dot{S}_r(x)$ despite the noise initially present in the observations. $x \in X$ corresponds to the position of observations over time, $X = \{0.01, 0.02, ..., 60.0\}$.

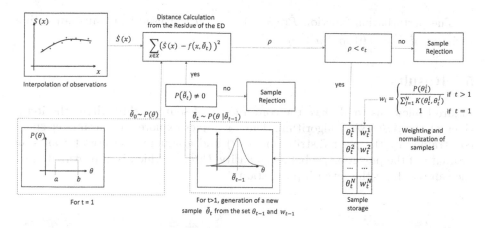

Fig. 3. General description of the method: The function $S(x)$ corresponds to a cubic spline interpolation performed on the observations of a state variable. The different steps during the iterations t of the *ABC-SMC* algorithm are represented. $f(x, \theta_t)$ represents a differential equation parameterized by θ_t.

The residue of the DE describing the system's dynamics is defined as follows:

$$\begin{cases} \dot{S}_{vy}(x) - \dfrac{F_{yf}}{m} \cos \delta_f + \dfrac{F_{yr}}{m} - v_x S_r(x) = 0 \\ \dot{S}_r(x) - \dfrac{l_f}{I_z} F_{yf} \cos \delta_f - \dfrac{l_r}{I_z} F_{yr} = 0 \end{cases} \tag{4}$$

An approximation of the solution for this system of DEs is given for:

$$\hat{\theta} = \arg \min_{\theta \in \Theta} \rho(\theta)$$

with:

$$\rho(\theta) = \frac{1}{|X|} \sum_{x \in X} \left(\left(\dot{S}_{vy}(x) - \frac{F_{yf}}{m} \cos \delta_f + \frac{F_{yr}}{m} - v_x S_r(x) \right)^2 + \left(\dot{S}_r(x) - \frac{l_f}{I_z} F_{yf} \cos \delta_f - \frac{l_r}{I_z} F_{yr} \right)^2 \right)$$

The value of $\rho(\theta)$ corresponds to a metric of adequacy between the parameters θ and the solution of the DE. $\rho(\theta)$ is used as a distance within the *ABC-SMC* algorithm to reject parameters that are unlikely to correspond to a solution of the DE system. To constrain the search for solutions of the DE, a uniform prior $P(\theta) = unif(a, b)$ is placed on each of the 7 parameters to be estimated:

Parameter	B	C	D	E	l_f	l_r	I_z
a	2.4	1.4	0.4	0.4	0.9	2.4	1.3
b	4	3	2	2	2.5	4	2.9

A number of iterations $T = 8$ is specified, with $N = 2000$ samples generated for each iteration. To filter the samples during each iteration, the following thresholds are used: $\epsilon_t = \{0.5, 0.3, 0.15, 0.05, 0.025, 0.02, 0.015, 0.0125\}$

The perturbation function $P(\theta|\tilde{\theta}_{t-1}) = \mathcal{N}\left(\mu, \sigma^2\right)$ used is a Gaussian with a variance of $\sigma^2 = 0.01$ and $\mu = \tilde{\theta}_{t-1}$

5   Results

Figure 4 allows us to observe the convergence of samples throughout the iterations of the *ABC-SMC* algorithm. These samples explore the parameter space constrained by the prior distribution $P(\theta)$ and gradually converge towards an estimate of the posterior probability density $P(\theta|\mathcal{D})$, which is approximated by the samples from the seventh population.

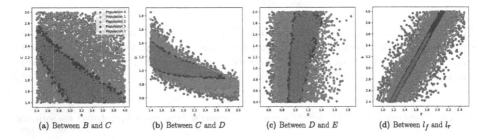

(a) Between B and C (b) Between C and D (c) Between D and E (d) Between l_f and l_r

Fig. 4. Pair diagrams showing the convergence of samples towards the posterior distribution.

The *ABC-SMC* approach is part of a comprehensive strategy thanks to the obtained posterior probability distribution. This distribution provides support for statistically analyzing the sensitivity of the system's various parameters concerning the observations used. To illustrate this approach, the results presented in this section first focus on analyzing the sensitivity of the parameters of the lateral dynamics model of a vehicle. To visualize the final distribution obtained, Fig. 5 show an estimation of the posterior probability density by kernel density based on the final samples obtained. Parameters showing high variance indicate low sensitivity of the system towards them, meaning these dimensions of the search space have limited influence on the likelihood of a sample being accepted or rejected. Figure 5 focuses on pairs of parameters showing correlations indicating system sensitivity to them. These graphs shed light on correlations between l_f and l_r, referring to the vehicle length and the center of gravity position. They also show correlations between parameters C and D, as well as between C and B. In addition to these correlations, parameter D shows a low standard deviation, which corresponds in the Pacejka model to the maximum force that can be generated by the tires on the ground. This sensitivity is consistent with the simulated data, representing curved trajectories at high speed. The estimation of E turns out to be distant from the one used to simulate the data. Figure 5c shows significant dispersion in the distribution of this parameter. This dispersion is explained by the fact that parameter E represents the curvature factor reached at after the grip peak of Pacejka model. A precise estimation of this curvature factor would require subjecting the vehicle to more aggressive solicitations. The analysis of the obtained probability distributions offers an interpretation of the global system parameters' observability. Among all the samples obtained in the last iteration of the *ABC-SMC* algorithm, the sample $\hat{\theta}$ that generated the minimal measure $\rho(\hat{\theta})$ is retained:

Parameter θ	B	C	D	E	l_f	l_r	I_z
$\hat{\theta}$	2.961	1.723	1.176	1.685	1.507	3.009	1.798
Standard deviation	0.348	0.291	0.095	0.427	0.153	0.269	0.327

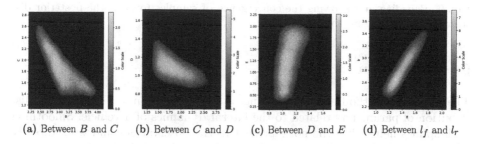

(a) Between B and C (b) Between C and D (c) Between D and E (d) Between l_f and l_r

Fig. 5. Pairwise plot resulting from kernel density estimation on the obtained samples.

Figures 6a and 6b allow for the visualization of the interpolation of observations made with the cubic spline. Integrating the system with parameter values $\hat{\theta}$ yields a solution very close to the trajectory modeled by the spline. The RMSE error is calculated between the trajectories obtained by integrating the system with $\hat{\theta}$ and the observations: concerning parameter $\hat{\theta}$ for v_y: RMSE $-$ 0.0398, and for r: RMSE $= 0.0020$; concerning parameter θ^* for v_y: RMSE $= 0.0397$, and for r: RMSE $= 0.0020$.

The RMSE error obtained following the integration generated by the parameters θ^* represents our baseline reference. This error corresponds to the additional noise added to each of the state variables. By comparing the RMSE error between the parameters $\hat{\theta}$ and that obtained with θ^*, we observe that the parameters estimated by our method generate an RMSE error similar to that of θ^* for both state variables. This demonstrates that our method was able to filter the noise from the observations and identify a set of parameters that could have generated the observed data.

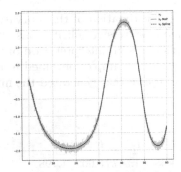

(a) Evolution of $v_y(x)$. In red, the trajectory generated by the parameterization $\hat{\theta}$, in blue for θ^*, and dashed line represents the interpolation $S_{vy}(x)$.

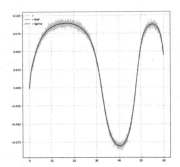

(b) Evolution of $r(x)$. In red, the trajectory generated by the parameterization $\hat{\theta}$, in blue for θ^*, and dashed line represents the interpolation $S_r(x)$.

Fig. 6. Pair diagrams showing the convergence of samples towards the posterior distribution. (Color figure online)

6   Conclusion

The proposed method allows obtaining an approximation of the posterior distribution of parameters related to the lateral dynamics of a vehicle. Smoothing with splines helps filter the noise present in the observations and work directly on an estimation of the derivatives of the system's state variables. This opens the way to using the Bayesian approach *ABC-SMC* using a distance metric based on the residual of the differential equation. By using this metric, we can avoid the computational times related to the integration of the system, favoring the generation of a larger number of samples with a stricter rejection threshold, which leads to better convergence towards the posterior probability density.

The incorporation of prior knowledge compensates for the small amount of available observations by restricting the parameter exploration space to be estimated. Obtaining a posterior probability distribution allows studying the sensitivity of the system and the nature of the relationships between its different parameters, thus providing interpretability of the obtained solutions.

We assume that the model used to generate the observations is similar to the model for which we are seeking to estimate the parameters. To strengthen the robustness of our approach, we plan to estimate the parameters of a system of differential equations while simultaneously taking into account discrepancies between the chosen model and the observations.

Acknowledgment. This research is financially supported by the Ministry of Defense through the Defense Innovation Agency (AID) and by the National Institute for Research in Computer Science and Automation (INRIA).

References

1. Sinha, N.K.: System identification - theory for the userâĂf: Lennart ljung. Autom **25**(3), 475–476 (1989)
2. Biegler, L.T., Damiano, J.J., Blau, G.E.: Nonlinear parameter estimation: a case study comparison. AIChE J. **32**(1), 29–45 (1986)
3. Wang, S., Xu, X.: Parameter estimation of internal thermal mass of building dynamic models using genetic algorithm. Energy Convers. Manage. **47**(13), 1927–1941 (2006)
4. Gutowski, N., Schang, D., Camp, O., Abraham, P.: A novel multi-objective medical feature selection compass method for binary classification. Artif. Intell. Med. **127**, 102277 (2022)
5. Eftaxias, A., Font, J., Fortuny, A., Fabregat, A., Stüber, F.: Nonlinear kinetic parameter estimation using simulated annealing. Comput. Chem. Eng. **26**(12), 1725–1733 (2002)
6. Schwaab, M., Biscaia, E.C., Jr., Monteiro, J.L., Pinto, J.C.: Nonlinear parameter estimation through particle swarm optimization. Chem. Eng. Sci. **63**(6), 1542–1552 (2008)
7. Ascher, U.M., Petzold, L.R.: Computer Methods for Ordinary Differential Equations and Differential-Algebraic Equations, 1st edn. Society for Industrial and Applied Mathematics, USA (1998)
8. Raissi, M., Perdikaris, P., Karniadakis, G.: Physics-informed neural networks: a deep learning framework for solving forward and inverse problems involving nonlinear partial differential equations. J. Comput. Phys. **378**, 686–707 (2019)
9. Barber, D., Wang, Y.: Gaussian processes for Bayesian estimation in ordinary differential equations. In: International Conference on Machine Learning (2014)
10. Calderhead, B., Girolami, M., Lawrence, N.: Accelerating bayesian inference over nonlinear differential equations with gaussian processes. In: Advances in Neural Information Processing Systems, vol. 21. Curran Associates, Inc. (2008)
11. Rasmussen, C.E., Williams, C.K.I.: Gaussian Processes for Machine Learning. Adaptive Computation and Machine Learning, MIT Press, Cambridge (2006)
12. Chib, S., Greenberg, E.: Understanding the metropolis-hastings algorithm. Am. Stat. **49**(4), 327–335 (1995)
13. Casella, G., George, E.I.: Explaining the Gibbs sampler. Am. Stat. **46**(3), 167–174 (1992)
14. Marin, J.-M., Pudlo, P., Robert, C.P., Ryder, R.J.: Approximate Bayesian computational methods. Stat. Comput. **22**, 1167–1180 (2012)
15. Toni, T., Welch, D., Strelkowa, N., Ipsen, A., Stumpf, M.P.: Approximate Bayesian computation scheme for parameter inference and model selection in dynamical systems. J. R. Soc. Interface **6**, 187–202 (2009)
16. Liepe, J., Kirk, P., Filippi, S., Toni, T., Barnes, C.P., Stumpf, M.P.H.: A framework for parameter estimation and model selection from experimental data in systems biology using approximate Bayesian computation. Nat. Protoc. **9**(2), 439–456 (2014)
17. Secrier, M., Toni, T., Stumpf, M.P.H.: The ABC of reverse engineering biological signalling systems. Mol. BioSyst. **5**, 1925–1935 (2009)
18. Lillacci, G., Khammash, M.: Parameter estimation and model selection in computational biology. PLOS Comput. Biol. (2010)
19. Pacejka, H. Bakker, E., Pacejka, W., Hans, B., Afsar, M.: Tyre and Vehicle Dynamics. Elsevier Science (2006)

Assessing Distance Measures for Change Point Detection in Continual Learning Scenarios

Collin Coil[ID] and Roberto Corizzo[(✉)][ID]

American University, Washington DC 20016, USA
{cc6121a,rcorizzo}@american.edu

Abstract. Detecting relevant change points in time-series data is a necessary task in various applications. Change point detection methods are effective techniques for discovering abrupt changes in data streams. Although prior work has explored the effectiveness of different algorithms on real-world data, little has been done to explore the impact of different distance measures on change detection performance. In this paper, we modify the architecture of a change point detection workflow to assess the impact of distance measure choices on change detection accuracy and efficiency in continual learning scenarios, where the goal is detecting transitions between tasks or concepts. An experimental evaluation of 41 distance measure across several benchmark datasets demonstrated that the change detection accuracy depends on the distance measure selected. Furthermore, our analysis showed performance patterns for distance measures in the same family.

Keywords: Continual learning · Change point detection · Time-series data

1 Introduction

Change point detection has a critical relevance in real-world applications characterized by dynamic environments. Offline approaches analyze the probability distributions of data before and after a candidate change point using methods such as likelihood ratios, which confirm whether the distributions are significantly different [25]. Some offline methods can discover multiple change points through binary segmentation [20], and can include pruning [26] as well as tolerance to outliers [19]. Since these methods rely on predefined parametric models, their major drawback is limited flexibility in real-world scenarios [2]. Moreover, most of them are limited to the analysis of univariate data. Probabilistic methods include offline and online Bayesian methods where change point prediction is carried out by comparing probability values [27]. Non-parametric change detection methods typically leverage hypothesis testing via test statistics [24,28] or are based on deep neural networks [8,13].

© The Author(s), under exclusive license to Springer Nature Switzerland AG 2024
A. Appice et al. (Eds.): ISMIS 2024, LNAI 14670, pp. 260–270, 2024.
https://doi.org/10.1007/978-3-031-62700-2_23

Besides its direct exploitation for decision support, change detection can also be leveraged to provide auxiliary information in automated machine learning pipelines. Continual learning (also known as lifelong learning) can be described as a continuous process in which a series of different problems, defined as tasks (or concepts), are presented to a machine learning method over time [30]. In continual learning, the main goal is to learn models that are capable to acquire new skills, by adjusting to newly presented tasks (or concepts), while preserving previously learned information, to tackle both new challenges and the recurrence of previously occurred ones [35]. Continual learning approaches in the literature can be categorized in regularization-based strategies [32], dynamic architectures [29,31], and replay-based strategies [7,18]. Popular continual learning scenarios focus on classification problems and are known as task-incremental, class-incremental, and domain-incremental [34]. Both class-incremental and task-incremental scenarios provide the model with new, previously unseen classes to be incorporated by the model [4,11]. Domain-incremental scenarios provide new distributions of already known classes [22]. Emerging types of scenario include online continual learning [12], scenarios with recurring tasks [10], and task-agnostic or task-free scenarios [17]. In task-agnostic and task-free scenarios, change detection is fundamental to uncover transitions between tasks and trigger specific behaviors, such as model updates, or the selection of the most suitable sub-model from a pool of existing ones [17]. Initial attempts to tackle this challenging learning scenario have shown the effectiveness of the Wasserstein distance [15]. This distance measures the work required to transport the probability mass from one distribution state to another, which provides a smooth and significant representation of the distance between distributions [14]. For this reason, the Wasserstein distance is increasingly being used in machine learning methods and notably in neural network architectures [21,33]. However, one crucial limitation of the Wasserstein distance is its computational complexity, which can result in a limited time efficiency when analyzing large-scale data. Motivated by this issue, in this paper we investigate the trade-off between accuracy and efficiency of different distance measures for change detection in continual learning scenarios. To this end, we generalize a Wasserstein distance-based change detection workflow and assess the performance of 41 measures belonging to 9 families in practical continual learning scenarios. Our extensive evaluation encompassing 4 real-world datasets exposes which measures provide accurate results while being efficient in different cases.

2   Method

We adopt WATCH, the change detection workflow defined in [14], and generalize it to support any distance measure of choice. The input of the algorithm is a time series of data samples $\mathbf{I} = (I_1, I_2, \dots, I_T)$, where each data sample is a d-dimensional vector: $I_t \in \mathbb{R}^d$. The algorithm presents the following parameters:

- κ – minimum number of points for the currently monitored distribution to trigger change point detection.

- μ – maximum number of points for the current distribution (memory size).
- ϵ – a ratio controlling how distant the samples can be from the current distribution to be considered a part of the same.
- ω – batch size controlling the number of samples processed at a time.

Our workflow allows us to learn a representation of a distribution D in an unsupervised manner with an initial set of data points. Subsequently, we incrementally store newly available samples in a buffer. Once the desired capacity of the buffer (κ) is reached, change detection is activated. We collect newly arriving points in non-overlapping batches of length ω, and calculate the distance between the multivariate mean $\overline{B}$ of every batch B and the multivariate mean $\overline{D}$ of a distribution D. If the distance is lower than the desired threshold γ (controlled by ϵ), we qualify B as belonging to D. The threshold is determined as:

$$\eta(D,\ \epsilon) = \epsilon \max_{B \in D} dist(\overline{B}, \overline{D}). \tag{1}$$

Subsequently, D is updated to include the new data points. On the contrary, if the distance is higher than γ, we assume that the current batch B to be part of a different distribution. We annotate this point in time as a change point and start creating a new distribution to replace D. Once D reaches its capacity μ, the oldest samples are removed to accommodate new ones. A pseudo-code of the algorithm is presented in Algorithm 1. Our approach allows us to realize change detection by simply comparing the multivariate means of two sets of points, and detect distributions of unknown types, without restricting to Normal distributions. Our approach creates the distribution D through sampling, then compares the mean $\overline{D}$ with the batch mean $\overline{B}$. Since the batch sizes we use are large ($\omega \geq 50$), the effectiveness of our approach is likely a result of approximating the mean of the sampling distribution, its normality from the Central Limit Theorem, and the low variance of the sampling distribution when the batch size is large.

To define continual learning scenarios, we resort to the characterization defined in [16] for continual/lifelong anomaly detection. The procedure leverages clustering functions to create normal and anomaly concepts (self-consistent sets of data points) based on normal and anomaly data, respectively. While training sets contain only normal data, evaluation sets contain both normal and anomaly data. To this end, the datasets in [16] present different scenarios: clustered anomaly concepts assigned to the closest normal concept (**A**), clustered anomaly concepts assigned randomly to normal concepts (**C**), and anomalies randomly assigned to normal concepts (**R**).

While the focus in [16] is on assessing the anomaly detection performance of different models when exposed to evolving normal concepts, in our work, we restrict to the analysis of a continuous sequence of normal concepts (training data) without the presence of anomalies, thus discarding evaluation data. We restrict our focus to the training data, so the three different scenarios can be seen as different orderings of data points within each normal concept. By performing experiments with all three scenarios, our goal is to investigate the accuracy,

Input   : $\mathbf{I} = (I_1, I_2, \ldots, I_T)$ Time series data

Result: R: Set of change points

Parameters: κ (min points dist.), μ (max points dist.), ϵ (threshold ratio), ω (batch size)

1   Set current distribution $D \leftarrow \emptyset$
2   $R \leftarrow \emptyset$

3   $\mathbf{B} \leftarrow$ Process $\mathbf{I}$ into sequence $(B_1, B_2, \ldots, B_{\frac{T}{\omega}})$ of non-overlapping batches B_i of size ω
4   **for** $B_i \in \mathbf{B}$ **do**
5   **if** $|D| < \kappa$ **then**
6         $D \leftarrow D \cup B_i$
7   **if** $|D| >= \kappa$ **then**
8   Determine η from Eq. (1) with D and ϵ
9   **end**
10   **else**
11         $v \leftarrow dist(\overline{B_i}, \overline{D})$
12   **if** $v > \eta$ **then**
13   Set change point $c \leftarrow i * \omega$
14           $R \leftarrow R \cup \{c\}$
15           $D \leftarrow \{B_i\}$
16   **else**
17   **if** $|D| < \mu$ **then**
18             $D \leftarrow D \cup B_i$
19   Determine η from Eq. 1 with D and ϵ
20   **end**
21   **end**
22   **end**
23   **end**

Algorithm 1: Pseudo-code of the change point detection workflow

efficiency, and stability of distance measures for change detection across these scenarios.

3   Experiments

3.1   Datasets

In our experiments, we adopt real-world datasets in relevant domains such as cybersecurity and smart grids.

- **NSL-KDD**: Network traffic data gathered during the DARPA Intrusion Detection Systems evaluation;
- **UNSW-NB15**: Hybrid network traffic from contemporary systems;
- **Energy**: Sensor-based observations collected from solar plants in Italy (17 plants, 2.5 years of data);
- **Wind**: Wind power production data from eolic/wind parks based on the Weather Research & Forecasting (WRF) model (5 plants, 2 years).

Each dataset presents a unique feature set, which allows us to assess the change detection accuracy in heterogeneous systems.

3.2   Metrics

In our experiments, we adopt popular metrics for change detection:

F1-Score: Change point detection algorithms may be evaluated in terms of classification between change-point and non-change point classes [6]. Usually, datasets are affected by class imbalance due to a low number of change points. To this end, robust metrics are *Precision*, i.e. the ratio of correctly detected change points over the number of all detected change points, and *Recall*, i.e. the ratio of correctly detected change points over the number of actual change points. The *F1-score* is the harmonic mean of *Precision* and *Recall*. It is common practice to consider a margin of error around the actual change points to allow for minor discrepancies. We use a margin of error equal to the batch size ω considered in the experiments.

Covering (Cover): The covering metric was presented in [3] and describes how well the predicted segments cover ground truth segments, corresponding to specific subsequences of contiguous data points. The metric is defined upon the Jaccard Index (also known as Intersection over Union), a statistic used for measuring the similarity/overlap of sample sets, and defined as: $J(A, A') = \frac{|A \cap A'|}{|A \cup A'|}$. The definition of the covering metric of partition G by partition G' introduced in [3] was adjusted to time series data in [6]:

$$Cover(G', G) = \frac{1}{T} \sum_{A \in G} |A| \max_{A' \in G'} J(A, A') \tag{2}$$

3.3   Setup

For the experiments, we chose the batch size ω so that it did not divide the change point indices. This was done to simulate trials on testing data where some batches would almost certainly contain data points from two concepts. Furthermore, we set the minimum number of points to trigger a change point detection κ to be an integer multiple of the batch size. We did not put a cap on the maximum number of points in a distribution μ.

   We left the threshold ratio ϵ for hyperparameter tuning. For this, we used Hyperopt [5], an automated hyperparameter tuning program. We defined out cost function to be the sum of cover and F1, and we used Hyperopt to select the threshold ratio ϵ to maximize the sum.

   We constrained this hyperparameter tuning in two ways. First, we limited the optimizer to run a maximum of 100 iterations or for a maximum of two hours. Second, we stopped tuning if the sum of cover and F1 exceeded 1.9. The first constraint ended tuning for trials that used distance measures that were either too sensitive to the threshold ratio ϵ, too computationally expensive, or ineffective. Few trials ended early due to sufficiently high performance.

   All experiments are implemented in Python v3.7.16, numpy v.1.21.6, and hyperopt v.0.2.7. We used the R implementation of the Wasserstein distance provided in [23], and called it from the Python code using the rpy2 v3.5.14

bridge. All experiments are run on a machine with an Intel Core i7-9750H CPU, GeForce RTX 2060 GPU, and 32 GB of RAM.

3.4   Analysis of Results

We assess performance of the change detection algorithm using the various distance metrics. We analyze the performance of the generalized algorithm with each distance measure against the unmodified WATCH performance, and examine the runtime for the best-performing distance measure in each family.

Cover and F1: Results in terms of Cover and F1 metrics are presented in Tables 1 and 2[1]. The tables are separated by family, and the members of the family are ordered by performance. An overview of each distance measure family, their characteristics, and which measures are metrics can be found in [9].

The tables show a wide range of performance across difference distance measure families. The L_1, L_p, and Intersection families had F1 and cover scores that were generally greater than other families. Although the inter-family variation in performance is notable, there is little intra-family variation as members of the same family tended to have similar performance. The results from the **C** scenario are very similar to the results from the **A** and **R** scenarios.

Some distance measures had lower cover and F1 scores, and this lower performance is combination of the measure being sensitive to the threshold ratio ϵ or generally being inferior. Members of the Vicissitude, Shannon's Entropy, and Squared L_2 families were most sensitive to ϵ.

WATCH generally had lower performance trials using other distance measures. For example, in the Energy dataset, the it has an average F1 of 0.326 and cover of 0.581 across the **A**, **C**, and **R** scenarios. In the same dataset, the members of the L_1 family had an average F1 of 0.988 and cover of 0.981. This is a 203.1% and 68.8% increase, respectively.

The notable exception to the lower performance of WATCH are the trials using the NSL-KDD dataset. Whereas other distance measures suffered lower performance on the NSL-KDD dataset than other datasets, WATCH had the opposite result. Furthermore, WATCH outperformed distance measures in other families on the NSL-KDD dataset.

Runtime: In addition to calculating the cover and F1 metrics, we assessed the runtime of the change point detection algorithm using each distance measure. We do this because the Wasserstein distance used in the original version of WATCH is computationally expensive. We present a subset of the results in Table 3.

The runtime for this algorithm depends on the computational expensiveness of the distance metric and the depth of the nested loops. The latter is determined by the performance of the algorithm since fewer detections leads to more steps needed in each loop. Therefore, the distance measures with the best cover and F1 performance (e.g., Kulczynski d and Czekanowski) having the lowest runtime

---

[1] Due to space limitations, we restricted our table to include only the **C** scenario. Code and additional experiments with the **A** and **R** scenarios are reported as an external appendix: https://github.com/CollinCoil/cpd-distances.

Table 1. Cover and F1 scores using various distance measures by family in order of performance

	Energy (C)		NSL-KDD (C)		UNSW (C)		Wind (C)	
L_1	Cover	F1	Cover	F1	Cover	F1	Cover	F1
Kulczynski d	0.988	1	0.746	0.895	0.98	1	0.996	1
Bray-Curtis	0.988	1	0.746	0.895	0.98	1	0.996	1
Gower	0.988	1	0.746	0.895	0.98	1	0.996	1
Lorentzian	0.988	1	0.738	0.872	0.98	1	0.996	1
Sorgel	0.988	1	0.721	0.865	0.98	1	0.996	1
Canberra	0.933	0.909	0.360	0.462	0.883	0.947	0.996	1
Intersection	Cover	F1	Cover	F1	Cover	F1	Cover	F1
Czekanowski	0.984	1	0.746	0.895	0.98	1	0.996	1
Tanimoto	0.984	1	0.721	0.865	0.972	1	0.996	1
Wave Hedges	0.984	1	0.531	0.71	0.883	0.947	0.996	1
Motyka	0.593	0.75	0.727	0.85	0.985	1	0.996	1
Nonintersection	0.299	0.429	0.61	0.727	0.855	0.9	0.995	0.909
L_p	Cover	F1	Cover	F1	Cover	F1	Cover	F1
Manhattan	0.988	1	0.746	0.895	0.98	1	0.996	1
Minkowski (p = 2)	0.984	1	0.641	0.773	0.98	1	0.996	1
Euclidean	0.984	1	0.462	0.647	0.972	1	0.996	1
Chebyshev (Max)	0.617	0.432	0.41	0.202	0.563	0.538	0.849	0.533
Chebyshev (Min)	0.372	0.12	0.068	0.095	0.223	0.333	0.463	0.17
Squared Chord	Cover	F1	Cover	F1	Cover	F1	Cover	F1
Squared-chord	0.871	0.69	0.626	0.857	0.972	1	0.996	1
Matusita	0.846	0.625	0.626	0.857	0.972	1	0.996	1
Hellinger	0.593	0.75	0.626	0.857	0.972	1	0.996	1
Bhattacharyya	0.299	0.429	0.253	0.205	0.482	0.438	0.203	0.333
Squared L_2	Cover	F1	Cover	F1	Cover	F1	Cover	F1
Squared χ^2	0.973	0.952	0.626	0.857	0.98	1	0.996	1
Divergence	0.675	0.476	0.531	0.71	0.883	0.947	0.996	1
Add. Sym. χ^2	0.974	0.952	0.531	0.579	0.698	0.526	0.954	0.8
Clark	0.397	0.571	0.495	0.556	0.883	0.947	0.996	1
Pearson χ^2	0.885	0.947	0.068	0.095	0.256	0.462	0.996	1
Max Sym. χ^2	0.974	0.952	0.302	0.444	0.357	0.2	0.411	0.571
Neyman χ^2	0.969	0.909	0.302	0.444	0.357	0.195	0.411	0.571
Combinations	Cover	F1	Cover	F1	Cover	F1	Cover	F1
ACC	0.988	1	0.746	0.895	0.98	1	0.98	1
Taneja	0.435	0.471	0.068	0.095	0.256	0.462	0.996	1

Table 2. Cover and F1 scores using various distance measures by family, continued

Shannon's Entropy	Energy (C)		NSL-KDD (C)		UNSW (C)		Wind (C)	
	Cover	F1	Cover	F1	Cover	F1	Cover	F1
K Divergence	0.66	0.636	0.562	0.75	0.75	0.889	0.703	0.667
Topsøe	0.952	0.952	0.068	0.095	0.256	0.462	0.996	1
Jensen-Shannon	0.952	0.952	0.068	0.095	0.256	0.462	0.996	1
Jensen	0.952	0.952	0.068	0.095	0.256	0.462	0.996	1
Jeffreys	0.435	0.471	0.068	0.095	0.256	0.462	0.996	1
KL Divergence	0.69	0.737	0.068	0.095	0.256	0.462	0.575	0.75
Vicissitude	Cover	F1	Cover	F1	Cover	F1	Cover	F1
Vicis-Wave Hedges	0.892	0.947	0.579	0.462	0.3	0.19	0.411	0.571
Vicis Sym. χ^2	0.614	0.56	0.551	0.316	0.43	0.3	0.441	0.444
Uncategorized	Cover	F1	Cover	F1	Cover	F1	Cover	F1
Penrose Shape	0.988	1	0.59	0.8	0.972	1	0.996	1
Google	0.496	0.667	0.799	0.919	0.98	1	0.996	1
Pearson Distance	0.377	0.051	0.557	0.197	0.202	0.125	0.494	0.286
	Cover	F1	Cover	F1	Cover	F1	Cover	F1
WATCH	0.633	0.353	0.612	0.512	0.586	0.36	0.339	0.102

Table 3. Mean runtime of trials using the three variants of each dataset. One metric from each family was selected. The fastest mean runtime for each dataset is in bold.

Metric	Runtime (s)			
	Energy	NSL-KDD	UNSW	Wind
Kulczynski d	17.352	**40.06**	17.888	22.514
Czekanowski	17.024	41.351	17.226	22.818
Manhattan	15.769	51.383	17.308	21.912
ACC	16.111	49.751	18.21	23.892
Squared-chord	**14.999**	56.455	**16.775**	**21.738**
Penrose Shape	16.466	71.524	19.013	23.446
Squared χ^2	18.31	71.861	17.836	22.598
K Divergence	46.887	95.399	44.355	130.316
Vicis-Wave Hedges	20.813	59.961	105.243	285.277
WATCH	222.304	637.036	463.18	311.539

is unsurprising. In the UNSW trials, WATCH returned more change points than trials using K divergence, so it should have run faster. However, WATCH was slower, indicating its computational expensiveness of calculating the Wasserstein distance is the driving factor in its runtime.

4   Conclusion

In this work, we assessed the impact of distance measures on the accuracy and efficiency of change point detection. We generalized the workflow in WATCH, a Wasserstein distance-based change point detection approach, to support a variety of distance measures in place of the Wasserstein distance. We find four notable results:

- The performance of the change detection algorithm depends heavily on the distance measure chosen;
- Distance measures in the same family tend to have similar performance;
- The WATCH algorithm can yield accurate change detection results when the Wasserstein distance is replaced with other distance measures;
- Distance measures from the L_1, L_p, and Intersection families generally had the highest performance across all datasets;
- Distance measures from the L_1, L_p, and Intersection families are generally easier to tune and are less sensitive to the threshold ratio hyperparameter ϵ.

The first two of these results align with the findings of [1], who examined the impact of distance measures on KNN classifier performance, suggesting distance measure choice effects a variety of machine learning algorithms. A natural extension of our study is integrating our generalized change detection algorithm in continual/lifelong learning anomaly detection frameworks. We also encourage further work assessing the impact of distance measure choice on other change point detection algorithms and machine learning algorithms for other tasks.

References

1. Abu Alfeilat, H.A., et al.: Effects of distance measure choice on k-nearest neighbor classifier performance: a review. Big Data **7**(4), 221–248 (2019)
2. Aminikhanghahi, S., Cook, D.J.: A survey of methods for time series change point detection. Knowl. Inf. Syst. **51**(2), 339–367 (2017)
3. Arbeláez, P., Maire, M., Fowlkes, C., Malik, J.: Contour detection and hierarchical image segmentation. IEEE Trans. Pattern Anal. Mach. Intell. **33**(5), 898–916 (2011)
4. Belouadah, E., Popescu, A., Kanellos, I.: A comprehensive study of class incremental learning algorithms for visual tasks. Neural Netw. **135**, 38–54 (2021)
5. Bergstra, J., Yamins, D., Cox, D.: Making a science of model search: hyperparameter optimization in hundreds of dimensions for vision architectures. In: International Conference on Machine Learning, pp. 115–123. PMLR (2013)
6. van den Burg, G.J.J., Williams, C.K.I.: An evaluation of change point detection algorithms. arXiv preprint arXiv:2003.06222 (2020)
7. Buzzega, P., Boschini, M., Porrello, A., Calderara, S.: Rethinking experience replay: a bag of tricks for continual learning. In: 2020 25th International Conference on Pattern Recognition (ICPR), pp. 2180–2187. IEEE (2021)
8. Ceci, M., Corizzo, R., Japkowicz, N., Mignone, P., Pio, G.: ECHAD: embedding-based change detection from multivariate time series in smart grids. IEEE Access **8**, 156053–156066 (2020)

9. Cha, S.H.: Comprehensive survey on distance/similarity measures between probability density functions. City **1**(2), 1 (2007)
10. Cossu, A., et al.: Is class-incremental enough for continual learning? Front. Artif. Intell. **5**, 829842 (2022)
11. De Lange, M., et al.: A continual learning survey: defying forgetting in classification tasks. IEEE Trans. Pattern Anal. Mach. Intell. **44**(7), 3366–3385 (2021)
12. De Lange, M., Tuytelaars, T.: Continual prototype evolution: learning online from non-stationary data streams. In: Proceedings of the IEEE/CVF International Conference on Computer Vision (ICCV), pp. 8250–8259 (2021)
13. Du, H., Duan, Z.: Finder: a novel approach of change point detection for multivariate time series. Appl. Intell. **52**, 2496–2509 (2022)
14. Faber, K., Corizzo, R., Sniezynski, B., Baron, M., Japkowicz, N.: WATCH: Wasserstein change point detection for high-dimensional time series data. In: 2021 IEEE International Conference on Big Data (Big Data), pp. 4450–4459. IEEE (2021)
15. Faber, K., Corizzo, R., Sniezynski, B., Baron, M., Japkowicz, N.: LIFEWATCH: lifelong wasserstein change point detection. In: 2022 International Joint Conference on Neural Networks (IJCNN), pp. 1–8. IEEE (2022)
16. Faber, K., Corizzo, R., Sniezynski, B., Japkowicz, N.: Lifelong learning for anomaly detection: new challenges, perspectives, and insights. arXiv preprint arXiv:2303.07557 (2023)
17. Faber, K., Corizzo, R., Sniezynski, B., Japkowicz, N.: VLAD: task-agnostic VAE-based lifelong anomaly detection. Neural Netw. **165**, 248–273 (2023)
18. Faber, K., Sniezynski, B., Corizzo, R.: Distributed continual intrusion detection: a collaborative replay framework. In: 2023 IEEE International Conference on Big Data (BigData), pp. 3255–3263. IEEE (2023)
19. Fearnhead, P., Rigaill, G.: Changepoint detection in the presence of outliers. J. Am. Stat. Assoc. **114**(525), 169–183 (2019)
20. Fryzlewicz, P.: Wild binary segmentation for multiple change-point detection. Ann. Stat. **42**(6), 2243–2281 (2014)
21. Gaujac, B., Feige, I., Barber, D.: Learning disentangled representations with the wasserstein autoencoder. In: Oliver, N., Pérez-Cruz, F., Kramer, S., Read, J., Lozano, J.A. (eds.) ECML PKDD 2021. LNCS (LNAI), vol. 12977, pp. 69–84. Springer, Cham (2021). https://doi.org/10.1007/978-3-030-86523-8_5
22. Gunasekara, N., Gomes, H., Bifet, A., Pfahringer, B.: Adaptive neural networks for online domain incremental continual learning. In: Pascal, P., Ienco, D. (eds.) DS 2022. LNCS, pp. 89–103. Springer, Cham (2022). https://doi.org/10.1007/978-3-031-18840-4_7
23. Hallin, M., Mordant, G., Segers, J.: Multivariate goodness-of-fit tests based on wasserstein distance. Electron. J. Stat. **15**(1), 1328–1371 (2021)
24. Haynes, K., Fearnhead, P., Eckley, I.A.: A computationally efficient nonparametric approach for changepoint detection. Stat. Comput. **27**(5), 1293–1305 (2017)
25. Hinkley, D.V.: Inference about the change-point in a sequence of random variables (1970)
26. Killick, R., Fearnhead, P., Eckley, I.A.: Optimal detection of changepoints with a linear computational cost. J. Am. Stat. Assoc. **107**(500), 1590–1598 (2012)
27. Knoblauch, J., Jewson, J.E., Damoulas, T.: Doubly robust Bayesian inference for non-stationary streaming data with beta-divergences. In: Advances in Neural Information Processing Systems, vol. 31, pp. 64–75 (2018)
28. Matteson, D.S., James, N.A.: A nonparametric approach for multiple change point analysis of multivariate data. J. Am. Stat. Assoc. **109**(505), 334–345 (2014)

29. Mignone, P., Corizzo, R., Ceci, M.: Distributed and explainable GHSOM for anomaly detection in sensor networks. Mach. Learn. 1–42 (2024)
30. Parisi, G.I., Kemker, R., Part, J.L., Kanan, C., Wermter, S.: Continual lifelong learning with neural networks: a review. Neural Netw. **113**, 54–71 (2019)
31. Pietroń, M., Żurek, D., Faber, K., Corizzo, R.: Ada-QPacknet–adaptive pruning with bit width reduction as an efficient continual learning method without forgetting. In: European Conference on Artificial Intelligence (ECAI), pp. 1882–1889 (2023)
32. Sharif Razavian, A., Azizpour, H., Sullivan, J., Carlsson, S.: CNN features off-the-shelf: an astounding baseline for recognition. In: Proceedings of the IEEE Conference on Computer Vision and Pattern Recognition Workshops, pp. 806–813 (2014)
33. Tolstikhin, I., Bousquet, O., Gelly, S., Schölkopf, B.: Wasserstein auto-encoders. In: 6th International Conference on Learning Representations (ICLR 2018). OpenReview.net (2018)
34. Van de Ven, G.M., Tolias, A.S.: Three scenarios for continual learning. arXiv preprint arXiv:1904.07734 (2019)
35. Wang, Y.H., Lin, C.Y., Thaipisutikul, T., Shih, T.K.: Single-head lifelong learning based on distilling knowledge. IEEE Access **10**, 35469–35478 (2022)

SPL^{index}: A Spatial Polygon Learned Index

Masoumeh Vahedi$^{(\boxtimes)}$ and Henning Christiansen

Roskilde University, Roskilde, Denmark
{vahedi,henning}@ruc.dk

Abstract. Effective indexing is crucial for any AI and big data analysis task involving huge datasets. Recently, machine-learned models for indexing have achieved much attention, and we apply such for spatial data, specifically huge collections of polygons. We propose an index structure SPL^{index} that organizes polygons into a tree of clusters with linear regression models for effective branching in search. It integrates an effective layout of polygon data to disk space that minimize disk access and amount of data to be kept in main memory. The approach is shown outperform the state-of-the-art R-tree for both range and point queries.

Keywords: Spatial Index · Learned Index · Spatial Polygon Data

1 Introduction

This paper concerns one of the foundations of all AI system, namely the storage and efficient retrieval of huge and complex datasets. This is also important for XAI as a prediction may be supported finding similar, previous cases.

Here we approach the special case of spatial polygon and point data, as used in GIS, location-based services and elsewhere. We suggest a new index structure for very efficient search in large polygon sets stored on disk, minimizing the amount of disk accesses, and our tests demonstrate significant improvements compared with state of the art.

Inspired by the influential work of [14], we build an index as a model SPL^{index} learned from data and thus reflecting properties of those data. We use a tree structure optimized by 1) abstracting clusters into their minimal bounding boxes (MBRs), 2) mapping these boxes to intervals on linear scale based on a Z-curve (explained below), and 3) linear regression models for effective branching. This allows to efficiently compare minimal bounding boxes of clusters and query polygons, and only involve cumbersome tests for overlap of box and polygon when strictly necessary. In this paper, we did not consider updates.

Traditional spatial indexes can be divided into: *Space-partitioning*, e.g., Quad-trees [7] and KD-trees [2], suitable for points but not complex objects, and *data-partitioning* such as R-trees [10] and its variants [1,12,23], effective for various spatial objects but deficient for real big datasets [27]. Learned indexes

© The Author(s), under exclusive license to Springer Nature Switzerland AG 2024
A. Appice et al. (Eds.): ISMIS 2024, LNAI 14670, pp. 271–281, 2024.
https://doi.org/10.1007/978-3-031-62700-2_24

have been used for point data only by, e.g., [3,17,22,27]. Our design is unique, integrating a scheme for storing complex objects on disk. Section 2 provides basic concepts and an overview of SPL^{index}. Section 3 explains our learned index, and Sect. 4 access table and disk layout. Query processing is described in Sect. 5. Section 6 is on experimental studies, comparing with R-trees, showing that SPL^{index} out-performs in terms of query response time and IO cost. Sections 7 and 8 comments on related work and concludes.

2   Preliminaries and Overview

Definition 1 (Polygon). *A polygon P is a closed two-dimensional shape delineated by line segments $(p_1, p_2), (p_2, p_3), ..., (p_{n-1}, p_n), (p_n, p_1), \geq 3$, where $p_1, ..., p_n$ is a list of points $\in R_{\geq 0}^2$ called its vertices. For simplicity, polygons should not have crossing lines, which implies natural definitions of point containment and overlap.*

Throughout this paper, we assume a given and potentially very large set of polygons $\mathbf{P} = \{P_1, P_2, ...\}$. A *cluster* is a set of polygons. Rectangles are assumed to be parallel to the axes and are characterized by their lower-left and upper-right corners, called its *minimum* and *maximum* points, respectively.

Definition 2 (Minimum Bounding Rectangle, MBR). *A minimum bounding rectangle, MBR of a polygon P, written $MBR(P)$, is the smallest rectangle that contains P. The notion is extended in the natural way for clusters.*

Definition 3 (Spatial Queries). *We consider two sorts of queries.*
Point query: *for point p, return the polygons of $\mathbf{P}$ containing p.*
Range query: *for rectangle R, return the polygons of $\mathbf{P}$ overlapping with R.*

Overview. Figure 1 gives an overview of the data structures used by SPL^{index}. First, the polygons of P are divided into clusters according to spatial proximity. Next, clusters are organized in a Hierarchical Z-Interval Tree (HZIT), optimized for spatial queries. For this, clusters are abstracted into their MBRs and further into Z-intervals (below), that provide a one-dimensional ordering. The polygon vertices are located on disk and accessed from the tree via a separate table describing the disk layout, organized in a way that reduces the number of disk accesses. To speed up the search, each tree node applies a linear regression model that predicts an approximate index for clusters that overlap with a given query.

3   The Polygon Index

For time and space efficiency, the index in main memory contains only *MBRs*, while polygon details are referred to disk. Clustering speeds up query evaluation, as whole clusters may be included/discarded based on a single MBR test.

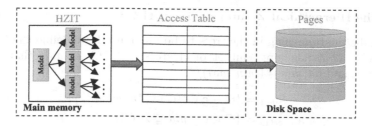

Fig. 1. SPL^{index} overview

3.1   Clustering of Polygons

We apply the BIRCH clustering algorithm [31] known its scalability and low memory usage; polygons are viewed as points in 4D space given by their minimum and maximum points. BIRCH is controlled by hyperparameters giving an upper limit for the number of elements in each cluster and a maximum diameter (largest distance between two points). These settings are important for obtaining a good balance between space and time consumption and depend on characteristics of the dataset **P**; we discuss this issue further in Sect. 6, below. We now assume a set of clusters $\mathbf{C} = \{C_1, C_2, \ldots\}$ produced by BIRCH for **P**.

3.2   Ranking Clusters by Z-Addresses

For efficient search, **C** is enumerated based on the Z-order curve [15] that provides a bijective mapping $Z \colon N_{\geq 0}^2 \to N_{\geq 0}$. It is extended for $\mathbb{R}_{\geq 0}^2$ as $Z((x, y)) = Z((\lfloor x \rfloor, \lfloor y \rfloor))$, called the Z-address of (x, y). For an MBR m given by points $p_{\min}, p_{\max}$, we define its Z-interval as $Z(m) = [Z(p_{\min}), Z(p_{\max})]$.

The Z mapping is monotonic w.r.t. the following domination relation:

$$(x, y) \text{ dominates } (x', y') \quad \Leftrightarrow \quad x \leq x' \wedge y \leq y'.$$

Monotonicity implies the following central properties:

$$\text{For any point } p \text{ and MBR } B \colon \quad p \in B \ \Rightarrow \ Z(p) \in Z(B).$$
$$\text{For any two MBRs } M_1, M_2 \colon \quad M_1 \subseteq M_2 \ \Rightarrow \ Z(M_1) \subseteq Z(M_2).$$

In other words, Z preserves point and MBR containment, although the implications in the other direction do not hold. In general, two clusters that are close in $\mathbb{R}_{\geq 0}^2$ may have arbitrary distant Z-intervals, but as argued by [19], on average, these intervals tend to be close. This assumption is crucial for the efficiency of our method as spatial queries look up polygons located within the same region.

Finally, the *rank* or *identifier* for a cluster $C \in \mathbf{C}$, denoted $ID(C)$, is given by its position in a sorting of the elements of **C**, according to the Z-address of the minimal points of their MBRs. We simplify the notation: for X being a polygon, a cluster, or a set of clusters, we define its Z-interval as $Z(X) = Z(MBR(X))$.

3.3   The Hierarchical Z-Interval Tree, HZIT

To describe the branching in the trees to be introduced, we define the following partitioning of a nonempty subset of clusters, $\mathbf{c} \subseteq \mathbf{C}$; assume that $Z(\mathbf{c}) = [min; max]$, and let $mid = (min, max)/2$.

$$\mathbf{c}_{left} = \{C \in \mathbf{c} \mid Z(C) \subseteq [min; mid[\, \}$$
$$\mathbf{c}_{overlap} = \{C \in \mathbf{c} \mid mid \subseteq Z(C)\}$$
$$\mathbf{c}_{right} = \{C \in \mathbf{c} \mid Z(C) \subseteq \,]mid; max]\}$$

HZIT is a tree consisting of *leaf* and *split* nodes. Each node is attributed by a linear regression model and a Z-interval, both explained below. A split node has 2 or 3 subtrees, and a leaf is also attributed by a set of cluster identifiers.

A node N *for* a subset $\mathbf{c} \subseteq \mathbf{C}$ is defined as follows; M is a hyperparameter that represents the maximum number of clusters in a leaf node; see Sect. 6.

- When $|\mathbf{c}| < M$, N is a *leaf node*. It is attributed by the set of cluster identifiers $\{ID(C) \mid C \in \mathbf{c}\}$, referred to as $IDs(N)$.
- Else, N is a *split node* with a subtree for each nonempty $\mathbf{c}_{left}, \mathbf{c}_{overlap}, \mathbf{c}_{right}$.
- Each node is attributed by
  1) its Z-interval $Z(N) = Z(\mathbf{c})$, and
  2) a linear regression model, referred to as $\mathscr{L}(N)$, learned from the set of pairs $\{(Z(C), ID(C)) \mid C \in \mathbf{c}\}$.

The entire HZIT is built recursively starting from a node for the entire set of clusters $\mathbf{C}$. A model $\mathscr{L}(N)$ is used in query evaluation to predict an approximate position of where to search for clusters that may overlap with a given MBR given by its Z-interval z; we use the notation $predict(\mathscr{L}(N), z)$.

4   Disk Space Layout and Access Table

Polygon vertex lists are stored on disk in a way that reduces the number of disk page requests during search. Clusters are layed out sequentially according to an in-order visiting of the leaf nodes of the HZIT, reflecting how they may be visited during search. Vertex lists for polygons inside a cluster appear in some order, giving an index $1, 2, \ldots$ They may be ordered to reduce the number of times a polygon continues into a next disk page, thus occasionally avoiding an extra disk page request. A traditional disk cache is assumed to hold pages recently read from disk; not explained further.

An *access table* is stored in main memory: given polygon P identified by cluster ID and index $\{1, 2, \ldots\}$ in that cluster, the table returns the tuple

$$(number\text{-}of\text{-}vertices\text{-}in\text{-}P, page\text{-}number, start\text{-}addr\text{-}in\text{-}page)$$

where vertices of P are stored starting at the indicated page and local address within that page. Finally, we assume tables that map each cluster ID id, and polygons identified by cluster ID plus index, (id, i), to their MBRs; we write $MBR(id)$ and $MBR(id, i)$ for those.

5 SPL^{index}-Based Query Processing

To process a **range query** given as a rectangle qr, we first find a set of clusters by a call search(N_{top}, $Z(qr)$) to the function defined below; N_{top} is the top node of the HZIT tree. We assume an error bound b to define a search interval around an approximate prediction given by one of the learned linear regression models. When mentioning a given polygon below, it refers to its vertex list, perhaps found in the disk cache, or perhaps involving a disk request.

> **function** search(N, z):
>   **if** $Z(N) \cup z = \emptyset$, **return** $\emptyset$;
>   **if** N is a leaf node,
>   **return** $\{id \in IDs(N) \mid Z(id) \cap z \neq \emptyset\}$
> $id := predict(\mathcal{L}(N), z);$ $result := \emptyset;$
>   **for** each child node N' of N
>   **if** $Z(N') \cap [id - b; id + b] \neq \emptyset$,
> $result := result \cup search(N', z);$
>   **return** $result$;
>
> **function** rangeQuery(N, qr):
>   // FILTERING:
> $pred\text{-}IDs := $ search(N, $Z(qr)$);
>   // OPTIMIZED REFINEMENT:
> $result := \emptyset;$ // FOR COLLECTING POLYGONS TO BE RETURNED
>   **for** all $id \in pred\text{-}IDs$
>   **if** $MBR(id) \subseteq MBR(qr)$,
> $result := result \cup all\text{-}vertex\text{-}lists\text{-}in\text{-}cluster(id);$
>   **else**
>   **for** all polygons p of $cluster(id)$,
>   **if** $overlap(MBR(qr), MBR(p), p)$
> $result := result \cup \{p\};$
>   **return** $result$;
>
> **function** overlap($MBR(qr)$, $MBR(p)$, p):
>   **if** $MBR(qr) \cap MBR(p) \neq \emptyset$,
>   **if** $MBR(qr) \cap poly \neq \emptyset$, **return** $True$;   // *** (SEE BELOW)
>   **return** $False$;

The call to the *overlap* function executes very fast when determined from the MBRs alone, or it can be time consuming when the test needs to process the vertex list; we use here a standard algorithm [21].

A modification of the rangeQuery function returning polygon indexes (of form (id, i)) rather than vertex lists, will run much faster as only the line marked *** needs access to cache or disk.

A **point query** qp is executed as a range query for a degenerate rectangle with minimum and maximum points equal qp.

6   Experimental Studies

6.1   Experimental Settings

For comparison, all experiments were implemented in Python 3.9 and run on a RedHat Enterprise Server 6.3 with Linux core 2.6.32, 64 GB memory and a 5 GHz Intel Core(TM) i9 processor, and 3.7 TB SSD storage.

Datasets. We experimented with the following real and synthetic datasets. (1) OSM: four real datasets extracted from OpenStreetMap including **Land**[1] with 755,769 land polygons, **Water**[2] with 53,351 polygons of water bodies, **Park**[3] with 11M polygons with a total size of 9.3 GB, represent map features for the whole world. Furthermore, we tested the robustness of our index by generating synthetic data; **Build** with 50M random polygons with a total size of 26 GB. **Uni**, **Corr**, and **Zipf** with uniform, bivariate Gaussian, and zipfian spatial distributions, respectively, containing 5M polygons in $\mathbb{R}^2$.

Baseline. We compare our SPL^{index}[4] to the most popular existing spatial disk-based index R-tree [10], using an open-source Python library[5].

Hyper-parameter Settings. Table 1 lists the parameter settings. These parameters have been set after experimenting with a particular dataset. Developing a theory for parameter settings from dataset analysis is a future research.

Evaluation Metrics. We use three evaluation metrics. (1) Index construction costs for a given dataset include creation time and storage size for both R-tree and SPL^{index}. (2) IO cost measures the number of disk pages loaded into memory when a query is processed. (3) Query response time is the time duration of processing a query, which equals CPU time plus IO time.

Queries. We compare SPL^{index} and R-tree on both range and point queries. In each single test, we run 100 random queries and report the average query cost. To check the effect of the linear regression, we tried to remove it, and observed an increase in execution time by a factor of several hundreds.

Table 1. Parameters settings

Parameter	Setting
Disk page size (DP)	4096
Vertex size (P)	16 bytes
MBR size (MS)	8 × 2 × 2 bytes
Page address size (PS)	4 bytes
R-tree fanout	DP/(MS+PS)
Maximum Capaciry (M)	DP/(MS+PS)
Error bound (b)	0.05
Range query selectivity	0.05%, 0.1%, 1%

Table 2. Number of nodes

Dataset	R-tree	SPL^{index}
Water	15798	79
Land	42578	126
Park	663742	11248
Build	938857	23561
Uni	94728	43697
Corr	99449	6192
Zipf	70770	1349

---

[1] https://osmdata.openstreetmap.de/data/land-polygons.html.
[2] https://osmdata.openstreetmap.de/data/water-polygons.html.
[3] https://spatialhadoop.cs.umn.edu/datasets.html.
[4] https://github.com/MasoumehVahedi/SPLindex.
[5] https://github.com/Toblerity/rtree.

Index Construction Costs. We observe the index construction time and storage size for our largest dataset, Build. The result shows that building index for SPL^{index} incurs, on average, 60% less time than R-tree. In addition, SPL^{index} occupies approximately 40% smaller space than R-tree. This is understandable as SPL^{index} has fewer nodes overall (see Table 2).

6.2 Range Query Performance

Query IO Cost. As shown in Fig. 2, SPL^{index} clearly performs better with range queries for IO consumption. In most scenarios, SPL^{index} saves more than 20–60% IO consumption compared to R-tree. However, at 0.05% selectivity, SPL^{index} incurs 10% higher IO costs than R-tree for the Water dataset. At 0.1% on Water, SPL^{index} and R-tree have comparable IO costs.

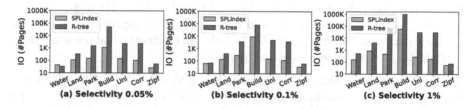

Fig. 2. Average IO cost per range query

Query Response Time. Figure 3 illustrates that SPL^{index} outperforms R-tree in terms of query response time for all selectivities on real datasets. Moreover, for the three synthetic datasets across all selectivity levels, SPL^{index} achieves approximately 40–60% faster time than R-tree. As IO cost in SPL^{index} is much smaller, it contributes to a shorter overall query response time.

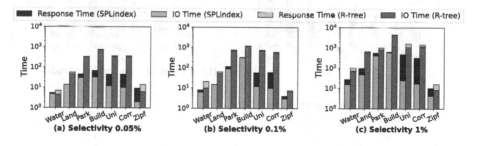

Fig. 3. Average response time (milliseconds) per range query

Scalability. Figure 4 shows the effect of dataset cardinalities on the efficiency of range queries on the Build dataset and its subsets $\{10M, 20M, 30M, 40M, 50M\}$, with selectivity at 1%. Clearly, both methods' time increase almost linearly as the dataset size increases. SPL^{index} is still faster than R-tree.

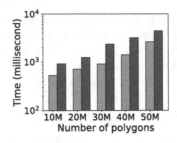

Fig. 4. Scalability: Response time vs. data cardinality

6.3 Point Query Performance

Query IO Cost. As shown in Fig. 5(a), SPL^{index} reduces IO costs 50%-60% compared to the R-tree for real-world datasets, and to approximately 80% for synthetic datasets. As expected, SPL^{index}'s design efficiently load relevant pages, avoiding the load of unnecessary disk pages.

Query Response Time. As illustrated in Fig. 5(b), SPL^{index} is 50%–60% faster than R-tree in all real-world datasets. Additionally, SPL^{index} outperforms R-tree by approximately 70%–80% for synthetic datasets. This makes sense because, in SPL^{index} query, fewer pages and nodes are examined compared to R-tree that eventually leads to shorter query response time.

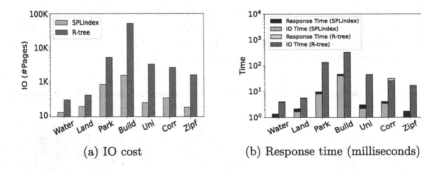

(a) IO cost   (b) Response time (milliseconds)

Fig. 5. Average response time and IO cost per point query

7 Related Work

The principle of learned index structures was introduced in [14]. A number of studies worked on learned one-dimensional index structure, including [4,6,8, 11,16]. Several works extended the idea of learned index for multi-dimensional point data [5,13,18,28], and [3,17,20,22,27] for spatial point data. Similarly to SPL^{index}, they aim to reduce index space and accelerate query time, but only for point data and not complex geometries, e.g., polygons. More recently, GLIN [26] was introduced as a learned index model for spatial range queries to complex geometries. However, the key difference with our approach is that GLIN is optimised for memory-based indexes for dynamic scenarios, while SPL^{index} considers disk-based model and minimize IO costs, which is essential for very large datasets. Other recent approaches that do not use machine learning models are [9, 24, 25, 29, 30] also consider IO optimization.

8   Conclusion and Future Work

This paper proposes SPL^{index}, a Spatial Polygon Learned Index, designed for disk-based systems. SPL^{index} is based on four main ideas: 1) dividing polygons into clusters, 2) mapping clusters to one-dimensional ordering using Z-address, 3) a hierarchical model that predicts cluster IDs for a given spatial queries, and 4) locating polygons on disk layout and accessing via a table for optimal disk access. Our experimental results on synthetic and real data demonstrate that SPL^{index} outperforms R-tree in terms of query time and IO costs for range and point queries; if the setting is changed to return polygon identifiers (id, i) (and not the actual vertices), disk usage is reduced drastically and we will likely observe an even greater difference.

An important unresolved issue is the lack of a proper method for optimal setting of hyperparameters for a dataset, but statistics about a distribution in space, sizes and shapes of polygons seems to be a good first step. In the future, we aim to extend SPL^{index} to handle updates, complex polygons with intersections and holes, and other queries like spatial joins.

Acknowledgement. This research is partly funded by Independent Research Fund Denmark (No. 1032-00481B). The authors are grateful to Prof. Hua Lu for initiating this project and for valuable advice.

References

1. Beckmann, N., Kriegel, H.P., Schneider, R., Seeger, B.: The R*-tree: an efficient and robust access method for points and rectangles. In: Proceedings of the 1990 ACM SIGMOD International Conference on Management of Data, pp. 322–331 (1990)
2. Bentley, J.L.: Multidimensional binary search trees used for associative searching. Commun. ACM **18**, 509–517 (1975)
3. Davitkova, A., Milchevski, E., Michel, S.: The ML-Index: a multidimensional, learned index for point, range, and nearest-neighbor queries. In: EDBT, pp. 407–410 (2020)
4. Ding, J., et al.: ALEX: an updatable adaptive learned index. In: Proceedings of the 2020 ACM SIGMOD International Conference on Management of Data, pp. 969–984 (2020)
5. Ding, J., Nathan, V., Alizadeh, M., Kraska, T.: Tsunami: a learned multi-dimensional index for correlated data and skewed workloads. ArXiv abs/2006.13282 (2020)

6. Ferragina, P., Vinciguerra, G.: The PGM-index: a fully-dynamic compressed learned index with provable worst-case bounds. Proc. VLDB Endow. **13**(8), 1162–1175 (2020)
7. Finkel, R.A., Bentley, J.L.: Quad trees a data structure for retrieval on composite keys. Acta Informatica **4**, 1–9 (1974)
8. Galakatos, A., Markovitch, M., Binnig, C., Fonseca, R., Kraska, T.: FITing-Tree: a data-aware index structure. In: Proceedings of the 2019 International Conference on Management of Data, SIGMOD Conference 2019, Amsterdam, The Netherlands, 30 June–5 July 2019, pp. 1189–1206. ACM (2019)
9. Georgiadis, T., Mamoulis, N.: Raster intervals: an approximation technique for polygon intersection joins. Proc. ACM Manag. Data **1**(1), 1–18 (2023)
10. Guttman, A.: R-trees: a dynamic index structure for spatial searching. In: Proceedings of the 1984 ACM SIGMOD International Conference on Management of Data, pp. 47–57 (1984)
11. Higuchi, S., Takemasa, J., Koizumi, Y., Tagami, A., Hasegawa, T.: Feasibility of longest prefix matching using learned index structures. SIGMETRICS Perform. Eval. Rev. **48**(4), 45–48 (2021)
12. Kamel, I., Faloutsos, C.: Hilbert r-tree: an improved rtree using fractals. In: VLDB, vol. 94, pp. 500–509. Citeseer (1994)
13. Kraska, T., et al.: SageDB: a learned database system. In: Conference on Innovative Data Systems Research (2019)
14. Kraska, T., Beutel, A., Chi, E.H., Dean, J., Polyzotis, N.: The case for learned index structures. In: Proceedings of the 2018 International Conference on Management of Data, pp. 489–504 (2018)
15. Lee, K.C., Zheng, B., Li, H., Lee, W.C.: Approaching the skyline in Z order. In: VLDB, vol. 7, pp. 279–290 (2007)
16. Li, P., Hua, Y., Jia, J., Zuo, P.: FINEdex: a fine-grained learned index scheme for scalable and concurrent memory systems. VLDB Endow. **15**(2), 321–334 (2021)
17. Li, P., Lu, H., Zheng, Q., Yang, L., Pan, G.: LISA: a learned index structure for spatial data. In: Proceedings of the 2020 ACM SIGMOD International Conference on Management of Data, pp. 2119–2133 (2020)
18. Nathan, V., Ding, J., Alizadeh, M., Kraska, T.: Learning multi-dimensional indexes. In: Proceedings of the 2020 ACM SIGMOD International Conference on Management of Data (2019)
19. Orenstein, J.A., Merrett, T.H.: A class of data structures for associative searching. In: ACM SIGACT-SIGMOD-SIGART Symposium on Principles of Database Systems (1984)
20. Pandey, V., van Renen, A., Kipf, A., Sabek, I., Ding, J., Kemper, A.: The case for learned spatial indexes. ArXiv abs/2008.10349 (2020)
21. Park, S.C., Shin, H., Choi, B.K.: A sweep line algorithm for polygonal chain intersection and its applications. In: Kimura, F. (ed.) Geometric Modelling. ITIFIP, vol. 75, pp. 309–321. Springer, Boston, MA (2001). https://doi.org/10.1007/978-0-387-35490-3_21
22. Qi, J., Liu, G., Jensen, C.S., Kulik, L.: Effectively learning spatial indices. Proc. VLDB Endow. **13**(12), 2341–2354 (2020)
23. Sellis, T., Roussopoulos, N., Faloutsos, C.: The R+-tree: a dynamic index for multi-dimensional objects. Technical report (1987)
24. Singla, S., Eldawy, A., Diao, T., Mukhopadhyay, A., Scudiero, E.: The raptor join operator for processing big raster+ vector data. In: Proceedings of the 29th International Conference on Advances in Geographic Information Systems, pp. 324–335 (2021)

25. Teng, D., Baig, F., Sun, Q., Kong, J., Wang, F.: IDEAL: a vector-raster hybrid model for efficient spatial queries over complex polygons. In: 2021 22nd IEEE International Conference on Mobile Data Management (MDM), pp. 99–108. IEEE (2021)
26. Wang, C., Yu, J.: GLIN: a lightweight learned indexing mechanism for complex geometries. arXiv preprint arXiv:2207.07745 (2022)
27. Wang, H., Fu, X., Xu, J., Lu, H.: Learned index for spatial queries. In: 2019 20th IEEE International Conference on Mobile Data Management (MDM), pp. 569–574. IEEE (2019)
28. Yang, Z., et al.: Qd-tree: learning data layouts for big data analytics. In: Proceedings of the 2020 ACM SIGMOD International Conference on Management of Data, SIGMOD 2020, pp. 193–208. ACM, New York (2020)
29. Yousfi, H., Mesmoudi, A., Hadjali, A., Matallah, H., Lahfa, F.: Efficient R-tree exploration for big spatial data. In: Kacprzyk, J., Balas, V.E., Ezziyyani, M. (eds.) AI2SD 2020. AISC, vol. 1418, pp. 865–874. Springer, Cham (2022). https://doi.org/10.1007/978-3-030-90639-9_70
30. Zardbani, F., Mamoulis, N., Idreos, S., Karras, P.: Adaptive indexing of objects with spatial extent. Proc. VLDB Endow. 16, 2248–2260 (2023)
31. Zhang, T., Ramakrishnan, R., Livny, M.: BIRCH: an efficient data clustering method for very large databases. ACM SIGMOD Rec. 25(2), 103–114 (1996)

Recommendation Systems
and Prediction

Action Rules Discovery: Leveraging Attributes Correlation Based Vertical Partitioning

Aileen Benedict[1]([✉]) [iD] and Zbigniew W. Ras[1,2] [iD]

[1] Computer Science Department, University of North Carolina at Charlotte, 9201
University City Blvd., Charlotte, NC 28223, USA
abenedi3@uncc.edu
[2] Polish-Japanese Academy of Information Technology, Institute of Computer
Science, 02-008 Warsaw, Poland

Abstract. This paper tackles the challenge of extracting actionable
insights from large datasets to enhance the knowledge bases of recommendation systems. We introduce a novel vertical dataset partitioning
method utilizing attribute correlation clustering, enabling efficient parallel action rule discovery and significant processing time reduction. Our
method's effectiveness is demonstrated by comparing it with traditional
random-based partitioning, focusing on precision, coverage, lightness,
rule yield, and efficiency. Employing a rule-based generation approach
with the RSES tool, we analyze rule quality, quantity, and computational
demand. The results reveal promising strategies for action rule discovery with large-scale datasets, with potential applications across various
domains like e-commerce and healthcare, offering valuable insights for
large-scale dataset analysis.

Keywords: Action Rules · Recommendation Systems · Dataset
Partitioning

1 Introduction

In the domain of data-driven intelligent systems and recommendation engines,
extracting actionable insights through action rules stands as a pivotal task guid-
ing decision-making processes. Yet, this endeavor presents challenges, particularly with large datasets, where the generation of action rules can be both
time-consuming and computationally intensive.

Action rules, introduced by Ras and Wieczorkowska [6], describe how objects
transition between states based on a decision attribute. In the context of our
study, action rules play a pivotal role in identifying actionable insights from
data. An action rule, derived from or directly discovered within a dataset, specifies a sequence of attribute transformations leading to a desired outcome. It is
formulated as follows:

A. Appice et al. (Eds.): ISMIS 2024, LNAI 14670, pp. 285–295, 2024.
https://doi.org/10.1007/978-3-031-62700-2_25

$$r = [[a_1 \wedge g_2 \wedge (B, b_1 \rightarrow b_2) \wedge (H, h_1 \rightarrow h_2)] \Rightarrow (D, d_1 \rightarrow d_2)], \qquad (1)$$

where a_1 and g_2 represent stable attributes, which are invariant properties of the data items. Stable attributes, such as 'customer segment' or 'product category', do not change over time and are integral to defining the context within which action rules are applied. On the other hand, $(B, b_1 \rightarrow b_2)$ and $(H, h_1 \rightarrow h_2)$ denote flexible attributes—attributes subject to change through intervention or over time, like 'customer satisfaction level' or 'product pricing'. The goal of an action rule is to outline the conditions under which modifying certain flexible attributes leads to a targeted change in the decision attribute $(D, d_1 \rightarrow d_2)$, such as transitioning a customer's status from 'passive' to 'promoter'.

Action rules help to form the foundations for a category of recommendation systems known as knowledge-based recommendation systems. These systems have been applied across various fields, including business (refer to [8]), healthcare (refer to [5]), music (refer to [4]), and art (refer to [3]).

We introduce a novel approach to distributed action rule discovery tailored for large datasets, advancing beyond traditional horizontal partitioning and random attribute grouping techniques. Our strategy employs attribute correlation-based clustering for vertical partitioning, enhancing the efficiency and effectiveness of action rule extraction. This method, supported by the Python Ray library for parallel execution, promises improved rule generation speed and coverage without the inconsistency often encountered in random partitioning.

Despite the additional step of clustering, our approach aims for consistent, high-quality outcomes, reducing the need for multiple iterations unlike conventional methods. By comparing these partitioning strategies, this research illuminates the trade-offs involved, contributing to the optimization of action rule discovery in large-scale data environments. We will further articulate our methodology, experiments, and findings in the following sections, demonstrating the impact of our approach on the development of intelligent systems and recommendation engines.

2   Related Work

The efficient extraction of action rules from large datasets has presented considerable challenges to traditional models that analyze data in a non-distributed manner. Recognizing this limitation, efforts have been made to adopt a distributed approach to better address the complexities associated with larger volumes of data. This section provides an overview of such distributed approaches to action rule discovery, setting the stage for our proposed enhancements.

Vertical partitioning, a critical technique in database design, involves dividing a database table into smaller, more manageable segments based on attribute usage. This approach, pioneered by seminal works such as Navathe et al. (1984), aims to optimize database performance by improving data access patterns and reducing disk I/O operations [2]. Navathe et al. introduced algorithms for vertical partitioning that analyze transaction access patterns and attribute affinity, leading to optimized database designs that significantly reduce disk I/O costs and enhance transaction processing efficiency. Their methodologies have laid the groundwork for subsequent research in database optimization. While Navathe et al.'s algorithms focus on structured data in relational databases, our proposed method extends vertical partitioning concepts to action rule generation.

Bagavathi et al. [1] introduce a distributed approach to extract action rules from large datasets. Their method emphasizes the data distribution phase, suggesting partitioning data into smaller granules both horizontally (across rows) and vertically (across attributes). Rules extracted from these vertical pairwise disjoint granules, which cover distinct attributes, are concatenated, thereby capturing some of the knowledge from the initial dataset. Similarly, Tarnowska et al. [7] propose another distributed method for action rule discovery using both horizontal and vertical partitioning. They employ Spark for horizontal partitioning, while their vertical approach randomly separates attributes. This method notably reduces the time required to discover action rules. However, a limitation lies in their combined study of both partitioning methods without isolation, making it unclear whether the benefits arise from horizontal partitioning, vertical partitioning, or a synergy of both.

While the aforementioned approaches provide a foundation for distributed action rule discovery methods, our work seeks to advance this further. We identify a potential limitation in the random nature of vertical data partitioning present in existing methods. Our contribution focuses on introducing a more logical and structured approach to vertical data partitioning. We propose using feature correlations as a distance measure in data partitioning, specifically for action rule discovery. This is a technique that, to the best of our knowledge, has not yet been explored. By doing so, we aim to enhance both the efficiency and the quality of the actionable patterns extracted from distributed datasets, distinguishing our work from existing methods.

3 Methodology

This study introduces a new approach for distributed action rule discovery, integrating attribute correlation and vertical data partitioning to address datasets with numerous flexible attributes.

3.1 Attribute Correlation-Based Vertical Partitioning

Our proposed attribute correlation-based vertical partitioning method includes the following steps:

1. Calculate the correlations between the flexible classification attributes in the dataset.
2. Perform agglomerative clustering on all flexible attributes using correlations as the distance measure, resulting in a dendrogram.

3. Iterate through the dendrogram levels to obtain various clusters of flexible attributes. Extract action rules for each cluster of flexible attributes, extended by the same stable attributes.
4. Create all possible combinations of rules from each cluster, considering their support.
5. Determine the optimal vertical partition of the dataset by comparing the F-scores of the sets of action rules discovered for each partition.

For our experiments, we also return all of the results for each dendrogram level for analysis. This includes the various performance metrics as well as the list of rules generated.

Step 1. To calculate the correlations between the flexible classification attributes, Pearson's correlation coefficient is used for continuous attributes, Cramer's V for categorical attributes, and the Correlation Ratio for a mix of both types. These equations are well-known and not included here for brevity. We utilized the Dython Python library to calculate these associations based on attribute types.

Step 2. To perform agglomerative clustering, we must first compute the distance matrix. This is done using the following equation:

$$DistanceMatrix = 1 - \frac{|associations| + 1}{2} \tag{2}$$

Here, *associations* are the correlations calculated in Step 1. We then performed agglomerative clustering using the distance matrix with the SciPy library, employing a single linkage to create a dendrogram. From this dendrogram, we derived clusters of flexible attributes. The goal is to break down flexible attributes into smaller groups for more efficient analysis. From this, we must then explore each level of the dendrogram to see which level of clustered flexible attributes yields the best set of action rules. Each level of the dendrogram, now consisting of the remaining Steps 3 and 4, can be run in parallel. To do so, we utilize the Python Ray library.

Step 3. Next, action rules were generated for each cluster of attributes within a single dendrogram level. Once again, we can generate action rules for each cluster in parallel for efficiency. Our experiments incorporated a technique for action rule discovery based on the Rough Sets Exploration System (RSES) for comparative analysis. In the RSES-based approach, classification rules are initially generated from the dataset using the RSES tool. These rules, representing patterns or associations between attributes, are then parsed and structured for further analysis. They are categorized into bins based on their decision attributes, with a focus on transitions between specific decision classes, such as from "passive" to "promoter", used in datasets for our experiments.

To generate action rules, pairs of classification rules are compared, emphasizing stable and flexible attributes. The method ensures the compatibility of stable attributes between rule pairs while examining differences in flexible attributes to suggest potential actions. Action rules are formulated to recommend changes in attribute values hypothesized to lead to desired transitions in decision classes.

Step 4. The rules were then concatenated by taking combinations from each cluster and considering their support. The updated rule combination process enhances efficiency and precision by introducing a depth-controlled approach, coupled with confidence and support threshold-based pruning. Initially, the method filters the input lists of rules, ensuring that only those surpassing the predefined confidence and support thresholds proceed. It then embarks on a systematic exploration of rule combinations, anchored by the concept of stable attributes, which serve as a basis for grouping and efficiently combining rules.

For each level of depth, up to a specified maximum, the algorithm meticulously generates combinations of rules drawn from the clusters. This is achieved by calculating a unique key for each combination based on the involved rules' stable attributes, thus streamlining the combination process. The algorithm then proceeds to merge rules within each combination, applying a critical evaluation at the maximum depth. At this juncture, it assesses the merged rules against the confidence and support thresholds, a step that decisively influences the retention or pruning of rules.

Step 5. Finally, once we have the action rules for each possible level of the dendrogram, we calculate and compare the F-scores for each level. The level with the best score is determined to be the best level, containing the final set of action rules. For the sake of this study, we use F-score to determine the best level. However, one could explore other metrics to determine the best score, such as lightness, coverage, or number of rules, or return rules at all levels for analysis.

This process is iterative, ensuring a thorough exploration of potential rule combinations while meticulously discarding those that fail to meet the established thresholds. The result is a refined collection of combined rules, each representing a meaningful and actionable insight. By incorporating a depth-controlled exploration and leveraging threshold-based pruning, the method significantly enhances the quality of the resultant rule set. This novel approach not only ensures the generation of high-quality combined rules but also optimizes the overall efficiency of the rule combination process, making it well-suited for further analytical endeavors.

3.2   Random Vertical Partitioning

We compare random vertical partitioning to our proposed method, vertical partitioning based on attribute correlation. In random partitioning, attribute splits are entirely arbitrary, unlike our proposed method, which leverages attribute correlations for more structured grouping. This approach was introduced by Tarnowska et al. [7].

The process involves generating multiple sets of attribute groupings, similar to the steps outlined in Correlation-Based Vertical Partitioning. However, instead of performing agglomerative clustering and dendrogram iteration, we randomly shuffle the attributes and divide them into groups, akin to the layers of clusters in a dendrogram. Specifically, we vary the number of groups (N) from two to the total number of attributes (exclusive) to create diverse arrangements. The attributes are randomly shuffled within each set and divided into k groups, where k ranges from 0 to N − 1. This approach yields a variety of attribute groupings, enabling the exploration of different partitions of the dataset. Subsequently, we extract action rules for each group of flexible attributes in each partition and create all possible combinations, considering their support.

In random partitioning, splits are made without regard to attribute correlations or clustering, unlike our proposed method. For our experiments, we then collect all the results from each grouping level for analysis. This encompasses various performance metrics and the list of generated rules.

4 Data Preparation

4.1 Dataset Source

Our study focuses on assessing customer experience using the Net Promoter Score (NPS®), a metric indicating the likelihood of customers recommending a company. We utilized the NPS dataset, comprising feedback from heavy equipment repair services across 38 companies and 340,000 customers in the USA and Canada. This dataset enables comparative analysis and employs data preprocessing techniques similar to prior research. Each survey captures numerical ratings (0–10) on benchmarks such as job accuracy and likelihood to recommend. The "PromoterStatus" decision attribute classifies customers as promoters, passive, or detractors. Our objective is to enhance customer satisfaction and loyalty, transitioning passive customers to promoters through action rules. Surveys from four companies in 2015 were used for our study.

4.2 Data Cleaning

Due to inconsistencies in dataset semantics across 38 companies, which undermined rule extraction confidence, we limited our analysis to datasets from two companies, denoted as A and B. Initially, we conducted several data preprocessing steps: (1) eliminating columns surpassing a sparsity threshold, (2) identifying and removing correlated columns, (3) addressing null values, and (4) binning benchmarks.

We assessed benchmark column sparsity, discarding those with 75% or more null values. Correlation analysis followed, employing the Pearson correlation coefficient to detect and eliminate redundant features exhibiting one-to-one relationships. Null values were managed based on column context. Rows with null

PromoterScores, our decision attribute, were removed, while nulls in benchmark features were categorized as "No Response." Subsequently, benchmarks were binned: null values designated "No Response," scores 0–4 as "Low," 5–6 as "Medium," 7–8 as "High," and 9–10 as "Very High."

4.3 Dataset Description and Experimental Parameters

In our dataset descriptions and experimental parameters, Dataset A comprised 542 rows, and Dataset B comprised 1279 rows. Both datasets underwent preprocessing and shared three stable attributes: 'division,' 'survey type,' and 'channel type.' Additionally, Dataset A contained 11 flexible attributes, while Dataset B contained 10. These flexible attributes, derived from the discussed surveys, constituted the benchmarks within the datasets. The resultant attribute across both datasets was the 'promoter status,' with possible values being 'Detractor,' 'Passive,' and 'Promoter.' Our experiments across all datasets focused on identifying action rules to transition the consequent from 'Passive' to 'Promoter.' We applied a confidence threshold of 0.8 and a support threshold of 2 to filter the rules extracted from the datasets.

Examples of some of the flexible attributes with their abbreviations and descriptions are as follows:

– *BenchmarkAllDealerCommunication* (DC): Evaluates communication effectiveness within the dealer network.
– *BenchmarkAllLikelihoodtobeRepeatCustomer* (RC): Measures the likelihood of customer repeat business.
– *BenchmarkPartsEaseofCompletingPartsOrder* (ECP): Gauges the ease of completing parts orders.
– *BenchmarkPartsTimeitTooktoPlaceOrder* (TPO): Measures the time taken for order processing.
– *BenchmarkPartsOrderAccuracy* (OA): Scrutinizes the accuracy of order processing.

5 Experimental Setup

Our objective was to investigate and compare the effectiveness of different methodologies, namely vertical correlation partitioning (our proposed approach), random vertical partitioning, and no partitioning. In the case of no partitioning, action rules were generated using all available flexible attributes.

For evaluation, we employed various metrics, including the time required to extract action rules, the total count of generated rules, the number of distinct objects covered by these rules, and the average number of identical objects encompassed within the domains of all extracted action rules, referred to as "lightness."

The experiments were conducted on the UNC Charlotte Orion research cluster, which comprises compute nodes equipped with Intel Xeon CPUs. Each node

is equipped with 32 cores and 128 GB of RAM. Job scheduling, resource allocation, and monitoring were managed using the SLURM workload manager. We
implemented the experiments using Python 3.8.5, leveraging the Ray parallel
processing framework to accelerate the training of our machine learning models.
Each experiment was executed on a single compute node featuring 16 CPUs and
4 GB of memory per CPU. Utilizing the computational power of the cluster
allowed us to reduce the training time of our models.

5.1   Evaluation Metrics

To facilitate the comparison of methods, we evaluated several metrics, including
runtime (in seconds), the number of generated rules, precision, coverage, and
lightness. Runtime refers to the duration required for a method or process to
complete its execution. In our study, runtime is measured in seconds, indicating
the time taken for each method to run. This metric provides insights into the
efficiency and computational speed of the evaluated method. Ensuring computational efficiency is crucial, especially for large datasets, as it ensures timely and
practical results suitable for real-world applications.

The number of generated rules represents the count of distinct rules produced
by each method, reflecting the richness and complexity of the rule set derived
from the dataset. A greater number of rules offers a richer set of insights and
actionable items. The precision of a set of rules reflects the quality of generated
rules by considering both their support and confidence. Higher precision indicates
widely applicable and reliable rules, essential for decision-making processes. The
coverage of each action rule is the number of tuples or items encompassed by
the rule in the dataset. The coverage for a set of action rules is determined by
the unique support of all the rules within the set.

We define lightness as a ratio indicating how evenly the coverage is distributed across the set of rules. It represents the average number of action rules
applying to each object, distinguishing between random and correlation-based
partitioning. The lightness for a set of action rules is defined in Eq. (3), where

$$\text{Lightness} = \frac{\sum \text{Coverage of each action rule}}{\text{Coverage of entire set of action rules}} \tag{3}$$

6   Results and Discussion

The primary objective of our study is to evaluate the performance of our proposed correlation-based vertical partitioning method in comparison to both the
random vertical partitioning approach and a base method devoid of vertical partitioning. Table 1 presents a comparative analysis of these methods across two
distinct datasets. The methods under comparison include "None," representing
our baseline method without vertical partitioning, "Correlation," denoting our
correlation-based vertical partitioning method, and "Random," indicating the
random vertical partitioning method. Time is quantified in seconds. Notably,

for the "Random" method, the displayed results represent the average outcomes derived from five iterations.

For Dataset A, the base method required 2,543.433 s to generate 312 rules, while the correlation-based method needed only 393.558 s, consistently producing 129 rules across runs. The random method was notably faster per individual run on average, taking 164.738 s to generate 41 rules. However, it is important to note that the random method's results are derived from averages across multiple iterations, as it often necessitates several runs to achieve satisfactory outcomes. Specifically, for Dataset A, the cumulative time for all five iterations of the random method amounted to 823.689 s, which ultimately exceeds the time required by the single-run correlation-based method. For Dataset B, the total duration for five iterations of the random method was 173.808 s, with one iteration yielding no rules at all. In contrast, the correlation-based method's results were consistent, requiring only a single execution, whereas the random method's need for multiple iterations to obtain good results introduces variability and potentially increases the total computation time.

Table 1. Comparison Analysis of Vertical Partitioning Methods Across Datasets.

Dataset	Method	Time (seconds)	Rules	Precision	Coverage	Lightness
A	None	2543.433	312.0	0.939	0.672	28.171
	Correlation	393.558	129.0	0.995	0.623	12.816
	Random	164.738	41.0	0.939	0.525	4.923
B	None	49.077	22.0	0.855	0.418	5.099
	Correlation	39.924	19.0	0.859	0.418	4.324
	Random	34.762	4.2	0.681	0.280	1.618

One of the rules obtained from Dataset A using the correlation-based method is shown below.

$$\text{Rule:}[\text{ChannelType} = \text{Construction Heavy}] \wedge$$
$$[\text{RC} = [\text{High} \rightarrow \text{Very High}]] \wedge$$
$$[\text{OA} = [\text{High} \rightarrow \text{Very High}]]$$
$$\Rightarrow [\text{PromoterStatus} = [\text{Passive} \rightarrow \text{Promoter}]]$$

This rule had a confidence of 0.932 and a support of seven and reads as: *"if the channel type stable attribute equals "construction heavy" and experiences a change in the RC flexible attribute from "high" to "very high" and a change in the OA attribute from "high" to "very high", this implies a change in our PromoterStatus from "Passive" to "Promoter."* In other words, we want to increase a company's ratings for the likelihood of customer repeat business and for the accuracy of order processing.

Across all datasets, the correlation-based method consistently outperforms the random-based method in key performance metrics, achieving superior precision, coverage, and lightness. It approaches the performance levels of the base method yet remains more time-efficient. This consistency highlights the importance of considering the balance between efficiency and the quality of generated rules.

7   Conclusion

The objective of our study was to evaluate key performance metrics of our proposed correlation-based vertical partitioning method in comparison with both the random vertical partitioning approach and a base method with no partitioning. The results illustrate the trade-offs between execution times and the consistency and quality of generated rules. Although the random-based method exhibited shorter completion times, its inherent randomness emerged as a drawback, necessitating multiple iterations to attain satisfactory outcomes. Conversely, our correlation-based method, despite appearing more time-consuming initially, offers the advantage of yielding higher lightness of discovered action rules (more options for recommendations) and consistent results without the need for repeated iterations. Notably, the correlation-based method strikes a balance between efficiency and effectiveness. It outpaces the base method in speed while delivering more comprehensive results than the random method.

Future research could further refine this approach, seeking to enhance the overall efficiency of the vertical partitioning methods. This exploration has the potential to open up new pathways for optimizing action rule mining.

References

1. Bagavathi, A., Tripathi, A., Tzacheva, A.A., Ras, Z.W.: Actionable pattern mining-a scalable data distribution method based on information granules. In: 2018 17th IEEE International Conference on Machine Learning and Applications (ICMLA), pp. 32–39. IEEE (2018)
2. Navathe, S., Ceri, S., Wiederhold, G., Dou, J.: Vertical partitioning algorithms for database design. ACM Trans. Database Syst. (TODS) **9**(4), 680–710 (1984)
3. Powell, L., Gelich, A., Ras, Z.: How to raise artwork prices using action rules, personalization and artwork visual features. J. Intell. Inf. Syst. **57**, 583–599 (2021). https://doi.org/10.1007/s10844-021-00660-x
4. Ras, Z., Dardzinska, A.: From data to classification rules and actions. Int. J. Intell. Syst. **26**, 572–590 (2011). https://doi.org/10.1002/int.20485
5. Ras, Z.: Reduction of hospital readmissions. Adv. Clin. Exp. Med. **31**(1), 5–8 (2022)
6. Ras, Z.W., Wieczorkowska, A.: Action-rules: how to increase profit of a company. In: Zighed, D.A., Komorowski, J., Żytkow, J. (eds.) PKDD 2000. LNCS (LNAI), vol. 1910, pp. 587–592. Springer, Heidelberg (2000). https://doi.org/10.1007/3-540-45372-5_70

7. Tarnowska, K., Bagavathi, A., Ras, Z.: High-performance actionable knowledge miner for boosting business revenue. Appl. Sci. **12**(23), 12393 (2022). https://doi.org/10.3390/app122312393
8. Tarnowska, K., Ras, Z.: NLP-based customer loyalty improvement recommender system (CLIRS2). Big Data Cogn. Comput. **5**(1), 4 (2021). https://doi.org/10.3390/bdcc5010004

HalpernSGD: A Halpern-Inspired Optimizer for Accelerated Neural Network Convergence and Reduced Carbon Footprint

Katherine Rossella Foglia[1]([✉])(iD), Vittorio Colao[1](iD), and Ettore Ritacco[2](iD)

[1] Department of Mathematics and Computer Science, University of Calabria,
Via P. Bucci, 30B, 87036 Arcavacata di Rende, CS, Italy
{katherine.foglia,vittorio.colao}@unical.it
[2] Department of Mathematics, Computer Science and Physics, University of Udine,
Via delle Scienze 206, 33100 Udine, UD, Italy
ettore.ritacco@uniud.it

Abstract. This research aims at focusing attention on Halpern iteration, a technique that can be exploited to define optimizers in neural network settings to outperform numerous current state-of-the-art methods. More specifically, we introduce HalpernSGD, an innovative network optimizer that leverages Halpern's method to enhance the rate of convergence of the Stochastic Gradient Descent (SGD), reducing the carbon footprint associated with neural network training. HalpernSGD exhibits a quadratic rate of convergence compared to the one of SGD, without compromising the accuracy of the model. We compared it with SGD and Adam through experiments that demonstrate HalpernSGD's superior efficiency by significantly reducing the number of epochs required for convergence, thereby lowering energy consumption and carbon emissions. The study also identifies potential improvements in Adam's approach to stability and convergence, suggesting a future direction for developing combined optimizers.

Keywords: Halpern's iterative method · Optimizers · Gradient Descent · Green AI · Environmental Footprint · Carbon emission · Rate of Convergence · Metric Fixed Point Theory · Non-expansive mapping

1 Introduction

Artificial Intelligence (AI) is currently experiencing a remarkable phase of development and expansion. The rapid progress in AI is leading to the design and implementation of increasingly large architectures. These kind of models, often characterized by Deep Neural Networks (DNNs), exhibit enhanced capabilities in tasks such as machine vision (Segmentation [24,36], Object Detection [10,25], Scene Recognition [9,38]), natural language processing (Transformers [37], Large Language Models [8]), and realistic data generation (Variational

A. Appice et al. (Eds.): ISMIS 2024, LNAI 14670, pp. 296–305, 2024.
https://doi.org/10.1007/978-3-031-62700-2_26

Autoencoders [27], GANs [19], Diffusion Models [23], synthetic data [31,32]). However, the growing complexity of these architectures comes at a cost, particularly in terms of computational resources. This escalating computational demand raises environmental concerns, since the environmental footprint of AI, especially with the surge in computational intensity, results in a substantial increase in carbon dioxide emissions [15,35,41]. It is important to emphasize that numerous research efforts [13,14] highlight the overall reduction in greenhouse gas pollution associated with the utilization of AI, particularly in industrial settings. However, the energy consumption of complex neural architectures underscores the importance of addressing the sustainability aspects of the AI development. The literature proposes various methodologies to mitigate carbon emissions during the training and inference processes of the DNNs, with a focus on: (i) reduction of the size of the neural architectures even through transfer learning, tolerating quality loss, (ii) hardware optimization by exploiting devices such as GPUs or TPUs, (iii) utilization of cloud services in specialized data centers with high performance in energy consumption, (iv) selection of the geographical site of the calculus units according to their energy production costs and environmental contexts, (v) change of the loss function adding regularization terms, and (vi) data processing to minimize the volume of information to be analyzed. In this study, we propose a different approach to address the challenge of carbon emission reduction. Central to our proposal is the enhancement of optimization algorithms underlying neural network training. While Gradient Descent (GD) is a widely used iterative algorithm, there are in the literature several alternatives, with a better convergence rate, that reduce the number of steps to train a network, and consequently the carbon emission, without compromising the fitting quality. We propose HalpernSGD, a network optimizer based on the Halpern's iterative method [21] that extends the GD [6] with a quadratic boost in terms of convergence rate. We compared HalpernSGD with the Stochastic Gradient Descent (SGD) highlighting the dramatic reduction of number of epochs and volume of carbon emission, through an experimentation with statistical relevance over well-known real benchmark datasets. In addition, we conducted a comparative analysis between HalpernSGD and Adam [26], a widely used optimizer globally. This led us to wonder about the feasibility of defining a similar generalization for Adam, with the potential for enhanced convergence. The rest of the paper is organized as follows: Sect. 2 provides a general overview of the current literature related to suggestions for carbon emission reduction; Sect. 3 delves into the theoretical foundations of the iterative methods, focusing on the GD, the Halpern's algorithm and Adam; Sect. 4 describes the experimental settings we used for the comparison and provides the interpretation and a critical analysis of the experimental results; and we conclude our study with final remarks and future perspectives in section Sect. 5.

2   Related Work

Academic interest in energy efficiency during the training of deep neural networks is increasing. This is evidenced by studies such as [28,40,41], which provide a comprehensive overview of the challenges and opportunities in sustainable AI. These studies underscore the pressing need to integrate energy efficiency and carbon footprint reduction into the research and development of new machine learning technologies, aiming for greener and more sustainable solutions. There are different ways to approach these challenges, and in particular there are approaches that are much more developed than others. In [41], the authors discuss the emergence of the Green AI movement, advocating green choices in software, hardware, and data centers to address environmental concerns related to AI. In order to reduce carbon footprints, this movement emphasizes the importance of location and hardware choices in model training. Green Learning is an eco-friendly alternative to traditional deep learning [28] that, focusing on lightweight models, aims to minimize the carbon footprint by reducing energy consumption in both cloud data centers and edge devices. According to [40], carbon emissions are a crucial evaluation metric in green deep learning. The authors discuss, as strategies to improve energy efficiency, a more efficient use of data, architecture optimization, efficient training and inference techniques. These techniques include smart initialization with transfer learning, batch normalization, progressive training, and hyper-parameter optimization. These advanced methodologies are important for balancing model accuracy with environmental impact and facilitating more energy-efficient training processes. Moreover, in favor of more environmentally friendly and sustainable strategies, in [22] and [20], the process of knowledge distillation is illustrated, which allows information to be transferred from a large and complex model to a smaller and more manageable one, facilitating the reduction of energy consumption and hardware resources. These studies demonstrate the potential for research in this area to significantly reduce the environmental impact of artificial intelligence. However, there are still unexplored approaches that leave ample room for future research. Our contribution stands out for adopting an innovative approach to optimize network parameters. Our methodology focuses directly on the optimizer instead of modifying or regularizing the loss function, as commonly seen in the literature. Our proposal is a distinctive optimizer due to its convergence properties, which enable a significant reduction in the number of iterations required to achieve convergence. This approach implies significant energy savings and a notable reduction in carbon emissions, demonstrating its potential for enhancing energy efficiency and promoting environmental sustainability in DNN training.

3   HalpernSGD

One of the major tasks in training neural networks can be described as the optimization problem:

$$\min_{\theta \in \mathbb{R}^d} f(\theta) \tag{1}$$

where $f : \mathbb{R}^d \to \mathbb{R}$ represents the loss function that describes the approximation error made by the neural network, which we minimize with respect to the parameters θ. Classically, the typical approach to problem (1) consists in using the Gradient Descent algorithm, given by the iterative procedure:

$$\begin{cases} \theta_0 \in \mathbb{R}^d \\ \theta_{m+1} = \theta_m - \eta \nabla f(\theta_m). \end{cases} \tag{2}$$

The sequence $\{\theta_m\}$ is known to converge to a minimum point of a convex and differentiable function f, under mild hypotheses on the parameter η and f itself [7]. The popularity of this approach originates from the fact that the gradient step $\nabla f(\theta_m)$ can be accurately calculated by the backpropagation algorithm [33]. On the other hand, the performances of the GD algorithm can be unpredictable poor without a further second order analysis as well as the convergence itself cannot be guaranteed outside the class of convex functions. To deal with the above mentioned difficulties, the classical approach is to focus onto the choice of the parameter η in Algorithm (2), usually termed as learning rate. Several discussions and experiments have been carried forward and studies are still being conducted on the role of the learning rate in the convergence speed and on the weakening of assumptions on the function f. In this direction, the state of the art is represented by the Adam algorithm [26], in which the coefficient η is replaced by a function over the momenta of the gradient and its squared value. Although the Adam algorithm and its variants are commonly used in various contexts, we focus on the fact that their effectiveness is not certain even under good assumptions [26]. In this context, we observe an imperfection in the original paper [26]. Indeed, it states the following: **Lemma 10.3.** Let $g_t = \nabla f_t(\theta_t)$ and $g_{t,i}$ as its i-th element. Defining $g_{1:t,i} \in \mathbb{R}^t$ as $g_{1:t,i} = (g_{1,i}, g_{2,i}, \dots, g_{t,i})$ and assuming that $\|g_t\|_\infty$ is bounded by a constant $G_\infty > 0$ then it holds that $\sum_{t=1}^{T} \left(g_{t,i}^2 / t \right)^{1/2} \leq 2 G_\infty \|g_{1:T,i}\|_2$. We observed that this result is not true in general, in fact setting $g_t = t^{-1}$, $G_\infty = 1$, and letting $T \to \infty$, we have $\zeta(1.5) \leq 2\sqrt{\zeta(2)}$, i.e. $2.61 \leq 2.56$ (where ζ is the Riemann zeta function), which is false and it directly represents a counter example for the Lemma 10.3. It is worth noting that this lemma is used in the convergence analysis of several optimizers that are variants of the Adam algorithm (e.g. [18]). Moreover, Adam can have some oscillating problems, in fact in [5] the authors discuss the existence of 2-limit cycles for the Adam optimizer. Those come out even for simple cases, such that the one of the scalar quadratic objective function: $f(w) = \frac{1}{2} c w^2$, with $c > 0$. Indeed, it corresponds to a situation where, after two iterations of the Adam optimizer, the state of the system returns to its previous state, which implies the optimizer is oscillating between two points in the parameter space without converging to a minimizer for f. Therefore, it would be interesting to delve into and improve algorithms where there is convergence to a solution, i.e. when the sequence $\{\theta_m\}$ generated by the Gradient Descent properly converges to a minimum point. To better understand the behavior of the algorithm (2), it is useful to carry out an analysis in a more general setting. In particular, a natural generalization of the Gradient Descent method is found in the framework of the Metric Fixed Point Theory; more specifically in the convergence of iterations of a map satisfying certain metric-type conditions. The above mentioned connection finds its roots in a now classical result first claimed in [2]. To explicitly state it, we point out that a map $T : Dom(T) \subset \mathbb{R}^d \to \mathbb{R}^d$ is said to be non-expansive if it does not increase the distance between points, i.e. if $\|T\theta - T\bar{\theta}\| \leq \|\theta - \bar{\theta}\|$, for any $\theta, \bar{\theta} \in Dom(T)$. We also recall that a fixed point θ^* for T is a point which remains fixed under the action of the map, that is $T\theta^* = \theta^*$. By the Baillon-Haddad theorem [3], the map of GD, defined by $T\theta = \theta - \eta\nabla f(\theta)$, is non-expansive and its fixed points coincide with $\arg\min f$, if f is convex and differentiable, ∇f is L-Lipschitz and the learning rate $\eta \in (0, 2/L)$. Under the light of this theorem, the Algorithm (2) can be restated as:

$$\begin{cases} \theta_0 \in \mathbb{R}^d \\ \theta_{m+1} = \lambda\theta_m + (1 - \lambda)T\theta_m \end{cases} \tag{3}$$

where λ is a real parameter in $[0, 1)$ and T is a non-expansive map, and the convergence problem translates into proving that $\{\theta_m\}$ approximates a fixed point θ^* for T. It can be proven that Algorithm (3) converges in norm to a fixed point of the operator T, which translates to $\nabla f(\theta^*) = 0$, recalling the above statement [11]. In this context, Algorithm (3) is named the Krasnoselskii-Mann Algorithm and its convergence properties are well known as is the convergence speed that is currently stated in terms of rate of asymptotic regularity, that is by evaluating the operator residual norm $\varepsilon_m := \|\theta_m - T\theta_m\|$, as shown in [4,11]. The question of finding a bound for the rate of asymptotic regularity for the Krasnoselskii-Mann iteration had been a long-standing problem in Fixed Point Theory, lastly solved by Baillon and Bruck [1] by proving that the estimate $\varepsilon_m \leq \frac{\|\theta_0 - \theta^*\|}{\pi(\lambda(1-\lambda))^{1/2}} \frac{1}{m^{1/2}}$ holds for in the general setting of Banach spaces and for any non-expansive map T. We highlight that, under this setting, the convergence rate (i.e., the residual norm) is $\varepsilon_m \leq \mathcal{O}(1/\sqrt{m})$. Another interesting approach had been proved to be effective in this context, that is the so-called Halpern iterative procedure, given by:

$$\begin{cases} \theta_0 \in \mathbb{R}^d \\ \theta_{m+1} = \lambda_m u + (1 - \lambda_m)T\theta_m \end{cases} \tag{4}$$

where $u \in \mathbb{R}^d$ is a fixed element, termed anchor, and $\{\lambda_m\}$ is a sequence of parameters in $(0, 1)$, whose properties will be later examined. Halpern iterations find their roots in [21] and represents a classical approach for approximating fixed points of non-expansive mappings. In recent years, this method has attracted great attention [16,29,42] due to the higher rate of asymptotic regularity $\varepsilon_m = \mathcal{O}(1/m)$ [30,34], that is quadratic with respect to the one of the Krasnoselskii-Mann iteration, and consequently to the one of GD. A full investigation on lower bounds for the rate of asymptotic regularity of Halpern iterations had not been carried out, while a preliminary lower bound for the error sequence $\|\theta_m - \theta^*\|$ has been established in [12]. Turning our attention to the step-dependent coefficients $\{\lambda_m\}$, it had been proved in [39] that the following conditions are sufficient to ensure the convergence of the whole sequence $\{\theta_m\}$ to the solution θ^*:

$$(i)\ \lim_{m\to\infty} \lambda_m = 0, \quad (ii)\ \sum_{m=1}^{\infty} \lambda_m = \infty, \quad (iii)\ \lim_{m\to\infty} \frac{\lambda_m - \lambda_{m-1}}{\lambda_m} = 0. \tag{5}$$

These conditions are mild enough to include the choice $\lambda_m = \mathcal{O}(1/m)$, which appears to guarantee good rate of asymptotic regularity. For a general non-convex function, the convergence of Algorithm (2) cannot be guaranteed under a constant learning rate. Thus, an inductive step of type $\theta_{n+1} = (I - \eta_n \nabla f)\theta_n$, with $\eta_n \to 0$, is required to ensure the convergence in a wider setting.

4   Experimental Evaluation

We conducted a comparative analysis between Halpern's iterative method and Gradient Descent through their stochastic variants, given the necessary use of mini-batches in neural network learning. This variation leads to the definition of two algorithms, HalpernSGD and the well-known SGD (Stochastic Gradient Descent). Since both benefit from the use of variable learning rates (see Sect. 3), we decided to employ the sequence $\eta_m = \eta_0/\sqrt{m}$, where m is the update step (not to be confused with epochs), and η_0 is the initial learning rate, which, through empirical analysis aimed at discovering the best values for both the optimizers, we set to 1.15 in our experimentation. In addition, we set in HalpernSGD $\lambda_m = 0.02/m$. Along with HalpernSGD and SGD, we also included the Adam algorithm as a baseline for comparison, which, as discussed in Sect. 3, is a variant of SGD. We conducted our experiments on three real well-known benchmark datasets:

- **Iris**[1] contains 150 samples from 3 species of iris flowers. Each sample includes 4 features: sepal length, sepal width, petal length, and petal width.
- **Optical Recognition of Handwritten Digits (Orhd)**[2] focuses on hand-written digit recognition (0–9), featuring $1,797$ images. Originally, the images were 32×32 bitmaps, that were divided into non-overlapping blocks of 4×4 and the number of on pixels were counted in each block. This generated an input matrix of 8×8 where each element is an integer in the range $[0, 16]$.
- **Mnist**[3] is another datasets of hand-written digits containing $70,000$ gray-scale images (28×28) of handwritten digits (0–9) for digit recognition tasks. In this case each pixel is considered as an input feature.

The best empirically-found learning rates for Adam are respectively $0.001, 0.001$, and 0.01. To ensure a fair comparison among optimizers, we utilized a single shared neural network, as depicted in the Fig. 1. The network is composed by a linear Input Layer and a linear Output Layer, whose sizes depend on the used dataset (respectively, 4 and 3 for Iris, 64 and 10 for Orhd, and 784 and 10 for Mnist), and three linear Hidden Layers, with sizes 128, 64 and 32. The activation function for the hidden layers is the Leaky ReLU, while the output layer is equipped with a softmax function to generate the prediction probabilities. The results of our experiments are summarized in Table 1. To determine if

[1] https://archive.ics.uci.edu/dataset/53/.
[2] https://archive.ics.uci.edu/dataset/80/.
[3] http://yann.lecun.com/exdb/mnist/.

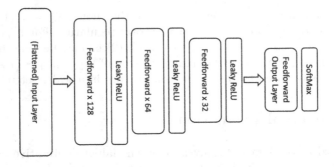

Fig. 1. The architecture of the neural network exploited for the experimentation.

Table 1. Comparative analysis for HalpernSGD, SGD and Adam.

	Optimizer	Loss	Epochs	Accuracy	Emission (μg)
Iris	SGD	0.65 ± 0.10	304 ± 257	0.93 ± 0.11	0.55 ± 0.54
	HalpernSGD	0.63 ± 0.04	223 ± 170	0.96 ± 0.02	0.39 ± 0.35
	Adam	0.63 ± 0.11	302 ± 239	0.93 ± 0.11	0.75 ± 0.62
Orhd	SGD	1.62 ± 0.05	1020 ± 427	0.85 ± 0.06	6.61 ± 2.95
	HalpernSGD	1.63 ± 0.05	230 ± 41	0.84 ± 0.05	1.45 ± 0.29
	Adam	1.60 ± 0.05	384 ± 222	0.86 ± 0.05	2.59 ± 1.48
Mnist	SGD	1.65 ± 0.03	2165 ± 356	0.81 ± 0.03	495.85 ± 75.404
	HalpernSGD	1.62 ± 0.04	888 ± 184	0.84 ± 0.04	201.13 ± 41.01
	Adam	**1.52 ± 0.01**	**116 ± 32**	**0.94 ± 0.01**	**28.02 ± 8.18**

there is a statistically significant difference between the compared optimizers we
exploited the 5×2 cv paired t-test [17], with a relevance of 95%. In the table,
values highlighted in **bold** are those that outperform HalpernSGD with statistical significance, while values shown in gray are the ones that underperform
HalpernSGD with statistical significance. Finally, those in normal formatting are
the ones for which there is no statistical difference compared to HalpernSGD.
During the cv paired t-test, each training set is further split into the actual
training set and a validation set, with a ratio of 20 : 80. The validation set was
used to define stopping criterion for the training that terminates if there is no
improvement greater than 10^{-4} in the loss on the validation set for at least 30
epochs, or if the number of epochs reaches 10, 000. The optimizers are compared
according to their best loss value on the validation set and the test global accuracy (all the datasets are balanced) to highlight their predictive capabilities, the
number of epochs to observe the convergence rate, and their CO2 emission in
micro-grams to understand their ecological impact. The latter parameter was
computed exploited the tool CodeCarbon[4]. Experiments were conducted on a

---

[4] https://codecarbon.io/.

MacBook Pro M1, with 32 GB of RAM, without exploiting GPU acceleration. Table 1 shows the average and the standard deviation for each of these features. As we can see, HalpernSGD significantly outperforms SGD in terms of number of epochs and low carbon emission for all the experiments (except for the number of epochs for Iris), without compromising the predictive quality, since there is no significant difference in terms of validation loss and test global accuracy, confirming what we stated in Sect. 3. Focusing on Mnist, Adam results to be the best contender, while there is no significant improvement in the other two cases. However, as seen, Adam may have problems with oscillation and convergence, hence, we wonder if it is possible to equip it with Halpern, given its good properties.

5   Conclusion and Future Work

In this paper, we have investigated the literature about iterative methods for fixed point searching in non-expansive maps. Exploiting the Baillon-Haddad theorem, we analyzed their involvement in the loss function minimization problem for neural networks. These methods enable the derivation of optimization algorithms for neural network training with quadratic convergence rates compared to the traditional Gradient Descent. We developed the HalpernSGD optimizer, grounded in the Halpern iteration, and compared its performance against Gradient Descent and Adam, over well-known real benchmark datasets. Experimental results indicate that HalpernSGD and GD have similar fitting performance, but the former achieves convergence in significantly fewer iterations, resulting in significantly lower carbon footprint. However, Adam remains the optimizer with the superior performance. We acknowledged the quality and the convergence speed of the Adam algorithm, so we better examined it. We found out an imperfection in the original paper and mentioned problems of oscillation and non-convergence arising in some specific situations. All these issues make us wondering about a future work that could aim at mixing the Adam's performance with the Halpern's stability and convergence properties. The aim is to define a set of brand new optimizers that are able to speed up the training process of neural networks, ensuring significantly reduced emissions of CO2.

Acknowledgemnt. This publication was partially funded by the PhD program in Mathematics and Computer Science at University of Calabria, Cycle XXXVIII with the support of a scholarship financed by DM 351/2022 (CUP H23C22000440007), based on the NRPP funded by the European Union.

References

1. Baillon, J., Bruck, R.E.: The rate of asymptotic regularity is $O(1/\sqrt{n})$. In: Lecture Notes in Pure and Applied Mathematics, pp. 51–82 (1996)
2. Baillon, J.B., Haddad, G.: Quelques propriétés des opérateurs angle-bornés et n-cycliquement monotones. Israel J. Math. **26**, 137–150 (1977)

3. Bauschke, H.H., Combettes, P.L.: The Baillon-Haddad theorem revisited. J. Convex Anal. **17**(3&4), 781–787 (2010)
4. Berinde, V., Takens, F.: Iterative Approximation of Fixed Points, vol. 1912. Springer, Heidelberg (2007). https://doi.org/10.1007/978-3-540-72234-2
5. Bock, S., Weiß, M.: Non-convergence and limit cycles in the adam optimizer. In: Tetko, I.V., Kůrková, V., Karpov, P., Theis, F. (eds.) ICANN 2019. LNCS, vol. 11728, pp. 232–243. Springer, Cham (2019). https://doi.org/10.1007/978-3-030-30484-3_20
6. Bottou, L.: On-line learning and stochastic approximations, pp. 9–42. Cambridge University Press (1999)
7. Boyd, S.P., Vandenberghe, L.: Convex Optimization. Cambridge University Press, Cambridge (2004)
8. Brown, T.B., et al.: Language models are few-shot learners. In: Proceedings of the 34th International Conference on Neural Information Processing Systems. Curran Associates Inc. (2020)
9. Chen, L., et al.: Deep neural network based vehicle and pedestrian detection for autonomous driving: a survey. IEEE Trans. Intell. Transp. Syst. **22**(6), 3234–3246 (2021)
10. Chen, S., Sun, P., Song, Y., Luo, P.: DiffusionDet: diffusion model for object detection. In: Proceedings of the IEEE/CVF International Conference on Computer Vision, pp. 19830–19843 (2023)
11. Chidume, C.: Geometric Properties of Banach Spaces and Nonlinear Iterations. Springer, London (2009). https://doi.org/10.1007/978-1-84882-190-3
12. Colao, V., Marino, G.: On the rate of convergence of Halpern iterations. J. Nonlinear Convex Anal. **22**(12), 2639–2646 (2021)
13. Das, K.P., Chandra, J.: A survey on artificial intelligence for reducing the climate footprint in healthcare. Energy Nexus **9**, 100167 (2023)
14. Delanoë, P., Tchuente, D., Colin, G.: Method and evaluations of the effective gain of artificial intelligence models for reducing CO2 emissions. J. Environ. Manag. **331**, 117261 (2023)
15. Dhar, P.: The carbon impact of artificial intelligence. Nat. Mach. Intell. **2**, 423–425 (2020)
16. Diakonikolas, J.: Halpern iteration for near-optimal and parameter-free monotone inclusion and strong solutions to variational inequalities. In: Conference on Learning Theory, pp. 1428–1451 (2020)
17. Dietterich, T.G.: Approximate statistical tests for comparing supervised classification learning algorithms. Neural Comput. **10**(7), 1895–1923 (1998)
18. Dubey, S.R., Chakraborty, S., Roy, S.K., Mukherjee, S., Singh, S.K., Chaudhuri, B.B.: diffGrad: an optimization method for convolutional neural networks. IEEE Trans. Neural Netw. Learn. Syst. **31**(11), 4500–4511 (2019)
19. Goodfellow, I., et al.: Generative adversarial nets. In: Advances in Neural Information Processing Systems, vol. 27. Curran Associates, Inc. (2014)
20. Gou, J., Yu, B., Maybank, S.J., Tao, D.: Knowledge distillation: a survey. Int. J. Comput. Vis. **129**, 1789–1819 (2021)
21. Halpern, B.: Fixed points of nonexpanding maps. Bull. Am. Math. Soc. **73**, 957–961 (1967)
22. Hinton, G., Vinyals, O., Dean, J.: Distilling the knowledge in a neural network. arXiv preprint arXiv:1503.02531 (2015)
23. Ho, J., Jain, A., Abbeel, P.: Denoising diffusion probabilistic models. In: Proceedings of the 34th International Conference on Neural Information Processing Systems. Curran Associates Inc. (2020)

24. Jain, J., Li, J., Chiu, M.T., Hassani, A., Orlov, N., Shi, H.: OneFormer: one transformer to rule universal image segmentation. In: Proceedings of the IEEE/CVF Conference on Computer Vision and Pattern Recognition, pp. 2989–2998 (2023)

25. Kaur, R., Singh, S.: A comprehensive review of object detection with deep learning. Digit. Sig. Process. **132**, 103812 (2023)

26. Kingma, D.P., Ba, J.: Adam: a method for stochastic optimization. In: International Conference on Learning Representations, ICLR (2015)

27. Kingma, D.P., Welling, M.: Auto-encoding variational bayes. In: International Conference on Learning Representations, ICLR (2014)

28. Kuo, C.C.J., Madni, A.M.: Green learning: introduction, examples and outlook. J. Vis. Commun. Image Represent. **90**, 103685 (2023)

29. Lee, S., Kim, D.: Fast extra gradient methods for smooth structured nonconvex-nonconcave minimax problems. In: Advances in Neural Information Processing Systems, vol. 34, pp. 22588–22600 (2021)

30. Lieder, F.: On the convergence rate of the Halpern-iteration. Optim. Lett. **15**(2), 405–418 (2021)

31. Liguori, A., Manco, G., Pisani, F.S., Ritacco, E.: Adversarial regularized reconstruction for anomaly detection and generation. In: IEEE International Conference on Data Mining, ICDM, pp. 1204–1209. IEEE (2021)

32. Manco, G., Ritacco, E., Rullo, A., Saccà, D., Serra, E.: Machine learning methods for generating high dimensional discrete datasets. WIREs Data Min. Knowl. Discov. **12**(2), e1450 (2022)

33. Rumelhart, D.E., Hinton, G.E., Williams, R.J.: Learning representations by back-propagating errors. Nature **323**, 533–536 (1986)

34. Sabach, S., Shtern, S.: A first order method for solving convex bilevel optimization problems. SIAM J. Optim. **27**(2), 640–660 (2017)

35. Strubell, E., Ganesh, A., McCallum, A.: Energy and policy considerations for modern deep learning research. In: Proceedings of the AAAI Conference on Artificial Intelligence, vol. 34, no. 09, pp. 13693–13696 (2020)

36. Sultana, F., Sufian, A., Dutta, P.: Evolution of image segmentation using deep convolutional neural network: a survey. Knowl.-Based Syst. **201**, 106062 (2020)

37. Vaswani, A., et al.: Attention is all you need. In: Advances in Neural Information Processing Systems, vol. 30 (2017)

38. Xie, L., Lee, F., Liu, L., Kotani, K., Chen, Q.: Scene recognition: a comprehensive survey. Pattern Recogn. **102**, 107205 (2020)

39. Xu, H.K.: Another control condition in an iterative method for nonexpansive mappings. Bull. Aust. Math. Soc. **65**(1), 109–113 (2002)

40. Xu, J., Zhou, W., Fu, Z., Zhou, H., Li, L.: A survey on green deep learning (2021)

41. Xu, Y., Martínez-Fernández, S., Martinez, M., Franch, X.: Energy efficiency of training neural network architectures: an empirical study. In: Hawaii International Conference on System Sciences (2023)

42. Yoon, T., Ryu, E.K.: Accelerated algorithms for smooth convex-concave minimax problems with $O(1/k^2)$ rate on squared gradient norm. In: International Conference on Machine Learning, pp. 12098–12109 (2021)

Integrating Predictive Process Monitoring Techniques in Smart Agriculture

Simona Fioretto[3]([✉]) [iD], Dino Ienco[2] [iD], Roberto Interdonato[1] [iD],
and Elio Masciari[3] [iD]

[1] CIRAD, INRIA, Univ. Montpellier, Montpellier, France
`roberto.interdonato@cirad.fr`
[2] INRAE, INRIA, Univ. Montpellier, Montpellier, France
`dino.ienco@inrae.fr`
[3] University of Naples Federico II, 08544 Naples, Italy
{`simona.fioretto,elio.masciari`}`@unina.it`

Abstract. Problems related to the environment are increasingly commonly known and consequently also technology is adapting to find suitable solutions. The ancestral technique of crop rotation was identified as a solution to address the problems related to pollution due to intensive food production (i.e. using fertilizers and pesticides). To ensure that this technique can actually improve food production, it is necessary to understand how modern technologies can support it; in particular the analysis of crop rotation can support farmers in decision making process and the optimization of farm management practices. The aim of this paper is to investigate how predictive process monitoring techniques can enhance crop rotation strategies by leveraging Agriculture 4.0 through real-time monitoring, resulting in more accurate and adaptive strategies. It is a position paper that proposes research questions for further study, which may help to develop the research area.

Keywords: Predictive Process Monitoring · Crop Rotation · Machine Learning

1 Introduction

The expected increase of the world population by about one third by 2050 leads directly to an increase in the demand for food [18]. Fertilisers and pesticides are employed to meet this ever-increasing demand, by preventing diseases that could cause a drastic reduction in production outputs. However, this new system of food production, without technological interventions or containment measures, will lead to an expected increase in pollution of 50–92% [18], causing environmental problems and not meeting the growing demand for food [4].

In this scenario, crop rotation can help in reducing the use of pesticides and fertilizers, meeting the needs of sustainable agriculture. Crop rotation is an ancestral agricultural technique that involves the successive cultivation of different crops in a specified order in the same field [1]. It allocates fields to different crops and selects the appropriate crop type [20]. Therefore, predictions about crop rotation sequences could support the farming system management in targeting production strategies. However, forecasting a suitable cropping plan is difficult due to the co-existence of many factors, such as weather, soil type, diseases, and pests, which strongly impact crop growth [7].

However, [20] suggests the use of detailed spatio-temporal data to improve the prediction of crop rotation sequences. Many proposed approaches leverage both spatial and temporal data, either alone or in combination. By modelling the crop sequence as a process, and integrating multi-dimensional and time-series data using event logs, it is possible to apply Predictive Process Monitoring (PPM) [2] to predict the crop rotation sequence. PPM is a subset of Process Mining (PM) techniques that use event log data to predict process behaviour [14,15].

This aim of this position paper is to present existing and emerging agricultural approaches, to identify related issues and to raise research questions. The structure of the paper is the following: in Sect. 2 we provide an overview of the current state of the art. In Sect. 3, after identifying the gaps that require further investigation, the research questions are discussed along with related comments. Finally, in Sect. 4, we present concluding remarks.

2   State of the Art

The aim of this section is to introduce Process Mining, Prediction Techniques, and Predictive Process Monitoring methods used in agriculture.

2.1   Process Mining Techniques

Process Mining (PM) is a research field that uses event logs to gain insights into business processes. PM includes *process discovery* (generating a model from an event log), *conformance checking* (comparing a process model with an event log of the same process) and *process enhancement* (improving a process model by using information from recorded processes in the event log) [21].

In [11] the authors utilise PM techniques to investigate user behaviour when employing decision support systems (DSS). The study specifically targeted farmers who used DSS for cultivar selection. The system assists farmers in selecting cultivars based on their characteristics, which can be used to identify potential features to add, when redesigning the system.

In [12] the authors combine PM and simulation-based technology to redesign processes. They use a bottom-up approach to identify processes and their performance from process data. Then, simulation on future performances of the process is employed to support the redesign phase. Finally, the approach is assessed on a web-based DSS in a Dutch agriculture organization, gaining insight into damages caused by parasites and the related financial consequences. Conclusions report that the redesign phase can benefit from potential performance gains.

The approaches analysed so far are summarised in Table 1. Therefore, it is clear that PM in the agricultural field is mainly used to identify the processes underlying the DSS used by farmers, using process discovery techniques.

Table 1. Process Mining methods

Study	Approach	Task
[11]	Process Discovery	User Behavior
[12]	Process Discovery and Simulation	BP Redesign

2.2 Prediction Techniques

The proposal [9] predicts the crop yield at the end of the cropping season on ten crops. They merge 2 data-sources, one for agricultural production data, and one for weather information data. Then, they compare regression methods, such as Multiple Linear Regression, M5-Prime Regression Trees (RT), Perceptron Multilayer Neural Networks, Support Vector Regression, and k-Nearest Neighbor, finding the best attribute subset for each method using a complete algorithm. The techniques were then evaluated using Root Mean Squared Error (RMSE), Root Relative Squared Error (RRSE), Correlation Factor (R), and Mean Absolute Error (MAE) metrics, showing M5 as the best tool.

The authors [17] predict crop yield under soil salinity effects, using vegetation indices and Stepwise Linear Regression (SLR) derived from satellite images. Firstly, they create crop pattern maps of the area and extract field types using object-based classification. Secondly, they predict crop yield and map soil salinity. Finally, they investigate the correlation between yield loss and soil salinity.

The authors [10] predict crop yield before sowing, considering rainfall and soil characteristics. First, they use historical rainfall data to forecast future rainfall quantities using ARIMA and Exponential Sequencing. Since rainfall data are seasonal, a time-series analysis approach takes into account the trend. Second, they use Recurrent Neural Network (RNN) to learn crop yield patterns based on soil properties. Finally, the RNN uses the previous predictions on rainfall and soil type to predict crop yield.

The study [19] proposes various data mining techniques for predicting crop yield in the agricultural domain based on rainfall data. After selecting the relevant features, the authors focus on using regression techniques to develop a predictor model. The dataset used contains information on crop type, season, year, crop coverage area, and production.

The authors in [16] propose a framework. First, they combine LoRa Wireless Sensor Networks (WSN) with short and medium area coverage to collect and organise data in time-series. Then, they apply ARIMA prediction model for predicting future complications in production and propose a notification system to support corrective actions and achieve a better crop yield. Finally, they do 3 experiments to analyze the proposed framework.

The proposal [8] considers pre-crop values and crop rotation matrices. The latter contains factors having a positive or negative effect on the next crop, mediated through the soil. First, they train an XGBoost Regressor to predict crop rotation matrices using Normalized Difference Vegetation Index (NDVI) around harvesting time. They used satellite image data combined with weather, soil, and land-use data to identify yield-enhancing crop combinations. Results show that the proposal can support farmers in adapting their crop rotation to local requirements and potential crops. However, its limitation lies in the use of the NDVI indicator for yield potential, as it measures plant land cover and also evaluates weeds.

The authors in [13] propose a two objective approach. First, they find the environmental factors that influence crop productivity and suitability. Then, they use six different tree-based ensemble learning models namely, Random Forest (RF), Gradient Boosting, AdaBoost, XGBoost, LightGBM, and CatBoost to predict crop suitability and productivity. The input data include environmental conditions, such as soil and climate measurements, and their impact on crop growth. The evaluation of the models on accuracy, precision, recall and F1-score, show that XGBoost and LightGBM outperform AdaBoost and RF.

The analysed approaches are summarised in Table 2. They are mainly focused on crop yield prediction, mostly using techniques for analysing and predicting time series data. Some studies also integrate Machine Learning (ML) and Artificial Neural Networks (ANN), either alone or in conjunction with ARIMA models.

Table 2. Prediction approaches

Study	Approach	Prediction Task
[9]	Regression Methods	Crop Yield
[17]	SLR	Crop Yield
[10]	ARIMA Exponential Sequencing RNN	Crop Yield Patterns
[19]	Regression Methods	Crop Yield
[16]	ARIMA	Deviations
[8]	XGBoost	Crop Rotation Matrices
[13]	Tree-based Ensemble Learning	Crop Suitability Crop Productivity

2.3 Predictive Monitoring Approaches

Predictive Process Monitoring (PPM) is a subset of PM techniques, which leverages event log data to predict the future behaviour of a process in execution by using prediction algorithms [2]. In fact, event log data can integrate multidimensional and time-series data, combining various factors affecting the crop growth.

In [5] the authors based on crop history data, use PM techniques and Markov Chains principles to predict and visualize the crop that will be planted in the following year (n + 1) for each field in the same geographic location. The methodology involved preparing the dataset and constructing Directly Follow Graphs (DFGs) to depict crop relationships through PM. Subsequently, a prediction model was developed using the weights derived from DFGs to generate an associated adjacency matrix. Finally, the transition matrix was estimated. Finally, the model performs predictions using the transition matrix, the SVP dataset and a coding dictionary. The evaluation of the model and presentation of the results are done using DFGs.

In [22] the authors model the crop growth as a process. More in details, they obtain crop harvest predictions by integrating static and dynamic information on crop growth into a single data input, i.e. the event log. First, they define the process model of crop growing process. Then, they use replay to merge various input data into one event log. Then, they extract prefixes from the obtained event log and group them into buckets, encoded as feature vectors with an encoding mechanism chosen according to agronomic knowledge. Finally they train statistical models for making explainable and transparent predictions. The experimentation on a real-life data case study shows promising results.

The authors in [6] leverage Long Short Time Memory (LSTM) networks to predict the most probable crop rotation scenario in a field. This is because crop rotation is a sequence-based problem. The paper employs the Seq2Seq architecture on LSTM cells auto-encoder to address the problem of predicting the most probable sequence of crops grown in a field, not only in year (n + 1), but also in year (n + x). The LSTM predicts both past and future scenarios. To propose plausible results, these predictions are compared with predictions from a probability-conditioned model. The results of the two models are integrated.

PPM approaches analysed so far in the context of agriculture are summarised in Table 3. Only a few approaches related to PPM are reported in the literature, addressing 3 different tasks. In addition, they also focus on different approaches, the first one relies on the application of PM to show the results based on relations between activities with the use of DGF, and the second and the third ones use SL methods and ANN respectively to make predictions.

Table 3. Predictive Monitoring Approaches

Study	Approach	Prediction Task
[5]	Markov Chains DGF	Crop Rotation
[22]	SL	Crop Harvest
[6]	ANN (LSTM)	Crop Sequence

The above approaches using PPM or PM techniques rely on the transformation of collected data into event log. Therefore, the application of PM based approaches can raise limitations related to data collection and transformation.

As a final point, the prediction approaches presented in Sect. 2.2 and Sect. 2.3 rely on datasets collected from multiple sources, depending on the prediction task and the availability of data in the field.

3 Research Challenges

Current research suggests that Agriculture 4.0, also known as the 4th agricultural revolution, is gaining momentum with promising results, also supported by digital transformation achieved by the introduction of new technologies. The literature review in Sect. 2 clearly shows an increasing tendency towards the application of PM techniques, Prediction Techniques, and PPM approaches.

Although not widely explored, PPM approaches could improve agricultural forecasting by integrating time series data with multi-dimensional data, thus enabling online forecasting. Furthermore, PPM facilitates what-if scenario analysis and the decision-making process. However, as previously mentioned, the application of PPM requires both historical and online event log data from multiple sources to be integrated. This integration is necessary to identify the features that influence the crop prediction task and to obtain more accurate predictions.

Research Questions. Current approaches in the literature raise questions about the potential benefits of the application of PPM techniques in precision agriculture. These observations trigger and motivate our following research questions:

RQ_1: Which prediction task should be faced? (e.g. if the goal is to achieve the maximum production, the prediction of the crop sequence is obviously more influential than the prediction of crop harvest)

RQ_2: Which approach should be used? (e.g., based on the prediction task and on the available data, what approach should be used in the field of predictive process monitoring? Are reported approaches showing better results than others? Is the rationale of the prediction process important?)

Remarks. This paper presents findings that open up fascinating possibilities for future research. Section 2 provides an overview of the current state of the art, indicating that PPM techniques have the potential to enhance crop predictions by integrating time-series data with multi-dimensional data. However, limited research studies suggest that further investigation is necessary. Since PPM predicts the future outcome of a process in execution [3], prediction results may assist farmers in decision-making processes, e.g. by re-adapting the decisions on crops based on possible predicted deviations.

The discussion of the above research questions could bring to important results, having a strong impact on the implementation of PPM in agriculture:

- First, by identifying the prediction task it is possible to detect potential future deviations (for instance a different crop production level than expected before it happens), provide results and predict probable outcomes of the crops (i.e., the crop yield, the crop harvest influencing the income), or to recommend solutions (i.e., predict the crop sequence or crop rotation);

– Second, by identifying the possible prediction models to be used, it is possible to conduct a comparative analysis of the models in order to choose the one which fits the objectives of the research (i.e., the one respecting the computational requirements and the interpretability of the model)

4   Conclusion

In this paper, the issue of food increasing demand is addressed, including concerns regarding environmental impacts. Among the available solutions, crop rotation is widely identified as an ancestral technique that can address the aforementioned problem. Along with the use of advanced technologies, it can effectively support farmers in field management. Therefore, predictive process monitoring is identified as an emerging approach currently used to address this problem. After providing a brief overview of the proposed approaches, we identified the issues related to the application of predictive monitoring in the research field. These issues are mainly related to the collection of input data. This, in turn, raises the research question to be addressed as future work, which is mostly attributable to identifying the predictive task to be addressed and the prediction model to be assessed.

References

1. Castellazzi, M., Wood, G., Burgess, P., Morris, J., Conrad, K., Perry, J.: A systematic representation of crop rotations. Agric. Syst. **97**(1), 26–33 (2008). https://doi.org/10.1016/j.agsy.2007.10.006. https://www.sciencedirect.com/science/article/pii/S0308521X07001096
2. Di Francescomarino, C., Ghidini, C.: Predictive process monitoring. In: van der Aalst, W.M.P., Carmona, J. (eds.) Process Mining Handbook. LNBIP, vol. 448, pp. 320–346. Springer, Cham (2022). https://doi.org/10.1007/978-3-031-08848-3_10
3. Di Francescomarino, C., Ghidini, C., Maggi, F.M., Milani, F.: Predictive process monitoring methods: which one suits me best? In: Weske, M., Montali, M., Weber, I., vom Brocke, J. (eds.) BPM 2018. LNCS, vol. 11080, pp. 462–479. Springer, Cham (2018). https://doi.org/10.1007/978-3-319-98648-7_27
4. Dias, T., Dukes, A.E., Antunes, P.M.: Accounting for soil biotic effects on soil health and crop productivity in the design of crop rotations. J. Sci. Food Agric. **95**(3), 447–54 (2015). https://api.semanticscholar.org/CorpusID:7890471
5. Dupuis, A., Dadouchi, C., Agard, B.: Predicting crop rotations using process mining techniques and Markov principals. Comput. Electron. Agric. **194**, 106686 (2022)
6. Dupuis, A., Dadouchi, C., Agard, B.: Methodology for multi-temporal prediction of crop rotations using recurrent neural networks. Smart Agric. Technol. **4**, 100152 (2023)
7. Dury, J., Schaller, N., Garçia, F., Reynaud, A., Bergez, J.E.: Models to support cropping plan and crop rotation decisions. A review. Agron. Sustain. Dev. **32**, 567–580 (2012). https://api.semanticscholar.org/CorpusID:16687797
8. Fenz, S., Neubauer, T., Heurix, J., Friedel, J.K., Wohlmuth, M.L.: AI-and data-driven pre-crop values and crop rotation matrices. Eur. J. Agron. **150**, 126949 (2023)

9. González Sánchez, A., Frausto Solís, J., Ojeda Bustamante, W., et al.: Predictive ability of machine learning methods for massive crop yield prediction (2014)
10. Kulkarni, S., Mandal, S.N., Sharma, G.S., Mundada, M.R., Meeradevi: Predictive analysis to improve crop yield using a neural network model. In: 2018 International Conference on Advances in Computing, Communications and Informatics (ICACCI), pp. 74–79 (2018). https://doi.org/10.1109/ICACCI.2018.8554851
11. Mărușter, L., Faber, N.R., Jorna, R.J., van Haren, R.J.: A process mining approach to analyse user behaviour. In: International Conference on Web Information Systems and Technologies, vol. 2, pp. 208–214. SCITEPRESS (2008)
12. Mărușter, L., Van Beest, N.R.: Redesigning business processes: a methodology based on simulation and process mining techniques. Knowl. Inf. Syst. **21**, 267–297 (2009)
13. Nti, I.K., Zaman, A., Nyarko-Boateng, O., Adekoya, A.F., Keyeremeh, F.: A predictive analytics model for crop suitability and productivity with tree-based ensemble learning. Decis. Anal. J. **8**, 100311 (2023). https://doi.org/10.1016/j.dajour.2023.100311. https://www.sciencedirect.com/science/article/pii/S2772662223001510
14. Pasquadibisceglie, V., Appice, A., Castellano, G., van der Aalst, W.M.P.: PROMISE: coupling predictive process mining to process discovery. Inf. Sci. **606**, 250–271 (2022). https://doi.org/10.1016/J.INS.2022.05.052
15. Pravilovic, S., Appice, A., Malerba, D.: Process mining to forecast the future of running cases. In: Appice, A., Ceci, M., Loglisci, C., Manco, G., Masciari, E., Ras, Z.W. (eds.) NFMCP 2013. LNCS (LNAI), vol. 8399, pp. 67–81. Springer, Cham (2014). https://doi.org/10.1007/978-3-319-08407-7_5
16. dos Santos, U.J.L., Pessin, G., da Costa, C.A., da Rosa Righi, R.: AgriPrediction: a proactive internet of things model to anticipate problems and improve production in agricultural crops. Comput. Electron. Agric. **161**, 202–213 (2019)
17. Satir, O., Berberoglu, S.: Crop yield prediction under soil salinity using satellite derived vegetation indices. Field Crop Res **192**, 134–143 (2016)
18. Springmann, M., et al.: Options for keeping the food system within environmental limits. Nature **562**(7728), 519–525 (2018)
19. Surya, P., Aroquiaraj, I.L., et al.: Crop yield prediction in agriculture using data mining predictive analytic techniques. Int. J. Res. Anal. Rev. **5**(4), 783–787 (2018)
20. Upcott, E.V., Henrys, P.A., Redhead, J.W., Jarvis, S.G., Pywell, R.F.: A new approach to characterising and predicting crop rotations using national-scale annual crop maps. Sci. Total Environ. **860**, 160471 (2023)
21. Van Der Aalst, W.: Process mining: overview and opportunities. ACM Trans. Manag. Inf. Syst. (TMIS) **3**(2), 1–17 (2012)
22. Yang, J., Ouyang, C., Dik, G., Corry, P., ter Hofstede, A.H.M.: Crop harvest forecast via agronomy-informed process modelling and predictive monitoring. In: Franch, X., Poels, G., Gailly, F., Snoeck, M. (eds.) CAiSE 2022. LNCS, vol. 13295, pp. 201–217. Springer, Cham (2022). https://doi.org/10.1007/978-3-031-07472-1_12

Author Index

A. Appice et al. (Eds.): ISMIS 2024, LNAI 14670, pp. 315–316, 2024.
https://doi.org/10.1007/978-3-031-62700-2

Printed in the United States
by Baker & Taylor Publisher Services